T0298959

Lozi Mappings
Theory and Applications

Lozi Mappings
Theory and Applications

Zeraoulia Elhadj

Department of Mathematics
University of Tébessa, Algeria

CRC Press
Taylor & Francis Group
Boca Raton London New York

CRC Press is an imprint of the
Taylor & Francis Group, an **informa** business

A SCIENCE PUBLISHERS BOOK

CRC Press
Taylor & Francis Group
6000 Broken Sound Parkway NW, Suite 300
Boca Raton, FL 33487-2742

First issued in hardback 2019

© 2014 by Taylor & Francis Group, LLC
CRC Press is an imprint of Taylor & Francis Group, an Informa business

No claim to original U.S. Government works

ISBN-13: 978-1-4665-8070-1 (hbk)

Visit the Taylor & Francis Web site at
http://www.taylorandfrancis.com

and the CRC Press Web site at
http://www.crcpress.com

Foreword

It is undeniable that nature is extremely complicated and in many instances absolutely unpredictable. It is therefore reasonable to assume that mathematical models of such behavior must be correspondingly complicated. However, in the past few decades it has been realized that extremely simple mathematical models can produce complicated and unpredictable results, which we now call "chaos".

One of the early pioneers in the field of chaos was the meteorologist Edward Lorenz, who in the 1960s sought a simple system of ordinary differential equations whose solution was not periodic. Upon finding such a system, he discovered the accompanying sensitive dependence on the initial conditions that is the hallmark of chaos and thus explained the difficulty of making long-range weather predictions.

In the 1970s, the theoretical ecologist Robert May pointed out to the biological community the complicated dynamics that can ensue from a simple one-dimensional quadratic mapping called "logistic map", while Michel Hénon proposed a simple two-dimensional quadratic mapping that captures the essence of the dynamics of the Lorenz system and that now bears his name. The logistic map was shown to be conjugate to the tent map, which is a piecewise-smooth map whose analysis is therefore simpler than the logistic map. The mathematician René Lozi soon thereafter suggested a piecewise-smooth variant of the Hénon map that now bears his name and that has been extensively studied and applied to many problems throughout nature. It might be considered the simplest non-trivial mathematical model of chaos.

In this book, Prof. Zeraoulia Elhadj has collected most of the important theoretical results and practical applications of the Lozi map as well as many of the fundamental principles for the analysis of general chaotic systems. It is thus suited as an introduction to chaotic dynamics and is also a treatise on this particular mapping about which much has been written. The concepts and methods can be applied to more complicated mathematical models of nature and hold the promise of a more thorough understanding of the world in which we live.

October 2012

Prof. Julien Clinton Sprott
Department of Physics
University of Wisconsin
Madison, WI USA

Preface

This book is based on the research carried out in the past three decades on Lozi mapping, discovered in 1978. This map is the simplest 2-D piecewise linear system displaying the so-called *true chaos* because it is characterized by a homogeneous and topologically stable structure. The novelty of this book lies in the fact that it includes a complete description of existing theories and results concerning Lozi mappings.

The first chapter presents some definitions and relevant results about the statistical properties and classification of chaotic attractors. Chapter 2 discusses the reality of chaos in the famous Hénon mappings. In general, this type of chaos is characterized by the occurrence of *dangerous homoclinic tangencies* that makes the corresponding attractor as the unified limit set of the whole attracting set of trajectories including a subset of both chaotic and stable periodic trajectories which have long periods and weak and narrow attraction basins and stability regions. This is a result of the effect that this attractor is *holed* by a set of basins of attraction of different periodic orbits. Also, the most interesting results concerning different shapes of the Hénon mappings are presented with some details in this chapter. Chapter 3 focusses on a self-contained introduction to chaos via the Lozi mappings. The second section of this chapter presents a rigorous proof that the Lozi map is truly chaotic, where almost results in this part are given by unified notations. In Chapter 4, is discussed the most interesting results concerned with maps obtained from the idea of generalizing the simple Lozi map to different forms. Chapter 5 is concerned with the real and mathematical applications of Lozi mappings in engineering, computers, communications, medicine and biology, management and finance and consumer electronics, where the potential application types of chaos are: control, synthesis, synchronization, information processing, etc.

Acknowledgements

I would like to thank the following people:

- Prof. René Lozi (Laboratoire de Mathématiques, Université de Nice Sophia Antipolis. France) for his valuable suggestions.

- Prof. Julien Clinton Sprott (Department of Physics, University of Wisconsin, Madison, USA) for his valuable suggestions and help.
- Prof. Zin Arai (Creative Research Initiative "Sousei", Hokkaido University, Kita-ku, Sapporo, Japan) for offering some of his papers and figures.
- James Yorke, Zheng Wei-Mou, Linda Stojanovska, Tamas Tèl, Michal Misiurewicz, Yongluo Cao, Akira Shudo, Yutaka Ishii, Garyfalos Papaschinopoulos, Aubin Arroyo, Artur Oscar Lopes, Ulrich Hoensch, Mukul Majumdar, Mark Pollicott, Stefano Luzzatto, Lorenzo J Diaz Casado and Izzet Burak Yildiz for offering some of their papers concerned the topics in this book.
- American Physical Society, American Institute of Physics, Institute of Physics (IOP) Publishing, Taylor & Francis Publishing, Hindawi Publishing Corporation, and Europhysics Letters (EPL), SPIE Digital Library, Elsevier Limited, Springer Science + Business Media, American Mathematical Society, Progress of Theoretical Physics Publishing (Yukawa Hall, Kyoto University Kyoto, Japan), Wiley-Blackwell Publishing, Hokkaido Mathematical Journal (Department of Mathematics, Hokkaido University, Japan) for their kind permission to reuse some of their copyrighted materials (figures).

Lozi Mappings: Theory and Applications is devoted to setting a practical guide to studying chaotic systems.

January 2013

Prof. Zeraoulia Elhadj
Department of Mathematics
University of Tébessa
12002, Algeria

Contents

List of Figures

Introduction

René Lozi

When more than two years ago, Prof., Zeroualia Elhadj informed me of his willingness to write a book on what is known as the "Lozi map" since the Misiurewicz's communication in the congress organized by the New York Academy of Science, 17–21 December 1979, I warned him that the task was not easy because hundreds of articles have been published on this topic in the past 30 years. These papers were scattered in various fields of research, not only in mathematics (dynamical systems), but in physics, computer science, electronics, chemistry, control science and engineering, etc. Nevertheless, he eventually collected and scrutinized more than one thousand papers before completing this outstanding book *Lozi Mappings: Theory and Applications*. The outcome of his enquiry is tremendous. Every aspect of the mathematical properties of this map of the plane (and its generalizations) is analyzed. The results are classified and systematized. Moreover, in order to make easy the comprehension for a fresh reader, the book begins with a comprehensive review of hyperbolicity, ergodicity and chaos. Once the background is clearly posed the reality of chaos in the Hénon mappings is examined followed by a survey on the Lozi mappings.

Responding to the kind invitation of Prof. Zeraoulia Elhadj to write an introduction, I take this opportunity to introduce some personal views not only on the matter of chaotic systems, but also on the current evolution of mathematics and some aspects of the life of one researcher in mathematics.

In life it is not so easy to recall a particular day. Thirty-five years after the pinpoint moment I had the idea to substitute the quadratic term in the Hénon map by an absolute value. I can remember the exact date because it took place during the presentation of the thesis of A. Intissar on 15 June 1977 around 11 am (I recently checked the date). In those days the Department of Mathematics of the University of Nice (later called the University of Nice-Sophia Antipolis) was a small community and every one attended the presentation of each Ph.D. thesis. Hence I was not very concerned with the talk but on the contrary I was thinking about the strange structure of the Hénon map that my colleague Gérard Iooss had told me about a few days before, at the "International Conference on Mathematical Problems in Theoretical Physics" organized by the University of Roma, Italy (June 6–15), which we had attended together. The opening talk of this conference given by David Ruelle (Dynamical Systems and Turbulent Behavior) emphasized the importance of such a simple discrete model in the study of turbulence (this is not recognized today). At that time I occupied the position of "Attaché de Recherches" at CNRS (Centre National de la Recherche Scientifique) after my Ph.D. thesis on numerical analysis of bifurcation problems (the first thesis on bifurcation theory in France, presented on 25 April, 1975). I was mainly interested in discretization problems and finite element methods, in which nonlinear functions are approximated by piecewise linear ones. I tried to apply my background to the quadratic map introduced by Michel Hénon a few months ago, in order to obtain a better amenable map for

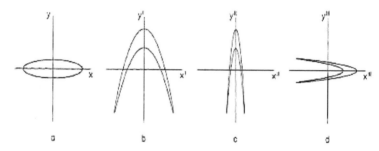

Figure 1: The initial area *a* mapped by T' into *b*, then by T'' into *c*, and finally by T''' into *d*.

analytical treatment. In Fig. 1 is shown a clear explanation of the folding and stretching process which led him to the formula of the map.

The area *b* on the figure is bounded by two parabolas generated by the formula: $T' : x' = x, y' = y +1-ax^2$ applied to the initial area *a*. Drawing on a sheet of paper the shape of this area, I embedded it on another area bounded by four line segments which eventually reminded me of the graph of the absolute value function. I then substituted $L' : x' = x, y' = y + 1 - a \, |x|$ to T'. Soon after the end of the presentation I went to my office situated on the upper floor of the seminar room to test the idea on the Hewlett-Packard 9820 calculator linked to the HP 9862 plotter I used to promote computer science for teachers at the Institute of Research in Educational Mathematics (IREM). Even if the parameter value giving the "classical" Hénon strange attractor (i.e., $a = 1.4$, $b = 0.3$) provides also a strange attractor for this new mappings, after a few tests I shifted it to $a = 1.7$, $b = 0.5$ in order to obtain a more striking picture of the strange attractor studied in this book. Back to lunch which celebrated the completion of the thesis, I showed the figure to Gérard Iooss and also to Alain Chenciner who encouraged me later to publish the formula (the genuine article comes from the presentation I gave during a conference on dynamical systems in July 1977 in Nice). In the following days I was convinced that few weeks would be enough to explain the structure of such an attractor basically composed of line segments. But the task proved more difficult than expected (mainly because contrary to as Michal Misiurewicz did, I did not limit the extent of the parameter value for the study). In the next few years I attended two meetings on the iteration theory: the first one on 21–23 May, 1979 at La Garde Freinet (a small town in the south of France) where Michel Hénon was also present and where Michal Misiurewicz, after some questions at the end of my talk (the purpose of which was the computation of homoclinic points of the map), came to the blackboard to give some clues of the forthcoming results of the New York meeting.

The second meeting was at a summer school in physics in July 1979 in Cargèse (Corsica), in the proceedings of which I eventually published the article entitled: "Strange attractors: a class of mappings of $\mathbb{R}^2$ which leaves some Cantor Set invariant". In this paper I used the genuine non-differentiable map in order to prove the existence of one homoclinic point for a smooth version of the Lozi map and then applying a theorem of Stephen Smale, I proved the existence of an invariant Cantor set. After that took place the congress organized by the New York Academy of Science where I am proud to have shook hands with Edward Lorenz, the father of strange attractors and I listened with a mix of anxiety and curiosity the first proof of existence of a strange attractor for an analytically given map of the plane. After Misiurewicz's work, hundreds of papers were published on countless aspects of this strange attractor, as has been in this book.

Today if we go back in thought in the late 1970s, in some aspects, life was very different from now. There was no personal computer (M. Hénon used one of the only two computers of the university of Nice, an IBM 7040, in order to plot the figure of his original paper), no internet, no wireless phone. Also, communications between researchers were done through slow post office mail, travels by air were very expensive, limiting personal contact between researchers in the west countries. Moreover, the Berlin Wall was still standing. James Yorke who coined the term "chaos" in his famous paper with his student Tien-Yien Li "Period three implies Chaos" in 1975, was unaware of Alexander Sharkovkii's theorem published in Russian in 1964 displaying more penetrating results on periodic orbits (however essential notion as sensitive dependence on initial conditions is only introduced in the paper of Li and Yorke). The technological progress in the last thirty years has been dramatic in all aspects of life. In contrast, the progress of mathematics is very slow. Near my entire professional life of mathematician has been needed to see published results I expected proved in few months. New results, as for example: "Topological entropy for the Lozi maps can jump from zero to a value above 0.1203 as one crosses a particular parameter and hence it is not upper semi-continuous in general" (I. B. Yildiz), or "Certain Lozi-like maps have the orbit-shifted shadowing property" (A. Sakurai) are continuously being published every year.

A tentative title first chosen for the book by Dr. Elhadj Zeraoulia was "The power of chaos" which reflects a part of what one can observe in numerous publications of the last few years. If on one hand, new theoretical important results are still regularly found as mentioned above, on the other hand applications of Lozi map are soaring—in engineering, computers, communications, control, medicine and biology and especially in the domain of evolutionary algorithms. In the last chapter of the book, due to the limitation of the number of pages, only some real-world applications are given. However, they allow the reader to get an idea of the power that holds the mastery of chaos generated by the Lozi mappings.

In the scope of evolutionary algorithms the use of chaotic sequences instead of random ones was introduced ten years ago by Caponetto et al. Several traditional chaotic maps in 1- or 2-dimensions are used. The difference between them is based on the two main differences: the shape of the invariant measure and the robustness with respect to numerical computation. The Lozi map is recognized to show better performance due to its shadowing property. These algorithms are particularly efficient in global optimization problems. Since 1975, when I was a young assistant professor, I faced disappointment for years in the search of a classical algorithm for solving such problems. What is funny is to notice how a map I introduced for an entirely different purpose, is routinely being now used to solve this problem.

Finally thanks to the work of Prof. Elhadj Zeroualia, I can go back and look at how a very small idea has blossomed into an area that still fascinates me: mathematics.

October 2012   **Prof. René Lozi**
Laboratory of Mathematics J. A. Dieudonné
University of Nice-Sophia Antipolis, France

Comprehensive Review of Hyperbolicity, Ergodicity, and Chaos

This chapter presents some definitions and relevant results about the statistical properties and classification of chaotic attractors. Chaos can be defined as follows: a chaotic strange attractor is a non-trivial attracting set that contains a dense orbit with at least one positive Lyapunov exponent. Generally, chaos refers to the existence of a *Smale horseshoe* with a hyperbolic structure. In Section 1.1 several criterions for measuring chaos in dynamical systems are presented and discussed in some details. This includes: the Lebesgue measure, the physical (or Sinai-Ruelle-Bowen) measure, the Hausdorff dimension, the topological entropy, Lyapunov exponents, ergodic theory and its importance in quantifying and understanding behavior of chaotic attractors. Different correlations are defined such as autocorrelation function (ACF) discussed in Section 1.1.7 and the decay of correlations presented in Section 1.1.9.1 and the Central Limit Theorem in Section 1.1.9.2. A hyperbolicity test is given in Section 1.1.8 in order to clarify the nature of chaos in a Hénon map. A classification of strange attractors of dynamical systems is given in Section 1.2. This classification contains the so-called *hyperbolic attractors, Lorenz-type attractors and quasi-attractors.*

1.1 Entropies and statistical properties of chaotic attractors

In this section, we will discuss briefly some statistical properties of chaotic attractors such as different types of entropies, ergodic theory, the rate of decay of correlations, the central limit theorem and other probabilistic limit theorems. The proof of these properties are based on Markov approximations which is an alternative to the conventional *Perron-Frobenius operator techniques* applied to dynamical systems.

1.1.1 The Lebesgue measure

The Lebesgue measure is a way of assigning a length, area or volume to subsets of Euclidean space. These sets are called *Lebesgue measurable*. For a dynamical system the invariant measure describes the distribution of the sequence formed by the solution obtained for any typical initial state. The mathematical formulation is given in the following definition:

Definition 1 *(a) A box in $\mathbb{R}^n$ is a set of the form $B = [a_1, b_1] \times [a_2, b_2] \times ... \times [a_n, b_n]$, where $b_i \geq a_i$.*
(b)   The volume of the box B is vol $(B) = (b_1 - a_1) \times (b_2 - a_2) \times ... \times (b_n - a_n)$.
(c)   The outer measure m^* (A) for a subset $A \subset \mathbb{R}^n$ is defined by:

$$m^* (A) = \inf \{Sum_{j \in J} vol (B_j) : \text{with the } \textbf{property B}\} \qquad (1.1)$$

Property B: *The set $\{B_j : j \in J\}$ is a countable collection of boxes whose union covers A.*
(d)   The set $A \subset \mathbb{R}^n$ is Lebesgue measurable[1] if

$$m^* (S) = m^* (S \cap A) + m^* (S \backslash A) \text{ for all sets } S \subset \mathbb{R}^n. \qquad (1.2)$$

(e)   The Lebesgue measure is defined by $m(A) = m^*(A)$ for any Lebesgue measurable set A.

Note that $m^* (S \backslash A) = 0$ in (1.2) if the set A is a topologically transitive attractor of a dynamical system.

1.1.2 The physical (or Sinai-Ruelle-Bowen) measure

The physical or Sinai-Ruelle-Bowen (SRB) measure was defined and proved for the first time for the so-called *Anosov diffeomorphisms* introduced in Section 1.2.1, and also for hyperbolic diffeomorphisms and flows studied in Sinai (1972), Ruelle (1976), Ruelle & Bowen 1975(a-b). The common definition of the SRB measure is given for discrete time systems by:

Definition 2 *Let A be an attractor for a map f, and μ be an f-invariant probability measure of A. Then μ is called SRB measure for (f, A) if one has for any continuous map g, i.e.*

$$\lim_{n \to \infty} \frac{1}{n} \sum_i g(f^i(x)) = \int g d\mu \qquad (1.3)$$

---

[1]We have $m^* (S \backslash A) = 0$ if set A is a topologically transitive attractor of a dynamical system.

where $x \in E \subset B(A)$ The basin of attraction of A, with m (E) > 0, where m is the Lebesgue measure.

The invariant measure means that the "*events*" $x \in A$ and $f(x) \in A$ have the same probability. This measure is defined as follows:

Definition 3 *(a) A probability measure μ in the ambient space M is invariant under a transformation f if $\mu(f^{-1}(A)) = \mu(A)$ for all measurable subsets A.*

(b) The measure μ is invariant under a flow f_t if it is invariant under f_t for all t.

(c) An invariant probability measure μ is ergodic if every invariant set A has either zero or full measure, or equivalently, μ is ergodic if it cannot be decomposed as a convex combination of invariant probability measures, that is, one cannot have $\mu = a\mu_1 + (1 - a)\mu_2$ with $0 < a < 1$ and μ_1, μ_2 are invariant.

A widely used definition of the concept of the SRB measure is given by:

Definition 4 *(a) An f-invariant probability measure μ is called SRB if and only if:*

(i) μ is ergodic,

(ii) μ has a compact support, and

(iii) μ has absolutely continuous conditional measures on unstable manifolds.

(b) The set Λ carries a probability measure μ if and only if $\mu(\Lambda) = 1$.

(1) This definition was used [Kiriki & Soma (2008)] to investigate the coexistence of homoclinic sets with/without SRB measures in Hénon maps (2.1) below.

(2) The existence of an invariant measure was proved theoretically [Sinai (1970–1979), Ruelle (1976–1978), Bunimovich & Sinai (1980), Eckmann & Ruelle (1985)] for hyperbolic and *nearly* hyperbolic systems.

(3) The results about existence and smoothness of these measures can be found [Jarvenpaa & Jarvenpaa (2001), Sanchez-Salas (2001), Jiang (2003), Bonetto, et al. (2004), Gallavotti, et al. (2004), Bonetto, et al. (2005), Amaricci, et al. (2007)].

(4) For the non-existence of such a measure, one can see for example the case studied [Hu & Young (1995)].

1.1.3 The Hausdorff dimension

It is defined as follows:

Definition 5 *Let A be an attractor for a map f. If $S \subset A$ and $d \in [0, +\infty)$, the d-dimensional Hausdorff content of S is defined by:*

$$C_H^d(S) = \inf \left\{ \sum_i r_i^d : \text{There is a cover of S with balls with radii } r_i > 0 \right\}. \quad (1.4)$$

The Hausdorff dimension of A is defined by:

$$D_H(A) = \inf \{d \geq 0, C_H^d(S) = 0\} \tag{1.5}$$

Definition 5 implies that the Hausdorff dimension of the Euclidean space $\mathbb{R}^n$ is n, for a circle $\mathbb{S}^1$ it is 1 and for a countable sets it is 0. The Hausdorff dimension of the *middle third Cantor set* is $\frac{\ln 2}{\ln 3} = 0.630\ 93$ and $\frac{\ln 3}{\ln 2} = 1.585\ 0$ for the *Sierpinski triangle*.

1.1.4 The topological entropy

The topological entropy of a dynamical system is a non-negative real number that measures the complexity of the system. This entropy can be defined in various equivalent ways, one of them is given in the following definition:

Definition 6 *(a) Let A be a compact Hausdorff topological space. For any finite cover C of A, let the real H(C) be defined by:*

$$H(C) = \log_2 \inf_j \{j = \text{The number of elements of C that cover A}\}. \tag{1.6}$$

For two covers C and D, let C ∨ D be their (minimal) common refinement, which consists of all the non-empty intersections of a set from C with a set from D, and similarly for multiple covers. For any continuous map f : A → A, the following limit exists:

$$H(C, f) = \lim_{n \to +\infty} \frac{1}{n} H(C \vee f^{-1} \vee ... \vee f^{-n+1}C) \tag{1.7}$$

Then, the topological entropy h(f) of f is the supremum of H(C, f) overall possible finite covers C, i.e.,

$$h(f) = \sup_C \{H(C, f), C \text{ is a finite cover of A}\} \tag{1.8}$$

(b) The topological dimension (Hurewicz and Wallman, 1984) of a space E is either −1 (if E = ∅) or the last integer k for which every point has arbitrarily small neighborhoods whose boundaries have a dimension less than k.

In a topological sense, a dynamical system is called *chaotic* if its topological entropy is positive. A more appropriate definition for the topological entropy can be formulated by using the notion of (n, ε)-separated sets and interval arithmetic. Basically, the method of calculation uses the number of periodic orbits of such a map in a specific range.

Definition 7 *(a) Let f = X → X be a map. A set E ⊂ X is called (n, ε)-separated if for every two different points x, y εE, there exists $0 \leq j < n$ such that the distance between $f^j(x)$ and $f^j(y)$ is greater than ε. Let us define the number $s_n(\varepsilon)$ as the cardinality of a maximum (n, ε)-separated set:*

$$s_n(\epsilon) = max \{card\ E : E\ is\ (n,\ \epsilon)\text{-}separated\} \tag{1.9}$$

The number H (f) defined by:

$$H(f) = \lim_{\epsilon \to 0} \limsup_{n \to \infty} \frac{1}{n} \log s_n(\epsilon) \tag{1.10}$$

is called the topological entropy of f.
  (b) *If f is an Axiom A diffeomorphism then*

$$H(f) = \limsup_{n \to \infty} \frac{\log C(f^n)}{n} \tag{1.11}$$

where C (f^n) is the number of fixed points of f^n (periodic orbits of f). See (Bowen, 1971).

Now, the formula:

$$H(f) = \frac{\log C(f^n)}{n} \tag{1.12}$$

can be used as the lower bound for the topological entropy $H(f)$ when the distance between the periodic orbits of length n is uniformly separated from zero, i.e., the formula for sufficiently big n. For example, the upper and lower bounds for the topological entropy of the Hénon map (2.1) was estimated based on the works given in Misiurewicz & Szewc (1980), Zgliczynski (1997(a), Galias (1998(b)), and Stoffer & Palmer (1999) using the interval technique and the following result:

Theorem 1 *The topological entropy H(h) of the Hénon map h (x, y) = (1−ax² + y, bx), given also by (2.1), located in the interval:*

$$0.3381 < H(h) \leq \log 2 < 0.6932 \tag{1.13}$$

We notice that there are also more recent methods which increase the lower bound of the topological entropy of the Hénon map upto 0.4646 as shown in Newhouse, et al. (2008). In fact, a rigorous lower bound for the topological entropy of planar diffeomorphisms was given in terms of the geometry of finite pieces of stable and unstable manifolds of hyperbolic periodic points.

1.1.5 Lyapunov exponent

The Lyapunov exponent of a dynamical system with evolution equation f_t in an n-dimensional phase space is a quantity that characterizes the rate of separation of infinitesimally close trajectories. The standard definition of the Lyapunov exponents for a discrete n-dimensional mapping is given by the following definition:

Definition 8 *Consider the following n-dimensional discrete dynamical system:*

$$x_{k+1} = f(x_k), \ x_k \in \mathbb{R}^n, \ k = 0, 1, 2, \ldots \tag{1.14}$$

where $f : \mathbb{R}^n \rightarrow \mathbb{R}^n$, is the vector field associated with system (1.14), let $J(x)$ be its Jacobian evaluated at x, also define the matrix:

$$T_r(x_0) = J(x_{r-1}) J(x_{r-2}) \ldots J(x_1) J(x_0). \tag{1.15}$$

Moreover, let $J_i(x_0, l)$ be the modulus of the ith eigenvalue of the lth matrix $T_r(x_0)$, where $i = 1, 2, \ldots, n$ and $r = 0, 1, 2, \ldots$
Now, the Lyapunov exponents of a n-D discrete time systems are defined by:

$$\lambda_i(x_0) = \ln\left(\lim_{r \rightarrow +\infty} J_i(x_0, r)^{\frac{1}{r}} \right), i = 1, 2, \ldots n \tag{1.16}$$

Some characteristics of the Lyapunov exponents can be summarized as follow: Lyapunov exponents can be considered as functions in the corresponding orbit $(x_r)_{0 \le r < \infty}$ as shown by relations (1.15) and (1.16). Generally, these exponents are not continuous functions (in the variable $(x_r)_{0 \le r < \infty}$) and they need some conditions of existence (in the sense that these exponents are bounded) of the solution under consideration. Also, the spectrum of Lyapunov exponents depends on the starting point x_0 and for conservative systems a volume element of the phase space will stay the same along a trajectory, i.e., $\sum_{i=1}^{n} \lambda_i = 0$. If the system is dissipative, then $\sum_{i=1}^{n} \lambda_i < 0$. The Lyapunov spectrum can be used to estimate the so called *Kaplan-Yorke dimension* D_{KY} defined as follows:

$$D_{KY} = k + \sum_{i=1}^{k} \frac{\lambda_i}{|\lambda_{k+1}|} \tag{1.17}$$

where k is the maximum integer such that the sum of the k largest exponents is still non-negative, and in this case D_{KY} is an upper bound for the *information dimension* of the system as shown in [Kaplan & Yorke (1987)]. Also, the sum of all the positive Lyapunov exponents gives an estimate of the *Kolmogorov-Sinai entropy* [Pesin (1977)]. Also, we note that the analytic calculation of Lyapunov exponents using the matrix $T_r(x_0)$ is not always possible. In fact, there is a large number of methods and algorithms that deal with numerical estimations [Shimada & Nagashima (1979), Benettin, et al. (1980), Sprott (2003)]. In some situation, it is important to determine the upper and the lower bounds for all the Lyapunov exponents of a given n-dimensional system. A recent result about this topic was given in [Li & Chen (2004)] for discrete mappings as follow:

Theorem 2 *If a system* $x_{k+1} = f(x_k)$, $x_k \in \Omega \subset \mathbb{R}^n$, *verify*

$$\|Df(x)\| = \|J\| = \sqrt{\lambda_{max}(J^T J)} \leq N < +\infty, \tag{1.18}$$

with a smallest eigenvalue of $J^T J$ *that satisfies:*

$$\lambda_{min}(J^T J) \geq \theta > 0, \tag{1.19}$$

where $N^2 \geq \theta$, *then, for any* $x_0 \in \Omega$, *all the Lyapunov exponents at* x_0 *are located inside* $[\frac{\ln\theta}{2}, \ln N]$. *That is,*

$$\frac{\ln\theta}{2} \leq l_i(x_0) \leq \ln N, \, i = 1, 2, \ldots, n, \tag{1.20}$$

where $l_i(x_0)$ *are the Lyapunov exponents for the map f.*

The range of examples covered by Theorem 2 for demonstrating positive Lyapunov exponents include some *coupled lattices maps* studied extensively in chemical reactions, pattern recognition, biology,...etc. One the these maps is the *L-dimensional globally coupled map* considered in [Ding & Yang (1997)] as follow: $x_{k+1}(i) = (1 - \varepsilon)f(x_k(i)) + \frac{\varepsilon}{L}\sum_{j=1}^{L} f(x_k(j))$. Theorem 2 implies that this map is chaotic if $l_1 > -\ln(1 - \epsilon)$, where l_1 is the first Lyapunov exponent. See [Ding & Yang (1997)] for more details. The second example is the 2-D piecewise linear map $f(x, y) = (x - ah(y), bx)$ studied in [Zeraoulia & Sprott (2009)] where a and b are the bifurcation parameters, $h(x) = \frac{2m_1 x + (m_0 - m_1)(|x+1| - |x-1|)}{2}$ is the characteristic function of the so-called *double scroll attractor* [Chua et al. (1986)], where m_0 and m_1 are respectively the slopes of the inner and outer sets of the original Chua circuit. Theorem 2 implies that this map is chaotic if $|a| > \max\left(\frac{1}{|m_1|}, \frac{1}{|m_0|}\right)$ and $|b| > \max$ $\left(\frac{|am_1|}{\sqrt{a^2 m_1^2 - 1}}, \frac{|am_0|}{\sqrt{a^2 m_0^2 - 1}}\right)$.

1.1.6 Ergodic theory

The ergodic theory was motivated by problems of statistical physics. The most important results in ergodic theory are the ergodic theorems of Birkhoff and von Neumann. Its central aspect is the investigation of the behavior of a dynamical system when it is allowed to run for a long period of time taking into account that the time average is the same for almost all initial points and that ergodic systems has stronger properties, such as mixing and equidistribution. The domains of applications of ergodic theory are, for example: the study of the problem of metric classification of systems, stochastic processes, study of the geodesic flow on Riemannian manifolds,

Markov chains, harmonic analysis, Lie theory and number theory....etc. The important concepts and theorems of ergodic theory are discussed in detail in [Kalikow (2009)]: Here, we give only some basic definitions and results: (1) Isomorphism: An *isomorphism* is a bijective map f such that both f and its inverse f^{-1} are homomorphisms, i.e., structure-preserving mappings. From the general setting of category theory point of view, an isomorphism is a morphism $f : X \to Y$ in a category for which there exists an "inverse" $f^{-1} : Y \to X$, with the property that both $f^{-1} \circ f = id_X$ and $f \circ f^{-1} = id_Y$, where id_X and id_Y are identity mappings of X and Y respectively. (2) Independent process: a stationary process on an alphabet in which all letters are completely independent of each other is called *an independent process.* (3) Ornstein isomorphism theorem: Two stationary independent processes are isomorphic if and only if they have the same entropy. (4) Ergodic process: An ergodic process or transformation is a process (or transformation) which cannot be written as a linear combination of two other processes (or transformations). (5) Birkhoff ergodic theorem: This fundamental result is described by Theorem 3 below and it means that the frequency of time that process spends in a given set is the measure of that set when an ergodic transformation is repeatedly applied to form an ergodic process. (6) Entropy: different types of entropies are described with some details in Section 1.1. For more details, see [Birkhoff (1931), von Neumann (1932(a-b)), Hopf (1939), Birkhoff (1942), Fomin & Gelfand (1952), Mautner (1957), Moore (1966), Arnold & Avez (1968),Walters (1982), Petersen (1990), Bedford (1991), Breiman (1992), Rosenblatt & Weirdl (1993), Anosov (2001)].

1.1.7 Autocorrelation function (ACF)

Let $x(t)$ be a process ($x(t)$ can be a solution of a dynamical system). Let $x_i = x(t_i)$ be the values of the process $x(t)$ at time $t_1, t_2, ..., t_N$, where N is the number of realizations of the $x(t)$ process. The autocorrelation function (ACF) was introduced in [Box & Jenkins (1976)] and it is used to detect non-randomness in data and to identify an appropriate time series model if the data are not random. Thus, it suffices to consider only the first (lag 1) autocorrelation, but if the purpose is the identification of an appropriate time series model, then the autocorrelations are usually plotted for many lags. In the case of constant location and scale, randomness, and fixed distribution A_0 have acceptable values and the univariate process Y_i can be modeled as:

$$Y_i = A_0 + E_i \tag{1.21}$$

where E_i is an error term. Model (1.21) is not valid if the randomness assumption is not valid. Generally, the necessary model is either a time series model or a time-independent variable non-linear model [Box & Jenkins (1976), pp. 28–32]. Checking randomness is done by autocorrelation plots by computing autocorrelations for data values at varying time lags. Thus, a model is random if these autocorrelations are near zero for any and all time-lag separations and the model is non-random if one or more of the autocorrelations is significantly non-zero. The evaluation of the ACF for chaotic system or for a chaotic process $x(t)$ can be done by a practical algorithm developed in [Anishchenko et al. (2004)] as follows:

1. Choose a sufficiently large number of realizations, N, and calculate the time-average value $\bar{x}$ using the formula:

$$\bar{x} = \frac{1}{N} \sum_{i=1}^{N} x(t_i) \tag{1.22}$$

2. Calculate the mean product $x(t)x(t + \tau)$ by averaging over time:

$$K_l(\tau) = \frac{1}{p} \sum_{i=1}^{p} x(t_i) \, x(t_i + k\Delta t), \, \tau = k\Delta t_i, \, k = 0, 1, ..., n - p \tag{1.23}$$

where $l = 1, ...,N$ is the number of realization. In this case, the limitation of a number of $x(t_i)$ values, $i = 1, 2, ...,N$ imply that the time-averaging converged if the number of averagings p is sufficiently large which gives the larger value of p, and the smaller value of the time $\tau_{max} = (n-p)\Delta t$, for the ACF estimation. Thus, the ACF must be computed on a very large time interval since, the rate of correlation splitting is not high in the regime being considered. Hence, the value of p must be not too large. Finally, to attain a high precision of the ACF calculation, the obtained data must be further averaged over N realizations, i.e.,

$$\psi(\tau) = \frac{1}{N} \sum_{l=1}^{N} K_l(\tau) - \bar{x}^2 \tag{1.24}$$

3. Normalize the ACF on its maximal value at $\tau = 0$, that is,

$$\Psi(\tau) = \frac{\psi(\tau)}{\psi(0)} \tag{1.25}$$

4. Plot $\ln \Psi(\tau)$ versus τ.

An example of the above algorithm can be found in [Anishchenko, et al. (2004)] using the Lorenz system with noise given by: $x' = \sigma(y - x) + \sqrt{2D}\xi(t), y' = rx - y - xz, z' = -bz + xy$. Here, $\xi(t)$ is a *Gaussian white noise source* with the mean value $\langle \xi(t) \rangle = 0$ and correlation $\langle \xi(t) \xi(t + \tau) \rangle = \delta(\tau)$, where $\delta()$ is the Dirac's function. For this system, we choose two different regimes, the first is a quasi-hyperbolic attractor obtained for $\sigma = 10$; $b = 8/3$, and $r = 28$ and the second is a non-hyperbolic attractor obtained for $\sigma = 10$; $b = 8/3$, and $r = 210$. If $\Psi_0(\tau)$ is the envelope of $\Psi(\tau)$, then it is better to plot $\Psi_0(\tau)$ versus τ to make figures more informative and compact as shown in Fig. 1.1.

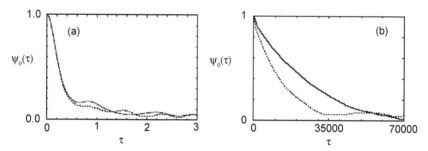

Figure 1.1: Envelopes of the normalized autocorrelation function $\Psi_0(\tau)$ for attractors in the Lorenz system with noise. (a) $r = 28$ and $D = 0$ (solid line), and $D = 0.01$ (dotted line); (b) $r = 210$, $D = 0$ (solid line), and $D = 0.01$ (dotted line). Reused with permission from Anishchenko, V. S., Vadivasova, T. E., Strelkova, G. I. and Okrokvertskhov, G. A., Math. Biosciences. Engineering (2004). Copyright 2004, American Institute of Mathematical Sciences Publishing.

1.1.8 Hyperbolicity tests

Detecting hyperbolic (see Section 1.2.1 below) and non-hyperbolic behaviors for chaotic systems is the subject of many works. For example, see [Abraham & Smale (1970), Newhouse (1979), Robinson (1983), Grebogi, et al. (1987-1988), Dawson, et al. (1994), Alligood, et al. (2006), Ott, et al. (1992), Kubo, et al. (2008), Lai, et al. (1996), Viana, et al. (2004), Pereira, et al. (2007), Hirsch & Pugh (1970), Newhouse & Palis (1971), Moser (1973), Palis & Takens (1993), Diamond, et al. (1995), Robinson (1999), Anishchenko, et al. (2000(ab)), Arai, (2007a), Mazure, (2008), Mazur & Tabor, (2008), Mazur, et al. (2008), Newhouse, (2004), Bedford & Smillie, (2006), Hruska, (2006(a-b))] and references therein. In this section, we present the method developed in [Lai, et al. (1993)] for calculating the angles between the stable and unstable directions of a given map: The angles between stable and unstable manifolds are used to detect hyperbolic and non-hyperbolic behaviors for chaotic systems. A whole picture of this procedure can be seen in what follows. Generally, this angle of a map f depending on two bifurcation parameters a and b is defined as follows:

Definition 9 (a) *We say s is the stable direction for a map f at x if there is a constant* $0 < K < 1$ *such that:*

$$\| Df^n(x)(s) \| \leq K^n \|s\| \text{ as } n \to \infty \tag{1.26}$$

(b) *We say u is the unstable direction at x if:*

$$\| Df^{-n}(x)(u) \| \leq K^n \|u\| \text{ as } n \to \infty \tag{1.27}$$

Let $\theta\,(x, a, b) \in [0, 2\pi]$ be the angle between s and u at x in the invariant set. In some cases, $\theta\,(x, a, b) = 0$ if s and u are identical. Let x_0 be an initial condition and

$$\theta_{\text{inf}}\,(x_0, a, b) = \inf_{i=0,1,2,\ldots,\infty}\,\theta\,(f^{\,i}\,(x_0), a, b) \tag{1.28}$$

be the lower bound of the angle of a trajectory $x_{n+1} = f\,(x_n, a, b)$. Due to the limitation of the number of iterates, one must replace (1.28) by:

$$\theta_m\,(x_0, a, b, N) = \min_{i=0,1,2,\ldots,N}\,\theta\,(f^{\,i}\,(x_0), a, b) \tag{1.29}$$

where $\theta\,(f^{\,i}\,(x_0), a, b)$ is the angle at a point, $\theta_m\,(x_0, a, b, N)$ is the minimum of the angle for $(N + 1)$ points of the trajectory. If the dependence on initial data was neglected, then one can write $\theta_{\text{inf}}\,(a, b)$ and $\theta_m\,(a, b, N)$ instead of $\theta_{\text{inf}}\,(x_0, a, b)$ and $\theta_m\,(x_0, a, b, N)$, i.e.,

$$\theta_{\text{inf}}\,(a, b) = \lim_{N \to \infty}\,\theta_m\,(a, b, N) \tag{1.30}$$

The numerical procedure given in [Lai, et al. (1993)] for calculating the angles is as follows:

Let $\theta\,(x) = \theta\,(x, a, b)$ be the angles between the stable and unstable directions of a map f for points x at fixed parameter values along a trajectory on the chaotic set.

Step 1: Calculate a single orbit on the chaotic set using direct iterations of the map f or by using the *PIM-triple method* described in [Hsu, et al. (1988), Nusse & Yorke (1989)].

Step 2: Calculate the stable and unstable directions for each point x on the chaotic set:

(a) Calculate the stable direction at point x.

(a-1) Iterate the point x forward under the map f, N times (It is possible to use $N = 100$, but $N = 20$ is quite adequate) to get a trajectory $f^1(x), f^2(x), \ldots, f^N(x)$.

(a-2) Consider a circle of radius $\varepsilon > 0$ on the point $f^N(x)$, and iterate this circle backward once. This gives an ellipse at the point $f^{N-1}(x)$ with the major axis along the stable direction of the point $f^{N-1}(x)$.

(a-3) Iterating this ellipse backwards N times and keeping the ellipse's major axis of order ε via some necessary normalizations. In this case, all the way back to the point x, and the ellipse becomes very *thin* with its major axis along the stable direction at point x. In practice, a unit vector at the point $f^N(x)$ was used instead of a small circle (for sufficiently large N, the unit vector we get at point x is a good approximation of the stable direction at x). Hence, all the steps in this case are (a-3-1) Iterate the unit vector backward to point x by multiplying by the Jacobian matrix $D f^{-1}$ of the inverse map f^{-1} at each point on the already existing orbit (the Jacobian matrix Df^{-1} was calculated by storing the inverse Jacobian matrix at every point of the orbit $f^i(x)$ ($i = 1, ..., N$) when one iterate forward the point x beforehand).

(a-3-2) Normalize the vector after each multiplication to the unit length. (b) Calculate the unstable direction at point x.

Step 3: Choose $0 \leq \theta(x) \leq 2\pi$ to be the smaller of the two angles defined by the two straight lines along the stable and unstable directions at x.

We notice that a computer programs which do these steps automatically can be found in [Nusse and Yorke, (1997)]. Generally, this *Dynamics* program can be helpful to mention.

For the Lozi map (2.1) the same algorithm can be applied. The result is shown in Fig. 1.3(a) where the distribution $P(\varphi)$ of the angle φ between the manifolds of a chaotic orbit of the Lozi map (2.1) is shown. We remark that there is some minimal and bounded values φ_{min} of the angle φ obtained for $a = 1.7$ for $b = 0.3$ and shown in Fig. 1.3(b). Indeed, for $1.5 \leq a \leq 1.8$ one has $\varphi_{min} > 400$ and is never equal to 0.

Generally, in 2-D invertible dissipative map, the following properties are characteristic for the *Lorenz-type attractors* [Anishchenko, et al. (1998)]:

(1) The transversality is a necessary condition for the attractor to be a Lorenz-type attractor.
(2) The probability density of the angle between separatrices of a chaotic orbit of the attractor is strictly equal to 0 in the neighborhood of a zero value of the angle, i.e., $p(\phi) = 0$ for $\phi \approx 0$ and $p(\phi) > 0$ for $0 < \phi < \phi_0$ (see Fig. 1.3(a)).
(3) The dependence of the largest Lyapunov exponent λ_1 is a smooth positive definite function on the parameter in the region where the attractor exists, i.e., $\lambda_1(a) > 0$, $a_1 < a < a_2$ as shown in Fig. 1.3(c)).
(4) The power spectrum of Lorenz-type attractors depends smoothly enough on the frequency and does not contain any pronounced peaks.

1.1.9 Correlations

In this section, we present some relations between the concept of correlations and the *Strong law of large numbers* (SLLN), the *correlations decay* and the *Central Limit Theorem* properties. Indeed, let $f : X \rightarrow X$ be a discrete map considered as a deterministic dynamical system. A *preserving probability measure, m,* on X verify the following property: for any measurable subset $A \subset X$ one has $m(A) = m(f^{-1}(A))$, where $f^{-1}(A)$ is the set of points mapped into

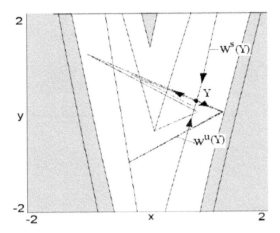

Figure 1.2: Stable and unstable manifolds of saddle points in the Lozi map (3.1) for $a = 1.7$ and $b = 0.3$. Reused with permission from Anishchenko, V.S., Vadivasova, T. E., Strelkova, G. I. and Kopeikin, A. S., Discrete Dynamics in Nature and Society (1998). Copyright 1998, Hindawi Publishing Corporation.

A. Note that the states $x_n \in X$ are not observable due to some errors effect. Thus, any function $F : X \to \mathbb{R}$ can be seen as a random variable on X called an *observable can be* used to generate a sequence $F_n = F(x_n)$. This means that for each n, the function $F_n = F \circ f^n$ is a random variable and the sequence $\{F_n\}$ is a stationary stochastic process since the measure m is invariant. Now, define the space average $\mu_F = m(F)$. Then we have the following standard results [Rosenthal (2000)]:

Theorem 3 *(a) (The classical Birkhoff ergodic theorem) for m-almost every initial state $x_0 \in X$ the time averages converge to the space average, i.e.,*

$$\frac{F_0 + F_1 + ... + F_{n-1}}{n} \to \mu_F = m(F) \tag{1.31}$$

(b) Birkhoff ergodic theorem can be stated using the partial sums of the observed sequence F_n given by $S_n = F_0 + F_1 + ... + F_{n-1}$ as

$$\frac{S_n - n\mu_F}{n} \to 0, \text{ i.e., } S_n = n\mu_F + O\left(\sqrt{n}\right). \tag{1.32}$$

if in addition each F_i has finite variance σ_i^2.

Proposition 1.1 *If the observable F is square integrable, i.e., $m(F^2) < \infty$, then the random variables F_n have finite mean value μ_F and variance σ_F^2 given by:*

$$\begin{cases} \mu_F = m(F) = \int_X F dm \\ \sigma_F^2 = m(F^2) - (m(F))^2 \end{cases} \tag{1.33}$$

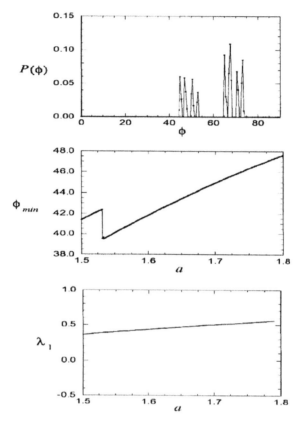

Figure 1.3: Calculation results of the characteristics for the Lozi attractor (3.1). (a) A distribution of the probabilities of the angle ϕ between stable and unstable manifolds for $a = 1.7$ and $b = 0.3$, (b) a plot of the minimal angle ϕ_{min} versus parameter a for $b = 0.3$ and (c) a dependence of the largest Lyapunouv exponent on a for $b = 0.3$. Reused with permission from Anishchenko, V.S., Vadivasova, T. E., Strelkova, G. I. and Kopeikin, A. S., Discrete Dynamics in Nature and Society (1998). Copyright 1998, Hindawi Publishing Corporation.

The quantity σ^2_F characterizes only one random variable F.

Now, to define the so called *Strong law of large numbers* (SLLN) related to the convergence property, we use the concept of *correlations* defined by the mean of the covariances $C_F(n)$ called *autocorrelations* given by:

$$C_F(n) = m(F_0 F_n) - \mu^2_F = m(F_k F_{n+k}) - \mu^2_F, \text{ for any } k \qquad (1.34)$$

The quantities $C_F(n)$ are called *correlations* even without normalization assumption, i.e., $\sigma^2_F = 1$. More generally, for any two square-integrable observables F and G the correlations are defined by:

$$C_{F,G}(n) = m(F_0 G_n) - \mu_F \mu_G = m(F_k G_{n+k}) - \mu_F \mu_G, \text{ for any } k. \qquad (1.35)$$

Now, if $\sigma^2_F = 1$, then $C_F(n)$ is the *correlation coefficient* between random variables F_k and F_{n+k}. If the system is chaotic, then for large n, the values of F_k, F_{n+k} are nearly independent, i.e., as n grows, the correlations decrease (the *decay* property).

1.1.9.1 The decay of correlations

There is a relation between correlations and mixing attractors. The decay of correlations is essential in the studies of *relaxation to equilibrium* in non-equilibrium statistical mechanics.

Definition 10 *The transformation f is said to be mixing if for any two measurable sets A, B ⊂ X one has:*

$$m\,(A \cap f^n\,(B)) \to m(A)\,m\,(B) \text{ as } n \to \infty. \tag{1.36}$$

In order to present a result about correlations decay for the class of squareintegrable observables, we need to define strongly (robust) and fragile chaotic attractors. Indeed, chaotic dynamical systems display two kinds of chaotic attractors: One type has *fragile chaos*, i.e., the attractors disappear with perturbations of a parameter or coexist with other attractors, and the other type has *robust chaos*, defined by the absence of periodic windows and coexisting attractors in some neighborhood of the parameter space. The existence of these windows in some chaotic regions means that small changes of the parameters would destroy the chaos, implying the fragility of this type of chaos. See [Zeraoulia & Sprott (2008)]. Also, sufficiently regular observables are those having sufficiently many *uniformly Hölder derivatives* (Recall Definition 11) in the flow direction. See [Dolgopyat (1998)] for more details. Thus, we have the following result:

Proposition 1.2 *(a) The map f is mixing if and only if correlations decay*

$$C_{F,G}\,(n) \to 0 \text{ as } n \to \infty. \tag{1.37}$$

(b) *The correlations decay rapidly if the system is strongly (robust) chaotic and the observables F and G are sufficiently regular.*

The relation between correlations and SLLN property can be seen as follow: Typical values of $S_n - n\mu_F$ are of order $\sqrt{n}$, i.e.,

$$S_n = n\mu_F + O\,(\sqrt{n}) \tag{1.38}$$

The approximation (1.38) can be obtained from the following relations:

$$
\begin{cases}
m[S_n = n\mu_F)]^2 = nC_F(0) + 2\sum_{k-1}^{n-1}(n-k)C_F(k) \\
\qquad \text{(the root-mean-square value)} \\
\sum_{n=0}^{\infty}|C_F(n)| < \infty \text{ (the correlations decay fast enough)}
\end{cases}
\tag{1.39}
$$

Thus, the following sum characterizes the entire process $\{F_n\}$ and it is defined by:

$$
\sigma^2 = \sum_{n=-\infty}^{\infty} C_F(n) = C_F(0) + 2\sum_{n=1}^{\infty} C_F(n)
\tag{1.40}
$$

is always non-negative and it is positive for generic observables F. Thus, we have that the mean square of $S_n - n\mu_F$ grows as follow:

$$
m[(S_n - n\mu_F)]^2 = n\sigma^2 + O(n).
\tag{1.41}
$$

1.1.9.2 The Central Limit Theorem

We have the following result:

Proposition 1.3 *(a) The observable F satisfies the Central Limit Theorem if the sequence $\frac{S_n\, n\mu_F}{\sqrt{n}}$ converges in distribution to normal law N $(0, \sigma^2)$, i.e., for every real $z \in (-\infty,\infty)$ one has*

$$
m\left(\frac{S_n\, n\mu_F}{\sqrt{n}} < z\right) \to \frac{1}{\sqrt{2\pi\sigma^2}} \int_{-\infty}^{z} \exp\left(\frac{-t^2}{2\sigma^2}\right) dt, \text{ as } n\to\infty.
\tag{1.42}
$$

(b) The central limit theorem holds if the correlations CF (n) decay fast enough, i.e., the asymptotics

$$
|C_F(n)| = O(n^{-(2+\varepsilon)}), \varepsilon > 0
\tag{1.43}
$$

As a standard example of strongly (robust) chaotic systems is the so called *angle doubling map* $f(x) = 2x$ modulo 1 of a circle and the *Arnold's cat* $(x, y) \to (2x + y, x + y)$ modulo 1 of the unit torus. For these maps correlations decay exponentially fast, i.e., $|C_{F,G}(n)| = O(n^{-na})$ for some $a > 0$, and Central Limit Theorem holds, if the observables F and G are Holder continuous, i.e.,

Definition 11 *A function F(x) is said to be Hôlder continuous if*

$$
|F(x) - F(y)| \le C(F)\,[\rho(x, y)]^\beta
\tag{1.44}
$$

where β > 0 is called the Hôlder exponent. More generally, let ξ be a partition of U into a finite number of domains separated by a finite number of compact smooth curves. Then let $H_\beta(\xi)$ be the class of functions that are Hôlder continuous (with the exponent β) within each of those domains. We say that such functions are piecewise Hôlder continuous (PHC) (with respect to the given partition ξ).

But, for continuous observables, correlations decay arbitrarily slowly and Central Limit Theorem fail. If chaos is fragile, then correlations decay polynomially, i.e., $|C_{F,G}(n)| = O(n^{-b})$ for some $b > 0$. Also, correlations decay exponentially fast and Central Limit Theorem holds for some examples of two dimensional maps called *generalized hyperbolic systems* described in Section 3.4.

More details about the subject of statistical properties of chaotic attractors, in particular, the rate of the decay of correlations, the central limit theorem and other probabilistic limit theorems can be found in [Ruelle (1968–1976), Sinai (1972), Bowen (1975(a-b)), Hofbauer & Keller (1982), Rychlik (1983), Denker (1989), Keller & Nowicki (1992), Liverani (1995), Chernov (1999), Young (1998(a)-1999), Benedicks & Young (2000), Szasz (2000), Markarian (2004), Balint & Gouezel (2006)].

1.2 Classification of strange attractors of dynamical systems

A common classification of strange attractors of dynamical systems was given on the basis of rigorous mathematical analysis and is not accepted as significant from the experimental point of view. Generally, in the present time, strange attractors can be classified into three principal classes [Anishchenko & Strelkova (1997), Plykin (2002)]: *hyperbolic, Lorenz-type or pseudo-hyperbolic,* and *quasi-attractors.*

1.2.1 Hyperbolic attractors

In this section, we give a short introduction to chaotic attractors with hyperbolic structure. Indeed, let $f: \Omega \ \mathbb{R}^n \to \mathbb{R}^n$ be a C^r real function that define a discrete map called also f and Ω is a manifold. Then, one has the following definition given in [Abraham & Marsden (1978)] :

Definition 12 *(a) A point x is a non-wandering point for the map f if for every neighborhood U of x there is $k \geq 1$ such that $f^k(U) \cap U$ is nonempty.*

(b) The set of all non-wandering points is called the non-wandering set of f.

(c) An f-invariant subset Λ of $\mathbb{R}^n$ satisfies $f(\Lambda) \subset \Lambda$.

(d) If f is a diffeomorphism defined on some compact smooth manifold $\Omega \subset \mathbb{R}^n$. An f-invariant subset Λ of $\mathbb{R}^n$ is said to be hyperbolic if there exists $0 < \lambda_1 < 1$ and $c > 0$ such that

(d-1) $T_\Lambda \Omega = E^s \oplus E^u$, *where* $\oplus$ *mean the algebraic direct sum.*

(d-2) $Df(x)E^s_x = E^s_{f(x)}$, *and* $Df(x)E^u_x = E^u_{f(x)}$ *for each* $x \in \Lambda$.

(d-3) $\|Df^k v\| \leq c\lambda^k_1 \|v\|$, *for each* $v \in E^s$ *and* $k > 0$.

(d-4) $\|Df^{-k}v\| \leq c\lambda^k_1 \|v\|$, *for each* $v \in E^u$ *and* $k > 0$.

where E^s, E^u are respectively, the stable and unstable sub-manifolds of the map f, i.e., the two Df-invariant sub-manifolds, and E^s_x, E^u_x are the two $Df(x)$-invariant sub-manifolds. For a compact surface, a result of Plykin imply that there must be at least three holes for a hyperbolic attractor and looks locally like a Cantor set (the stable direction) × interval (the unstable direction). Also, hyperbolic attractors have dense periodic points and a point with a dense orbit.

Definition 13 *A hyperbolic set* Λ *is locally maximal (or isolated) if there exists an open set* U *such that* $\Lambda = \bigcap_{n \in \mathbb{Z}} f^n(U)$.

The Smale horseshoe and the Plykin attractor are examples of locally maximal sets. In fact, if Λ is a hyperbolic set with nonempty interior for a map f, then f is Anosov if it is transitive, locally maximal and Ω is a surface. This type of attractors have the properties including: Shadowing property, structural stability, Markov partitions and the SRB measures. When $\Lambda = \Omega$, then the diffeomorphism f is called an *Anosov diffeomorphism* or *uniformly hyperbolic map*.

Definition 14 *The map f is uniformly hyperbolic or an Axiom A diffeomorphism if*

(a) *The non-wandering set* $\Omega(f)$ *has a hyperbolic structure,*

(b) *The set of periodic points of f is dense in* $\Omega(f)$*, i.e.,* $\overline{Per(f)} = \Omega(f)$*, the closure is the non-wandering set itself.*

Definition 14 implies that the homoclinic tangencies are excluded, and the resulting attractor is robust (structurally stable). Axiom A maps serve as models for the general behavior at a transverse homoclinic point, where the stable and unstable manifolds of a periodic point intersect, and they play a crucial role in the study of homoclinic bifurcations [Anosov (1969), Mãné (1982), Hayashi (1992)]. Note that the stable and unstable subspaces E^s and E^u are invariant and they depend continuously on the point x. In this case, not every manifold admits an Anosov diffeomorphism. In fact, the *hairy ball theorem* shows that there is no Anosov diffeomorphism on the 2-sphere. Also, we note that Anosov diffeomorphisms are rather rare and every known example of them is a generalization of automorphisms of a nul-manifold up to a topological conjugacy. For instant, the Smale horseshoe [Smale (1967)]. At this end, we note that all Anosov systems are expansive, i.e., there is a universal distance by which any two orbits will be separated at some time. Mathematically, there is a constant $\delta > 0$ such that for any points x and y, if $d(f^n(x), f^n(y)) \leq \delta$ for all n then $x = y$.

The most important properties of Anosov diffeomorphisms are:

1. (The *cone criterion*, or the Alekseev cone criterion) [Anosov (1969)]: At each point x the map requires the existence of two complementary closed sectors denoted by $C_\alpha^s(x)$ and $C_\alpha^u(x)$ in the tangent space $T_x\Omega$ that are strictly invariant in the following sense: There is a $\gamma \in (0, 1)$ such that:

$$Df(C_\alpha^u(x)) \subset C_{\gamma\alpha}^u(f(x)) \text{ and } Df^{-1}(C_\alpha^s(x)) \subset C_{\gamma\alpha}^s(f^{-1}(x)) \qquad (1.45)$$

where the cone $C_\alpha^s(x)$ is defined to be the set of vectors in the tangent space at x that make an angle less than α with E^s, and similarly for E^u. The stable and unstable subspaces are given by:

$$\begin{cases} E^s = \cap_{n \in \mathbb{N}} Df^{-n}(C_\alpha^s f^n(x)) \\ E^u = \cap_{n \in \mathbb{N}} Df^n(C_\alpha^u(f^{-n}(x))) \end{cases} \qquad (1.46)$$

2. The coexistence of highly complicated long-term behavior[2], sensitive dependence on initial conditions, and the overall stability of the orbit structure.

Examples of Anosov diffeomorphisms are those defined on the torus $\mathbb{T}^n$. The first example of a such situation are Anosov automorphisms obtained using symbolic model [Yoccoz (1993)]. This type of maps have the following properties: (a) the periodic points are dense, (b) there exists a point whose orbit is dense and (c) there are many ergodic invariant probability measures with full support. For example the map defined by the matrix $\begin{pmatrix} 1 & 1 \\ 1 & 0 \end{pmatrix}$ and its square $\begin{pmatrix} 2 & 1 \\ 1 & 1 \end{pmatrix}$ has the entropy at the value $\log \dfrac{3+\sqrt{5}}{2}$. The second example can be realized as a limit of the inverse spectrum of the expanding cycle map [Williams (1967)] given by: $\theta = m\theta \, [mod 1]$, $m \in \mathbb{R}$. This map displays an *expanding solenoid* or hyperbolic attractor where their method of construction is similar to that of minimal sets of limit-quasi-periodic trajectories. More generally, an example of the Anosov diffeomorphism is a mapping of an n-dimensional torus given by: $\bar{\theta} = m\theta$ modulo 1, $m \in \mathbb{R}$, where A is a matrix with integer entries other than 1, det $A = 1$, the eigenvalues of A do not lie on the unit circle, and $f(\theta)$ is a periodic function of period 1. For more details see [Ledrappier (1981), Mané (1985), Shub & Sullivan (1985), Burns, et al. (2001), Dedieu & Shub (2003), Ledrappier, et al. (2003)]. A recent result on this topic is the one given in [Anosov, et al. (2008)] for

---

[2] i.e., coexistence of many chaotic attractors in the same time.

the 2-torus case. In this work, the existence of an isomorphism between a deterministic dynamical system and a random process was proved by using an example of the circle expanding map, with a classification of hyperbolic toric automorphisms. Other examples of hyperbolic maps are: The *Blaschke product* given by: $B(z) = \theta_0 \left(\dfrac{z - a_1}{1 - \overline{z}a_1} \right) \left(\dfrac{z - a_2}{1 - \overline{z}a_2} \right) \cdots \left(\dfrac{z - a_n}{1 - \overline{z}a_n} \right)$, where $n \geq 2$, $a_i \in \mathbb{C}$; $|a_i| < 1$, $i = 1, ..., n$ and $\theta_0 \in \mathbb{C}$ with $|\theta_0| = 1$. The *Bernoulli map* given by: $\phi_{n+1} = 2\phi_n$ modulo 2π. or $\phi_{n+1} = 2\phi_n + const$ modulo 2π. See [Helstrom (1984), Gaspard (1992), Driebe (1999), Bertsekas & Tsitsiklis (2002), Dolgopyat & Pesin (2002)] for more details. The Arnold cat map [Anosov (1967)] given by:

$$\begin{cases} x_{n+1} = x_n + y_n, \text{modulo } 1 \\ y_{n+1} = x_n + 2y_n, \text{modulo } 1 \end{cases} \tag{1.47}$$

and its small perturbation map given by:

$$\begin{cases} x_{n+1} = x_n + y_n + \delta \sin 2\pi y_n, \text{modulo } 1 \\ y_{n+1} = x_n + 2y_n, \text{modulo } 1 \end{cases} \tag{1.48}$$

For more details, see [Sinai (1972), Farmer, et al. (1983), Arnold (1988), Bunimovich, et al. (2000)]. From the above presentation and examples, we can conclude that the hyperbolic attractors have the following properties:

1. They are the limit sets for which Smale's Axiom A is satisfied.
2. They are structurally stable.
3. Periodic orbits and homoclinic orbits are dense in the attractor and they are of the same saddle type (the same index), i.e., the same dimension for their stable and unstable manifolds.

In conclusion, we note that the previous properties are results of a rigorous axiomatic foundation that exploits the notion of hyperbolicity [Ott (1993), Katok & Hasselblatt (1995)]. In fact, hyperbolic chaos is often called true *chaos* because it is characterized by a homogeneous and topologically stable structure as shown in [Anosov (1967), Smale (1967), Ruelle & Takens (1971), Guckenheimer & Holms (1981)]. More details can be found in [Anosov (1967), Newhouse, et al. (1978), Sinai (1979), Devaney, Plykin (1984–1989), Shilnikov (1993), Ott (1993), Katok & Hasselblatt (1999), Hunt (2000), Kuznetsov, (2001), Anishchenko, et al. (2003), Hunt & MacKay (2003), Belykh, et al. (2005)].

1.2.2 Lorenz-type attractors

Almost all known Lorenz-type attractors are not structurally stable[3], although their homoclinic and heteroclinic orbits are structurally stable (hyperbolic), and no stable periodic orbits appear under small parameter variations, as for example in the case of Lorenz system [Lorenz (1963)] given by:

$$\begin{cases} x' = \sigma\,(y - x) \\ y' = rx - y - xz \\ z' = -bz + xy \end{cases} \qquad (1.49)$$

Hence, these attractors are closest in their structure and properties to robust hyperbolic attractors. Thus, Lorenz system (1.49) is called some times *quasi-hyperbolic* [Afraimovich, et al. (1977), Mischaikow & Mrozek (1995–1998)]. On the other hand, Lorenz-type attractors were considered as examples of *truly* strange attractors [Shil'nikov (1980), Williams (1977), Cook & Roberts (1970)] and there are a finite number of these attractors in the literature. For the Lorenz attractor (1.49) all trajectories are of saddle type and the variation of parameters does not create stable points or cycles as shown in [Bykov & Shil'nikov (1989), Afraimovich (1984–1989–1990)]. Also, the Belykh [Belykh (1982–1995)] and Lozi attractors (2.1) are examples of such situation as Lorenz-type attractors realized in two-dimensional maps. A Computer assisted proof of chaos for the Lorenz equations (1.49) was given in [Franceschini, et al. (1993), Mischaikow & Mrozek (1995), Galias & Zgliczynski (1998), Tucker (1999), Stewart (2000), Sparrow (1982)].

From the above notes, it follows that the Lorenz-like attractors can be defined as follow [Araújo & Pacifico, (2008)]:

Definition 15 *A Lorenz-like attractor is an attractor with the following characteristics:*

(1) robust, transitive attractor which is not hyperbolic.
(2) The origin (0, 0, 0) is accumulated by hyperbolic periodic orbits.
(3) The attractor has sensitive dependence to initial conditions (or chaotic).

More detailed and recent results can be found in [Komuro (1984), Morales, et al. (1998), Anishchenko, et al. (2002), Klinshpont, et al. (2005), Bautista & Morales (2006), Klinshpont (2006), Alves, et al. (2007), Araujo

---

[3] In term of C^k topology, the structural stability can be defined as follow: (a) A diffeomorphism f is C^r structurally stable, if for any C^r small perturbation g of f, there is a homeomorphism h of the phase space such that $h \circ f(x) = g \circ h(x)$, for all points x in the phase space. (b) A flow f is C^r structurally stable, if there exists a homeomorphism h sending trajectories of the initial flow f to the trajectories of any small C^r perturbation g. If these conditions are not satisfied by a diffeomorphism (flow), then the corresponding attractor is a non structurally stable set.

& Pacifico (2007), Arroyo & Pujals (2007), Araujo, et al. (2007), Araújo & Pacifico (2008)].

In conclusion, we summarize the properties of Lorenz-like attractors. Let Λ be an attractor of a Lorenz-like system in a compact boundaryless 3-manifold M. Let $C^1(M)$ be the set of C^1 vector fields on M endowed with the C^1 topology. Thus, in [Araújo & Pacifico (2008)], an overview about some recent results on the dynamics of Lorenz-like attractors are given:

1. There is an invariant foliation whose leaves are forward contracted by the flow.
2. There is a positive Lyapunov exponent at every orbit.
3. They are expansive and so sensitive with respect to initial data.
4. They have zero volume if the flow is C^2.
5. There is a unique physical measure whose support is the whole attractor and which is the equilibrium state with respect to the center-unstable Jacobian.

A number of concrete Lorenz-like systems were found and it was proved mathematically that they are robustly transitive and not hyperbolic and they are sensitive to initial conditions as shown in [Afraimovich, et al. (1977), Guckenheimer & Williams (1979), Robinson (1989), Rychlik (1990)]. A relatively new kind of attractor in 3-D is the so called the *contracting Lorenz attractor* presented in [Rovella (1993)]. This attractor contains a hyperbolic singularity with real eigenvalues but not robust. As a general result, it was proved in [Araujo, et al. (2005)] that the so-called *singular-hyperbolic attractor* of a 3-dimensional flow is chaotic in two different strong senses: if the flow is expansive or there exists a physical (or Sinai-Ruelle-Bowen) measure supported on the attractor whose ergodic basin covers a full Lebesgue (volume) measure subset of the topological basin of attraction. These results show that both the classical Lorenz system and the geometric Lorenz flows are expansive. Also, it was proved in [Tucker (2002(a))] that this flow exhibits a singular-hyperbolic attractor.

Another proof of the robustness of chaos in the classical Lorenz attractor was given in [Franceschini, et al. (1993)] where the chaotic attractors of the Lorenz system associated with $r = 28$ and $r = 60$ were characterized in terms of their unstable periodic orbits, eigenvalues, Hausdorff dimension and topological entropy. A general method for proving chaos in a set of systems called C^1-*robust transitive sets with singularities* for flows on closed 3-manifolds was given in [Morales, et al. (2004)]. The elements of this set are partially hyperbolic with a volume-expanding central direction and are either attractors or repellers. In fact, any C^1-robust attractor with singularities for flows on closed 3-manifolds always have a positive Lyapunov exponent at every orbit. Thus, any C^1-robust attractor resembles a geometric Lorenz attractor. Other than the Lorenz attractor, there are some works that focus on the proof of chaos in 3-D continuous systems, for example the set C^1 studied in [Morales, et al. (2004)].

For the classical Lorenz system, it was proved in [Guckenheimer & Williams (1979)] that the resulting attractor is robust in the sense that there is a structurally stable 2-parameter family of flows containing the geometric Lorenz flow, i.e., the geometric Lorenz flow is structurally stable of codimension 2. That is, any perturbation of the geometric Lorenz flow is topologically conjugate on a neighborhood of the attractor to a nearly member of this family. The proof of this result is based on the following things: (1) The construction of suspension [Smale (1967)] and inverse limits [Williams (1975)] in dynamical systems. (2) Results on the bifurcations of maps of the unit interval given in [Guckenheimer (1977)].

Also, it was shown rigorously in [Tucker (1999)] that the geometric model does indeed give an accurate description of the dynamics of the classical Lorenz system, i.e., it supports a strange robust attractor. The proof is based on a combination of *normal form theory* and *rigorous numerical computations*. The novelty and originality of this proof lies in the construction of an algorithm in C-program which, if successfully executed, proves the existence of the strange attractor.

Generally, expanding and contracting Lorenz attractors systems are examples of Lorenz-type attractors realized in 3-D continuous time systems. Indeed, in [Morales, et al. (2006)] the so called *contracting Lorenz attractor* (in the unfolding of certain resonant double homoclinic loops in dimension three) was obtained from the geometric Lorenz attractor by replacing the usual expanding condition at the origin by a contracting condition given in [Rovella (1993)]. Also, in [Morales, et al. (2005)], a generalization of the idea given in [Robinson (2000)] was studied, where the existence of Lorenz attractors was proved in the unfolding of resonant double homoclinic loops in dimension three. The method of analysis is based on two steps. First, the search of *attractors* instead of *weak attractors* used in [Robinson (2000)] and second, a large region was considered in the parameter space corresponding to flows presenting expanding Lorenz attractors.

1.2.3 Quasi-attractors

Quasi-attractors are the limit sets enclosing periodic orbits of different topological types (for example stable and saddle periodic orbits) and structurally unstable orbits. We note that most known observed chaotic attractors are quasi-attractors [Afraimovich & Shilnikov (1983), Lichtenberg & Lieberman (1983), Rabinovich & Trubetskov (1984), Schuster (1984), Neimark & Landa (1989), Anishchenko (1990-1995)], i.e., neither robust hyperbolic nor almost hyperbolic attractors. These attractors have the following properties:

1. They separatrix loops of saddle-focuses or homoclinic orbits of saddle cycles in the moment of tangency of their stable and unstable manifolds, because they enclose non-robust singular trajectories that are *dangerous*.
2. A map of *Smale's horseshoe-type* appear in the neighborhood of their trajectories. This map contains both non-trivial hyperbolic subset of trajectories and a denumerable subset of stable periodic orbits [Shilnikov (1963), Gavrilov & Shilnikov (1972-1973)] and Newhouse's theorem [Newhouse (1980)].
3. The quasiattractor is the unified limit set of the whole attracting set of trajectories including a subset of both chaotic and stable periodic trajectories which have long periods and weak and narrow basins of attraction and stability regions. This is a result of the effect that this attractor is *holed* by a set of basins of attraction of different periodic orbits.
4. The basins of attraction of stable cycles are very narrow.
5. Some orbits do not reveal themselves ordinarily in numeric simulations except some quite large stability windows where they are clearly visible.

Thus, quasi-attractors have a very complex structure of embedded basins of attraction in terms of initial conditions and a set of bifurcation parameters of non-zero measure. In this case, the homoclinic tangency of stable and unstable manifolds of saddle points in the Poincaré section is the principal cause of this complexity as shown in [Gavrilov & Shil'nikov (1972-1973), Afraimovich (1984–1989–1990)]. On the other hand, for the quasi-attractors the basins of attraction of co-existing limit sets is very narrow and it can have fractal boundaries. Hence, rigorous mathematical description of quasi-attractors are still an open problem, because almost non-hyperbolic attractors are obscured by noise. An example of such a situation is the attractor generated by Chua's circuit [Chua, et al. (1986)] described by the following equation:

$$\begin{cases} x' = \alpha(y - h(x)) \\ y' = x - y + z \\ z' = -\beta y \end{cases} \tag{1.50}$$

where

$$h(x) = m_1 x + \frac{1}{2}(m_0 - m_1)(|x+1| - |x-1|) \tag{1.51}$$

System (1.50)-(1.51) is associated with saddle-focus homoclinic loops. This type of attractor is more complex than the above two attractors, and it is not suitable for some cases of potential applications of chaos such as secure communications and signal masking. The second example of quasi-attractor is the Hénon map (2.1). See Section 1.5 for more details. Recent studies show a basic difference between properties of hyperbolic, Lorenz-type attractors and quasi-attractors [Anishchenko & Strelkova (1998)]. Homoclinic tangencies of the trajectories is the most common property of quasi-attractors and Lorenz-type ones, and the difference between them is that quasi-attractors contain a countable subset of stable periodic orbits. Classical methods can be used to prove chaos in the above types of attractors, namely, by finding a positive Lyapunov exponent, a continuous frequency spectrum, fast decaying correlation functions...etc. On the other hand, the chaotic behavior of Lorenz-type attractor can be studied using statistical methods introduced in Section 1.1. The reason is that they admit the introduction of reasonable invariant measures contrary to quasi-attractors who are hardly admit this property. Thus, a rigorous foundation is possible for studying characteristics of hyperbolic and Lorenz attractors. They are called *stochastic*. For the characterization of quasi-attractors, it is possible to use the *Autocorrelation function* (ACF) introduced in Section 1.1.7. A characterization of this method is the presence of a periodic component in the ACF and sudden peaks at certain characteristic frequencies in the spectrum which is a typical feature for these attractors. For the general case, some results (for example the co-existence of countable sets of saddle periodic orbits of distinct topological types) about homoclinic tangencies for 3D systems and 2D diffeomorphisms are also hold as shown in [Gonchenko, et al. (1993)]. This result implies that the complete theoretical analysis of these models is not possible. More details about topological classification of strange attractors can be found in [Vietoris (1927), van Dantzig (1930), Bourbaki (1942-1947), Cassels (1961), Arov (1963), Hewitt & Ross (1963), Novikov (1965), Smale (1967), Franks (1970), Ruelle & Takens (1971), Perov & Egle (1972), Aleksandrov (1977), Plykin (1977), Rand (1978), Plykin (1980), Watkins (1982), Plykin (1984), Ustinov (1987), Aarts & Fokkink (1991), Fokkink (1991), Plykin, et al. (1991), Pesin (1992), Plykin & Zhirov (1993), Zhirov (1994(a-b)-1995)), Gorodetski & Ilyashenko (1996), Anishchenko & Strelkova (1997), Zhirov (2000), Plykin (2002)].

2

Reality of Chaos in the Hénon Mapping

This chapter is concerned with discussions about reality of chaos in Hénon mappings. In general, this type of chaos is characterized by the occurrence of *dangerous homoclinic tangencies*. In this case the corresponding chaotic attractor can be seen as the unified limit set of the whole attracting set of trajectories. This set includes a subset of both chaotic and stable periodic trajectories which have long periods and weak and narrow basins of attraction and stability regions. This is a result of the fact that this attractor is *holed* by a set of basins of attraction of different periodic orbits. Now, the most interesting results concerned with different shape of the Hénon mappings are presented with some details in this chapter. Indeed, in Section 2.1 several methods on measuring chaos in the Hénon mappings are presented and discussed. In Section 2.2 the most interesting bifurcations phenomena in the Hénon mappings are given. In particular, those related with the existence of tangencies and unusual bifurcation phenomena are discussed in Section 2.2. Section 2.3 deals with the type of chaos in the Hénon mappings, i.e., these maps are quasi-attractors. Section 2.4 and Section 2.5 are concerned with the problem of period-doubling cascades for large perturbations of the Hénon families. Especially, a general theory is presented with many explanations, results and examples.

2.1 Measuring chaos in the Hénon map

In this section, we discuss the reality of chaos in the Hénon map [Hénon (1976)] given by:

$$h\left(x, y\right) = \begin{pmatrix} 1 - ax^2 + y \\ bx \end{pmatrix} \tag{2.1}$$

The Hénon attractor with its basin of attraction obtained for $a = 1.4$, $b = 0.3$ is shown in Fig. 2.1. The most important feature of the Hénon map (2.1) is the appearance of homoclinic tangencies between stable and unstable manifolds. This phenomenon is the principal cause of its complexity and non-robustness, i.e., the Hénon map (2.1) is a quasi-attractor. The map (2.1) can be considered as a simplified model[1] of the Poincaré map for the Lorenz model (1.49). Several form of the Hénon mappings were used to state the corresponding results, and the reader can deduce these results for the map (2.1) by using affine or linear transformations. The Hénon map (2.1) has several properties including:

1. It has two fixed points, invertible, conjugate to its inverse and its inverse is also a quadratic map. The parameter b is a measure of the rate of area contraction (dissipation).

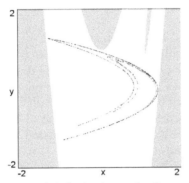

Figure 2.1: The Hénon attractor with its basin of attraction (in gray) obtained for $a = 1.4$, $b = 0.3$. The Hénon attractor with its basin of attraction (in gray) obtained for $a = 1.4$, $b = 0.3$.

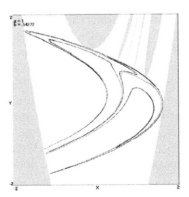

Figure 2.2: Attractor for the maximally complex Hénon map with $a = 1$ and $b = 0.54272$ along with its basin of attraction (in grey).

---

[1] A minimal normal form for modeling flows near a saddle-node bifurcation.

2. It is the most general 2-D quadratic map with the property that the contraction is independent of the variables x and y.

3. It reduces to the quadratic map when $b = 0$, which is conjugate to the well known *logistic map*.

4. It has bounded solutions over a range of a and b values, and a portion of this range (about 6%) yields chaotic solutions.

5. For $a = 1.4$ and $b = 0.3$, the Hénon map (2.1) has the chaotic attractor shown in Fig. 2.1 that has correlation exponent 1.25±0.02 [Grassberger & Procaccia (1983)] and capacity dimension 1.261 ± 0.003 [Russell, et al. (1980)] and the Lyapunov exponents are $\lambda_1 = 0.41922$ and $\lambda_2 = -1.62319$ [Grassberger & Procaccia (1984)].

6. A numerical algorithm to estimate the generalized dimensions for large negative q for the Hénon map (2.1) was given in [Pastor-Satorras & Riedi (1996)], because the standard fixed-size box-counting algorithms are inefficient for computing generalized fractal dimensions in the range of $q < 0$.

7. In [Sprott (2007)] the Hénon map $h_1 (x, y) = (1 - ax^2 + by, x)$ was studied numerically to find the location in the positive ab-plane with the highest value of λ_1 and Kaplan-Yorke dimension D_{KY}. The results are that the values rapidly and consistently converged to $a = 2$ and $b = 0$, i.e., the case of the logistic map with $\lambda_1 = \ln 2 = 0.69314....$ and $\lambda_2 = -\infty$ with a Kaplan-Yorke dimension of $D_{KY} = 1$. The parameters that maximize the Kaplan-Yorke dimension are $a = 1$ and $b = 0.54272$, for which the Lyapunov exponents are $\lambda_1 = 0.27142$ and $\lambda_2 = -0.88258$ with a Kaplan-Yorke dimension of $D_{KY} = 1.30753$. The corresponding Hénon-like attractor is shown in Fig. 2.2 with a bit more structure as befits its higher dimension, along with its basin of attraction with boundary that touches the attractor at numerous places. The method of analysis is based on a random search starting with the above values of a and b and exploring a Gaussian two-dimensional neighborhood in ab-plane with an initial fractional standard deviation of $\varepsilon = 0.1$, taking 2 $\times 10^7$ iterations at each set of parameters and calculating the Lyapunov exponents using the method presented in [Sprott (2003(d))]. In this case, if λ_1 is higher than any previously found value, then the search neighborhood was moved to those coordinates, and was increased by a factor of 1.1. Else, was reduced by a factor of 0.999 and the search continued until became negligibly small. If the optimum was found, then it was recalculated with many more iterations (typically 10^{11}).

It is clear that almost all results on the dynamics of the Hénon map (2.1) were obtained numerically. However, some analytical works can be found in the literature. For example in [Benedicks & Carleson (1991)] an excellent global characterization of such dynamics was given analytically.

In particular, consider the following question: *Is an orbit starting from a point would generically go into an attractive cycle?* If W^u (h) denotes the unstable manifold, then the main theorem for the answer to this question is stated as follows:

Theorem 4 *Let W^u be the unstable manifold of the Hénon map (2.1) at its fixed point in x, y > 0. Then for all c < log 2 there is a b_0 > 0 such that for all b ∈ (0, b_0) there is a set E (b) of positive one dimensional Lebesgue measures such that for all a ∈ E (b):*
  (i) *There is an open set U = U (a, b) such that for all z ∈ U:*

$$\text{dist} (h^n (z), cl (W^u)) \to 0, \text{ as } n \to \infty \tag{2.2}$$

  (ii) *There is a point z_0 = z_0 (a, b) ∈ W^u such that*
  (a) *The sequence $(h^n (z_0))^\infty_{n=0}$ is dense in W^u,*
  (b) *$|Dh^n (z_0) (0, 1)| \geq e^{cn}$.*

   Theorem 4 means that if a is approximately equal to 2 and sufficiently small b ($b = 0$ corresponds to the logistic map) then the Hénon map (2.1) admits a strange attractor. In other words, if a ∈ E (b), then the Hénon map (2.1) has the following characterizations: (a) The forward orbit is dense on the unstable manifold W^u for almost all starting points in a neighborhood of the origin. (b) The closure of the parameter values a for which h has an attractive cycle contains E (b). (c) There exist a Sinai-Bowen-Ruelle measures (recall Definition 2) which has absolutely continuous conditional measures with respect to the unstable foliation. (d) A partial theory of kneading sequences can be developed for a ∈ E (b) . The proof of Theorem 4 is based on the analysis of the one-dimensional map $x \to 1 - ax^2$ using a very long treatment. This proof is based on a modified version of the proof given in [Benedicks & Carleson (1985)] of Jakobson's Theorem. In order to obtain the chaotic behavior with simplified proof, the authors replace a property called *basic assumption* (BA) by a simple rule and similarly in the so called *binding condition* (BC). The main result using these assumptions is that the derivatives grow exponentially for a model case, and then they prove that the Hénon map behave in an analogous way. Some tools were defined in order to prove Theorem 4. For example, the concept of contractive direction closely tied to expansion (in all other directions) was introduced and analyzed. Also, the crucial points are defined which correspond to the origin for the 1-D mapping $x \to 1-ax^2$ with a full description of the related geometry of the unstable manifold and the choice of the parameters, i.e., the parameters of the map are $0 < a < 2, b > 0$ and all results are valid in the intervals $a_0 < a < 2, 0 < b < b_0$ and all estimates improve when a increases and b decreases. We note that the proof of SRB measure is given in [Benedicks & Young (1993)], and then again in [Benedicks & Young (2000)] with a result on the rate of mixing for Hénon maps. A more general theory can be found in [Mora & Vianna (1993)] as follow:

Theorem 5 *Let $(f_\mu)_\mu$ be a C^∞ one-parameter family of diffeomorphisms on a surface and suppose that f_0 has a homoclinic tangency associated to some periodic point P_0. Then, under generic (even open and dense) assumptions, there is a positive Lebesgue measure set E of parameter values near $\mu = 0$, such that for $\mu \in E$ the diffeomorphism f_μ, exhibits a strange attractor, or repeller, near the orbit of tangency.*

It is well known that the Lozi map (3.1) has good hyperbolic structure properties, and the ergodic and chaotic properties are also now well known for piecewise expanding maps [Young (1998)]. However, the Hénon map (2.1) lacks hyperbolicity. In this case, it would be interesting to understand the methods used for proving ergodicity and chaos in the Hénon family and similar maps. The main references for this subject are [Benedicks & Carleson (1991), Mora & Viana (1993), Benedicks & Young (1993-2000), Wang & Young (2001)]. It was shown in [Sinai (1973), Ruelle (1976)] that for a diffeomorphism f with an Axiom A attractor Λ, there is a unique f-invariant Borel probability measure μ on Λ. In fact, this idea of SRB measures can be generalized to arbitrary diffeomorphisms with abstract theorems, that is, these theorems contain no assertions of existence and without general methods on how to determine SRB measures for concrete dynamical systems. There are two main approaches to find SRB measures: The first, the *axiomatic approach*, seeks to relax the conditions that define Axiom A systems and to enlarging the set of maps with SRB measures. The second approach is called *phenomenological approach* and it is related with the concrete dynamical behaviors not necessarily related to Axiom A systems. This approach seeks underlying characteristics conducive to having SRB measures for the system under consideration. The construction of SRB measures on Axiom A attractors is not exactly the one used in the original works of Sinai, Ruelle or Bowen, but it is very close to the Sinai's construction given in [Pesin & Sinai (1982)]. The construction of SRB measures using the phenomenological approach for a particular class of strange attractors (described with a function f) is possible if f has the following properties:

1. The defining maps are strongly dissipative, i.e., $|\det(Df)| \ll 1$.
2. The attractors are chaotic with a single direction of instability.
3. Some unstable curves have *folds* similar to those in the Hénon maps.

Natural examples of this class of attractors include the Hénon family, Hénon-like attractors arising from homoclinic bifurcations, simple mechanical model with periodic forcing and strange attractors emerging from Hopf bifurcations. For this type of attractors, it was shown that there are small fractal sets near the critical points of the corresponding 1-D maps with the following property: *Away from these sets the map is uniformly hyperbolic, and upon approaching them, the stable and unstable directions of an orbit are confused, i.e., they are not uniformly hyperbolic, and all the problems are caused by the existence of certain well-defined, localized bad sets or sources of nonhyperbolicity.* Now, it was shown in [Jakobson (1981), Graczyk & Swiatek (1997), Lyubich (1997-2001)] that the logistic family $Q_a(x) = 1 - ax^2, x \in [-1, 1]$, $a \in [0, 2]$ has the following properties:

(i) There is an open and dense set A in parameter space such that for all $a \in A$, Q_a has a periodic sink to which the orbit of Lebesgue-a.e. point converges [Graczyk & Swiatek (1997)].

(ii) There is a positive Lebesgue measure set of parameters B such that for $a \in B$, Q_a has an invariant measure absolutely continuous wrt Lebesgue measure [Jakobson (1981)].

(iii) The union of A and B has full measure in parameter space [Lyubich (1997–2001)].

Item (ii) of this result is called *Jakobson's Theorem*. For the Hénon map h given by (2.1), when $b = 0$, h maps the x-axis to itself and reduces to the logistic family. Thus, by continuity, it follows that for $a \in (0, 2)$, h maps a rectangle into itself and has an attractor for small values of b. In [Benedicks & Carleson (1991)] the authors studied the map (2.1) for small values of b for $a \approx 2$, treating h as a small perturbations of Q_a. Hence, they developed some techniques for tracking the growth of derivatives along certain crucial orbits; see Theorems 4 and 5. Based on this result, it was shown in [Benedicks & Young (1993)] that for every $b > 0$ sufficiently small, there is a positive Lebesgue measure set $\Delta_b \subset (2 - \varepsilon, 2)$ such that for each $a \in \Delta_b$, the Hénon map (2.1) admits a unique SRB measure. We note that this is the first result about the construction of SRB measures for non-uniformly hyperbolic attractors. It seems that the proof of this result is also valid for the type of maps described above with properties (i), (ii) and (iii). The main idea of the proof is based on replacing the computation-based arguments in [Benedicks & Carleson (1991)] by a more conceptual approach, and to replace the formulas of the Hénon maps (2.1) by some qualitative, geometric conditions. We note that the full structure of the parameter space of the Hénon map (2.1) was the subject of several works such as in [Gallas & Jason (1993), Sprott (1996), Hansen & Cvitanovic (1998), Cao & Kiriki (2000), Endler & Gallas (2001), Cao & Mao, (2000) and references therein]. Figure 2.3 shows regions of unbounded, fixed point, periodic, and chaotic solutions in the *ab*-plane for the Hénon map (2.1), where we use $|LE| < 0.0001$ as the criterion for quasiperiodic orbits with 10^6 iterations for each point. There are several known methods about the rigorous proof of chaos in the Hénon mappings. Almost of these methods were discussed in [Zeraoulia & Sprott (2010)]. This includes:

1. Computer assisted proof of the horseshoe dynamics and topological entropies given in [Zgliczynski (1997(b)), Misiurewicz & Szewc (1980), Zgliczynski (1997(a)), Galias (1998(b)), Stoffer & Palmer (1999), Galias (2001), Tibor; et al. (2006)] for the 7th iterate of the classical Hénon map (2.1).
2. The standard numerical calculations given in [Gómez & Simó (1983)] for the existence of an infinite number of homoclinic and heteroclinic trajectories for the Hénon map (2.1) for $a = 1.4$, $b = 0.3$.
3. The linking of topological tools with a local hyperbolic behavior in [Galias & Zgliczynski (2001)].
4. The method of Wiener and Hammerstein cascade models was used in [Xu, et al. (2001)] and a method based on time series analysis [Bhattacharya & Kanjilal (1999)] were applied for detecting chaos in the Hénon map (2.1) with $a = 1.4$, $b = 0.3$.
5. Finally, a new chaos detector (just like the largest Lyapunov exponent and the different type of dimensions and entropies) was introduced in [Mcdonough, et al. (1995)] for detecting chaos in the Hénon map (2.1) also, for $a = 1.4$, $b = 0.3$.

2.2 Bifurcations phenomena in the Hénon mappings

The bifurcations phenomena in the Hénon map (2.1) was the subject of several works. In particular, the Hénon map in all its simplest forms has at most 2^n periodic points of period n [Moser (1960)]. Their existence was proved in [Hitzl & Zele (1985), Alligood & Sauer (1988), Galias (1998(b)), Michelitsch & Rössler (1998), Galias (1999) and references therein] using computer assisted methods, numerical and analytical treatments. More

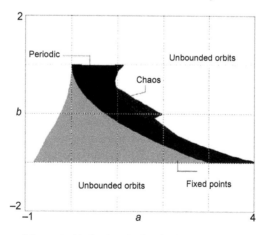

Figure 2.3: Regions of dynamical behaviors in the ab-plane for the Hénon map (2.1).

details can be found in the book [Zeraoulia & Sprott (2010)]. As an example of some unusual bifurcation phenomena observed in the Hénon map (2.1) is the occurrence of a new type of transients of high and varying periodicity for low-periodic orbits as shown in [Michelitsch & Rössler (1998)].

Second, the *landing phenomenon* presented in [Pei-Min & Bang-Chun (2004)], is a complex scenario, concerned both with the coexistence of attractors and the crises of chaotic attractors in the course of constructing domains of steady state solutions of the Hénon map in parameter space by numerical methods. In this case, a narrow domain of period-m solutions first co-exists with (lies on) a big period-n ($m < n$) domain. Then it enters the chaotic area of the big domain and becomes period-m windows and then the co-existence of attractors disappears. In other words, there is an interaction between the two domains in the course of landing: the chaotic area in the big domain is enlarged and there is a crisis step near the landing area.

The third example of unusual bifurcation phenomenon is the so called *sequence of global period doubling bifurcation* observed in [Murakami, et al. (2002)]. This sequence has some properties such as the existence of a region of bifurcated area in the phase space expands gradually when the control parameter increases over the critical threshold. Also, the relative prevalence of the bifurcated orbit is measured in terms of the relative ratio of the area covered by the bifurcated daughter orbit to that of the mother orbit.

The fourth phenomenon called *noisy parametric sweep through a period-doubling bifurcation* was observed in [Davies & Rangavajhula (2002)]. This phenomenon means that an explicit noise model is included, and the matched asymptotic expansions together with a center-manifold reduction in the vicinity of the bifurcation are used to describe trajectories sweeping up or down. The fifth phenomenon is the so called *entry and exit sets*[2] observed in [Petrisor (2003)] and it refers to regions through which any forward, respectively backward, unbounded orbit escapes to infinity. The method of analysis is based on the fact that the Hénon map can be considered as a discrete version of *open Hamiltonian systems*, that can exhibit chaotic scattering and then the proof that the right branch of the unstable manifold of the hyperbolic fixed point is the graph of a function. This function is the uniform limit of a sequence of functions whose graphs are arcs of the symmetry lines of the Hénon map, as a reversible map.

Since this book discusses ergodic properties, symbolic sequences, complexity, and strange attractors of the Lozi mappings, it might be worth contrasting results with what is known for Hénon maps and Lorenz-like maps. Indeed, in [Wang & Young (2001)] there is a detailed discussion on

---

[2] Hénon and Feit [Hénon (1976), Feit (1978)] have noted that for $b = 0.3$ and $a \in [-0.12, 2.67]$ then no attractors are observed for the Hénon map, numerically, all points seem to escape to infinity.

symbolic codings for the Hénon map (or Hénon-like maps) with proofs of some ergodic properties. Indeed, some conditions for the existence of strange attractors for known properties can be summarized as follows: (1) By letting dissipation go to infinity, the Hénon maps pass to the singular limit and give a 1-D maps. (2) Some of these 1-D maps show strong expanding properties, i.e., *Misiurewicz maps*. (3) Varying the parameters in these maps changes the dynamics effectively, i.e., transversality condition. (4) It is possible to reverse the process in (1) to return to the original Hénon or Hénon-like dynamics. By these considerations, the Hénon maps have several important statistical properties. The first result is the existence of non unique SRB measures for some *good* parameters with a complicated domain of attraction. For $a = 2$ and $b = 2$ the SRB measure is unique as shown in [Benedicks & Young (1999)]. The second result is that the correlations decay and the central limit theorem holds for these SRB measures. The third result is that the Hénon maps have a *coarse geometry of the attractor*, i.e., there is a sequence of neighborhoods such that each element of this sequence is the union of *monotone* branches[3]. The fourth result is about the coding of orbits on the attractor, i.e., there is a subshift of a full shift on finitely many symbols and a continuous surjection making the Hénon map under consideration equivalent to the mentioned subshift. Also, it is possible to define kneading sequences for critical points in the sense that every itinerary is represented by a unique sequence. An immediate result is the existence of *equilibrium states*. In particular every Hénon-like map admits an invariant Borel probability measure maximizing the corresponding entropy. The most important result for Hénon maps is the *homoclinic bifurcations*.

A classification of these bifurcations for the Hénon map $h_{k,b}(x, y) = (y - k + x^2, -bx)$ was given in [Sterling, et al. (1999)]. The method of analysis is based on the so-called *anti-integrable limit* discussed in [Aubry (1995), MacKay & Meiss (1992)]. Also, it was shown in [Sterling, et al. (1999)] that there exists a bound on the parameter range for which the Hénon map $h_{k,b}$ exhibits a complete binary horseshoe and a subshift of finite type. Indeed, let $x_t, t \in \mathbb{Z}$ be a sequence of points on an orbit of the Hénon map $h_{k,b}$, which can be rewritten as $x_{t+1} + bx_{t-1} + k - x_t^2 = 0$. The scaled coordinate $z = \epsilon x = k^{-\frac{1}{2}} x$ for $0 < k < \infty$ gives an implicit map (in the variable z) with parameter ϵ as follow: $\epsilon(z_{t+1} + bz_{t-1}) + 1 - z_t^2 = 0$. Thus, a period-$n$ orbit of the Hénon map $h_{k,b}$ is given by a sequence $z_0, z_1, ..., z_{n-1}$ that satisfies the implicit map formula with the condition $z_{t+n} = z_t$. Now, it was shown in [MacKay & Meiss (1992), Aubry (1995)] that the anti-integrable limit technique implies that the dynamics in discrete time can be represented by a relation $F(x, x') =$

---

[3] A monotone branch is a region diffeomorphic to a rectangle and bordered by two sub-segments.

0, where x and x' are points in some manifold. This last equation can be explicitly solved for $x' = h_{k,b}(x)$, giving a map $x_t = h_{k,b}(x_{t-1})$ on the manifold. If $F(x, x') = \epsilon G(x, x') + H(x)$, then the solutions of the implicit equation $F = 0$ when $\epsilon = 0$ correspond to an arbitrary sequences of points x_t that are zeros of H. It was shown in [Aubry (1995)] that: (a) the case $\epsilon = 0$ corresponds to an anti-integrable limit (AI) of the map $h_{k,b}$. (b) If the derivative of H is nonsingular, then the (AI) orbits can be continued for $\epsilon \neq 0$ to orbits of the map $h_{k,b}$ [MacKay & Meiss (1992), Aubry (1995)]. At the anti-integral limit, the equation $\varepsilon (z_{t+1} + bz_{t-1}) + 1 - z^2_t = 0$ reduces to $z^2_t = 1$. Hence, the orbits of this limit are arbitrary sequences in the following infinite set: $\Sigma = \{(-1,+1)^{\mathbb{Z}}\} = \{(-,+)^{\mathbb{Z}}\} = \{s : s_t \in \{-1,+1\}, t \in \mathbb{Z}\}$. If Σ is the set defined by: $\Sigma_F = \Sigma - \{s \in \Sigma : \exists t \in \mathbb{Z}, \text{ such that } s_{t-1} = s_{t+1} = -1\}$, then

(a) Orbits in the anti-integrable limit are bi-infinite sequences $s \in \Sigma$.
(b) The dynamics on $s \in \Sigma$ are given by the shift map, $\sigma : \Sigma \to \Sigma$ defined as $\sigma (...s_{-1}.s_0 s_1 s_2...) = ...s_{-1} s_0.s_1 s_2...$
(c) An orbit of the symbolic dynamics is periodic if the sequence s is periodic.
(d) A saddle-node bifurcation is a local bifurcation in which two fixed points *collide* and annihilate each other.
(e) A Pitchfork bifurcation is a particular type of local bifurcation. Pitchfork bifurcations, like Hopf bifurcations, are of two types–supercritical and subcritical.
(f) A period doubling bifurcation is a bifurcation in which the map switches to a new behavior with twice the period of the original map, i.e., the *Floquet multiplier* is -1.
(g) A rotational bifurcation occurs when the *winding number* of an elliptic orbit has the form $\omega = m/n$, where $m, n \in \mathbb{N}$. Thus, an orbit of least period n is denoted by the string of n symbols and a superscript ∞ to represent repetition as follow: $(s_0 s_1 s_2 s_{n-1})^{\infty} = ...s_{n-2} s_{n-1}.s_0 s_1...s_{n-1} s_0...$ From these considerations it was shown in [Sterling, et al. (1999)] that the map σ has the following properties:

(a) σ has two fixed points, $(+)^{\infty}$ and $(-)^{\infty}$, corresponding to the two fixed points of the Hénon map $h_{k,b}$.
(b) The two fixed points are born in a saddle-node bifurcation at $k = \dfrac{-(1+b^2)}{4}$, which we denote by sn $\{(+)^{\infty}, (-)^{\infty}\}$. To classify different homoclinic bifurcations for the Hénon $h_{k,b}$, it is necessary to state the following definitions and notions: (a) A *parent* refers to the orbit that is undergoing the bifurcation, if any. (b) The *type* is one of **sn**, **pf**, **pd**, or m/n, corresponding to a saddle-node, Pitchfork, period doubling, or rotational bifurcation, respectively.
(c) The set of orbits created in the bifurcation is listed as the *children*.

Thus, a classification of the periodic orbits up to period-6 and their bifurcations was given in [Sterling, et al. (1999)]. The bifurcations are denoted with the general template *parent → type (children)* as follow:

Parent	Type	Child	Child	k-values
	sn	$(-)^\infty$	$(+)^\infty$	-1
$(-)^\infty$	pd		$(+-)^\infty$	3
	sn	$(-+-)^\infty$	$(+-+)^\infty$	1
$(-)^\infty$	1/3	$(-+-)^\infty$		$\frac{5}{4}$
$(-)^\infty$	1/4	$(+--+)^\infty$	$(+-++)^\infty$	0
$(+-)^\infty$	pd		$(-+--)^\infty$	4
$(-)^\infty$	1/5	$(-+++-)^\infty$	$(++-++)^\infty$	$\frac{7-5\sqrt{5}}{8}$
$(-)^\infty$	2/5	$(--+--)^\infty$	$(-+-+-)^\infty$	$\frac{7+5\sqrt{5}}{8}$
	sn	$(+-+-+)^\infty$	$(+---+)^\infty$	$5.5517014^\otimes$
$(-)^\infty$	1/6	$(-+^4-)^\infty$	$(++-+^3)^\infty$	$\frac{-3}{4}$
$(+-+)^\infty$	pd		$(+-^4+)^\infty$	$\frac{5}{4}$
$(+-^4+)^\infty$	pf	$(++-+--)^\infty$	$(--+-++)^\infty$	3
	sn	$(--+-+-)^\infty$	$(--+-^3)^\infty$	$3.7016569^\boxtimes$
$(+-)^\infty$	1/3	$(--+-+-)^\infty$		$\frac{15}{4}$
	sn	$(+-+^3-)^\infty$	$(--+^3-)^\infty$	$5.6793695^\boxtimes$

We note that where there are two children, the one listed in the first column has negative residue just after *birth* (except for the **pf** case). The exact bifurcation values for the three approximations shown above in the table labeled by $\otimes$ and $\boxtimes$, respectively are the real roots of the polynomials $16k^5 - 108k^4 + 105k^3 + 27k^2 - 97k - 47$ and $16k^6 - 136k^5 + 213k^4 + 220k^3 + 126k^2 + 108k + 81$. Now, if α, ζ are two homoclinic orbits of the Hénon map $h_{k,b}$, and let W^s, W^u be respectively, the stable and unstable manifolds. Then, we have: (a) A segment of a manifold from a point to its iterate $W^s(\beta, h_{k,b}(\beta))$ called a *fundamental segment*. (b) The *transition time* is the number of iterates required for $\beta \in W^u(h_{k,b}^{-1}(\zeta), \zeta)$ to reach the stable segment $t_{trans}(\beta) = m$, if $h^m{}_{k,b}(\beta) \in W^s(h_{k,b}^{-1}(\zeta), \zeta)$. (c) The *type* of a homoclinic point is the number of iterates for which the stable initial segment $h^j_{k,b}(\beta)$ intersects with the unstable initial segment β, i.e., *type* $(\beta) = \sup \{j \geq 0, : W^s(p, h^j_{k,b}(\beta)) \cap W^u(p, \beta) \neq \varnothing\}$. (d) Primary homoclinic points have type 0. (e) Horseshoes have type 1.

Let us define the sets $U = W^u(\alpha, \zeta)$, and $S = W^s(h_{k,b}^{-1}(\zeta), \alpha)$; then it was proved in [Sterling, et al. (1999)] that (a) For every symbol sequence $s \in \Sigma$, there exists a unique orbit $z(\varepsilon)$ of the Hénon map $h_{k,b}$ such that $z(0) = s$ if

$|\varepsilon|\,(1 + |b|) < 2\sqrt{1 - \dfrac{2}{\sqrt{5}}} \approx 0.649839$. (b) There are no bifurcations in the

Hénon map $h_{k,b}$ when ε and b are in the range $|\varepsilon|\,(1 + |b|) < 0.649839$. (c) (Existence and uniqueness of Σ_F orbits): Suppose that $0 \le b \le 1$, then for every symbol $s \in \Sigma_F$, there exists a unique orbit $z\,(\varepsilon)$ of the Hénon map $h_{k,b}$

such that $z\,(0) = s$ if $0 \le \varepsilon \le \varepsilon_{max}$, where $\varepsilon_{max} = \dfrac{2}{1+b}\sqrt{\dfrac{-b^2 + 2b + 5 - 2\sqrt{5 + 4b}}{(1-b)(5-b)}}$. (d)

Assume there are exactly two primary homoclinic orbits, α and ζ, and the segments S and U defined above contain all of the homoclinic orbits.

Then for each homoclinic point in $\beta \in U$, one has $t_{trans}\,(\mu) = type\,(\beta)$. (e) Two homoclinic orbits β and γ cannot bifurcate unless they are double neighbors. (f) If two homoclinic orbits β and γ bifurcate, then they must have the same transition time t_{trans}. (g) The transition time of a homoclinic orbit never changes. (h) Two homoclinic orbits on U are neighbors in the complete horseshoe if and only if they are of the form $+^\infty - \cdot (s+) - +^\infty$ and $+^\infty - \cdot (s-) - +^\infty$. (l) The first homoclinic bifurcation of the invariant manifolds of the fixed point $(+)^\infty$ is $sn\,\{+^\infty - (+) - +^\infty, +^\infty - (-) - +^\infty\}$. For more details, see [Devaney (1984), Grassberger, et al. (1989), Davis, et al. (1991), Easton (1998), Sterling, et al. (1999)], and references therein.

2.3 Hénon attractor is a quasi-attractor

In this section, we shall present known results in the current literature that deal with the problem of the existence of homoclinic and heteroclinic trajectories for the Hénon mappings. This property implies that these maps have some features of the third class of chaotic attractors presented in Section 1.2.3, i.e., quasi-attractors. Indeed, the existence of an infinite number of homoclinic and heteroclinic trajectories was proved for the Hénon map (2.1) for $a = 1.4$, $b = 0.3$ in [Gómez & Simó (1983)] using the standard numerical calculations, and in [Galias & Zgliczynski (2001)] using a linking of topological tools with a local hyperbolic behavior. The existence of transversal homoclinic points for the Hénon map (2.1) was proved in [Kan, et al. (1995)] using numerical methods and an analytic method in [Marotto (1979), Misiurewicz & Szewc (1980), Brown (1995)]. Also, in [Kirchgraber & Stoffer (2006)] and for $b = -1$, $a \ge 0.265625$ by using the so called *shadowing techniques* in accordance with an old conjecture due to Devaney and Nitecki, claiming that the Hénon map admits a transversal homoclinic point in a region of interest. Namely, the following Theorem proved in [Devaney & Nitecki (1979)] states that the non-wandering set (recall Definition 12(a)) Ω $(H_{a,b})$ of the Hénon map $H_{a,b}\,(x, y) = (a + by - x^2, x)$ is topologically equivalent to the shift map of 2 symbols. This result indicates that the phenomena of the Hénon attractor are part of a bifurcation occurring in the creation of a horseshoe from nothing for $b \neq 0$.

Theorem 6 *(i) For $a < \dfrac{-(1+|b|^2)}{4}$ [4], then $(H_{a,b}) = \varnothing$.*

(ii) For $a > \dfrac{-(1+|b|^2)}{4}$ then, $\Omega\,(h)$ is contained in the square $S\,(a, b) = \{(x, y) \in \mathbb{R}^2,\ |x| \leq R(a, b)\},\ |x| \leq R(a, b)$, where $R(a, b) = \dfrac{1+|b|+\sqrt{(1+|b|^2)+4a}}{2}$.

(iii) For $a > 2\,(1+|b|^2)$, then $\Lambda = \cap_{n \in \mathbb{Z}} H^n_{a,b}\,(S)$ is a topological horse-shoe; for $b \neq 0$ there is a continuous semi-conjugacy of $\Omega\,(H_{a,b}) \subset$ onto the 2-shift.

(iv) For $a > \dfrac{(5+2\sqrt{2})(1+|b|^2)}{4}$, then, $\Lambda = \Omega\,(h)$ has a hyperbolic structure and is conjugate to the 2-shift.

The proof of Theorems 6 can be carried out by proving several Lemmas as in [Devaney & Nitecki (1979)]. Using the *graph transform method* in [Fontich (1990)] it was proved that the stable and unstable manifolds of the Hénon map $h_c\,(x, y) = (y, -x + 2y^2 + 2cy)$ intersect transversally for $c > 1$. Namely, *the following result:*

Theorem 7 *(a) For $c > 1$, the symmetric Hénon map h_c has a homoclinic point $p_c = (\mu_{c'} \mu_c) \neq (0, 0)$ on the intersection of its graph representation h_c^5 [5] with $y = x$.*
(b) The angle between the invariant manifolds at the homoclinic point $p_c = (\mu_{c'} \mu_c)$ is given by the function $\varphi : (1,\infty) \to \mathbb{R}$ defined by $\varphi\,(c) = 2\arctan\left(\dfrac{Dh_c(\mu_c)+1}{Dh_c(\mu_c)-1}\right)$.
(c) φ is an analytic function on $(1,\infty)$.

(d) For $c > 1.78$, we have $\varphi\,(c) > 2\arctan\left(1 - \dfrac{2}{\sqrt{1+\dfrac{16(c-1)^2}{2c-1}}}\right) > 0$.

The proof of Theorem 7 is based on the global graph transformation used to obtain estimates for certain invariant manifolds of the symmetric map under consideration. In [Tovbis (1998), Gelfreich & Sauzin (2001)] an investigation of exponentially small phenomena for $-1 < \rho << 0$ in the Hénon map given by $h\rho\,(x, y) = (x + \rho y, y + \rho x\,(1 - x))$ leads to the following

---

[4] This value is precisely the a-value at which the first fixed point of h^2 appears as H´enon remarked [Hénon (1976)], and statement (i) in [Devaney & Nitecki (1979)] comes from the Brouwer translation theorem [Brouwer (1912), Andrea (1965)].

[5] The function f_c is defined as the limit of a specified sequence of function $(f_k)_{k \in \mathbb{N}}$.

result proved in [Gelfreich (1991)] using the fact that the symmetry of the map implies that the intersection of separatrices with the horizontal axis is a homoclinic point:

Theorem 8 *The angle[6] between the stable and unstable separatrix at the first intersection of the separatrices with the horizontal axis of the map h_ρ is given asymptotically by* $\alpha = \dfrac{64\pi\, e^{\frac{-2\pi^2}{\rho}}}{9\rho^7}\, (|\Theta| + O(\rho)),\ \Theta \in \mathbb{C}.$

The last formula implies the exponentially small transversally of the homoclinic point for all small $\rho > 0$ if the factor $|\Theta|$ does not vanish. This fact implies that the Hénon map h_ρ has a homoclinic point for all small $\rho > 0$ as shown in [Gelfreich & Sauzin (2001)]:

Theorem 9 *In the case of the Hénon map h_ρ the splitting constant $|\Theta|$ does not vanish. More precisely, $\Theta\ i\mathbb{R}$ and $Im\Theta < 0$.*

The proof of Theorem 9 is based on a detailed study of the *Borel transform* of the formal separatrix of the parabolic fixed point of the map h_ρ. A proof of the existence of homoclinic orbits to a periodic orbit using shadowing theorems for $a = 1.4$, $b = 0.3$ for the Hénon map (2.1) was given in [Coomes, et al. (1997-2005)] using the *global Newton's method* and computer assisted methods. In [Pilyugin (1999), Kirchgraber & Stoffer (2004-2006)] the existence of a transversal homoclinic point for a Hénon map was proved using the shadowing Lemma:

Theorem 10 *(a) The Hénon map $h_{p,q}$ admits a transversal homoclinic point for all parameters p, q with $0 < |p| \le q \le \dfrac{1}{10}$, where $h_{p,q}\,(x, y) = \left(\dfrac{1}{q}[(1 + pq)\, x - x^2 - py], x\right)$. In terms of the classical parameters a, b this means:*

$|b| \in (0, 1]$ and $a \ge 20 + \dfrac{91}{20}b - \dfrac{19}{400}b^2.$

(b) The area and orientation preserving Hénon map $h_{p,q}$ admits a transversal homoclinic point for all parameters p, q with $p = q \in (0, \dfrac{1}{4}]$. In terms of the classical parameters a, b, this corresponds to $b = -1$ and $a \ge a_0 = \dfrac{17}{64} = 0.265625$[7].

---

[6] Called also the *splitting constant*, and for the map h_g one has that $|\Theta| \approx 2.474.10^6$ as shown in [Gelfreich & Sauzin (2001)].

[7] Theorem 10 generalizes one of the results in [Coomes, et al. (2001), Devaney & Nitecki (1979)] where they prove with computer assistance the existence of a transversal homoclinic point for $b = -1$ and $a = 1$ and for $b = -1$, $a \ge -0.866360$ respectively. Also, using the graph transforms method, it was shown in [Fontich (1990)] that the stable and unstable manifolds of the fixed point 0 intersect transversally for $a \ge -0.3916$.

Analytical methods for finding homoclinic point for the Hénon map (2.1) were given in [Marotto (1979), Misiurewicz & Szewc (1980)] for proving the existence of a transversal homoclinic point. The proof is based mainely on Marotto theorem [Marotto (1979)]:

Theorem 11 *The Hénon map (2.1) has transversal homoclinic point for all a >* 1.55 *and* $|b| < \varepsilon$, *for some* $\varepsilon > 0$.

However, for $a = 1.4$, $b = 0.3$, it was shown in [Misiurewicz & Szewc (1980)] that the Hénon map (2.1) has a transversal homoclinic point for the fixed point:

Theorem 12 *There exists a transversal homoclinic point for the fixed point P =* $\left(\dfrac{-0.7+\sqrt{6.09}}{2.8}, \dfrac{-0.7+\sqrt{6.09}}{2.8} \right)$ *of the Hénon map l* $(x, y) = (y, 1 - 1.4y^2 + 0.3x)$.

On the other hand, numerical analysis in [Hansen & Cvitanovic (1998)] suggested that the Hénon map $H_{a,b}(x, y) = (a + by - x^2, x)$ is not structurally stable in some parameter intervals such as $b = 0.3$ and $a \in [1.270, 1.420]$. This result implies that for a dense set of parameter values in this interval the Hénon family possesses a homoclinic tangency. Another rigorous computational method for finding homoclinic tangency and structurally unstable connecting orbits is proposed in [Arai & Mischaikow (2006)] based on several tools and algorithms, including the interval arithmetic, the subdivision algorithm, the *Conley index theory* [Mischaikow (2002)], and the computational homology theory:

Theorem 13 (a) *Fix any* b_0 *sufficiently close to 0.3. Then there exists a* $\in$ [1.392419807915, 1.392419807931] *such that the one-parameter family* f_{a,b_0} *has a generic[8] homoclinic tangency with respect to the saddle fixed point on the first quadrant.*

(b) *Fix any* b_0 *sufficiently close to* -0.3. *Then there exists a* $\in$ [1.31452..., 1.31452...] *such that the one-parameter family* f_{a,b_0} *has a generic homoclinic tangency with respect to the saddle fixed point on the third quadrant.*

Similar results depending on the analyticity of maps can be obtained as in [Fornæss & Gavosto (1992–1999)]. The comparison shows that the method presented in [Arai & Mischaikow (2006)] is rather geometric and topological, and it can be applied to a wider class of maps, i.e., a continuous family of C^2 diffeomorphisms is required for which one can compute the image of the maps using interval arithmetic. A good reference about homoclinic tangency for diffeomorphisms is [Gonchenko, et al. (2005(b))]. The uniform hyperbolicity of the Hénon mappings will be discussed in the following: First, there are many known results that give sufficient conditions

---

[8] The tangency in a one-parameter family is generic if the intersection of unstable and stable manifolds is quadratic, and the intersection is unfolded generically in the family.

for hyperbolicity, for example see [Hirsch & Pugh (1970), Newhouse & Palis (1971), Moser (1973), Palis & Takens (1993), Robinson (1999), Newhouse, (2004)]. Second, some of these results were applied to prove the well-known theorems for hyperbolicity of the set of bounded orbits in real and complex Hénon mappings using *complex methods* as in [Bedford & Smillie (2006)]. Thus, the following result was proved in [Newhouse (2004)]:

Theorem 14 *Consider the real or complex Hénon family* $H_{a,b}$ *with* $0 < |b| \leq 1$. *Let* $\Lambda_{a,b}$ *denote the set of points with bounded orbits. Assume that* $|a| >$

$\dfrac{(5+2\sqrt{2})(1+|b|^2)}{4}, 0 < |b| \leq 1$. *In the complex case or the real case with* $a > 0$, *we have that* $\Lambda_{a,b}$ *is a non-empty compact invariant uniformly hyperbolic set. In the real case with* $a < 0$, *the set* $\Lambda_{a,b}$ *is empty.*

In [Arai (2007(a))] another proof of the uniform hyperbolicity[9] was given for the Hénon map $H_{a,b}$ using rigorous computational method. If $\mathcal{R}(H_{a,b}) =$

$\dfrac{1+|b|+\sqrt{(1+|b|^2)+4a}}{2}$ is the *chain recurrent set* defined in [Arai (2007(a))] of

the Hénon family $H_{a,b}$, then the knowledge a priori of its size is needed to prove the following result:

Theorem 15 *There exists a set* $P \subset \mathbb{R}^2$,[10] *which is the union of 8943 closed rectangles, such that if* $(a, b) \in P$ *then* $\mathcal{R}(H_{a,b})$ *is uniformly hyperbolic. The set* P *is illustrated in Fig. 2.4 (shaded regions).*

Theorem 15 implies that on each connected component of P, no bifurcation occurs in $\mathcal{R}(H_{a,b})$ and hence numerical invariants such as the topological entropy, the number of periodic points, etc., are constant on it. On the other hand, Theorems 15 does not determine regions for non-hyperbolic parameter. More computations for the one-parameter area-preserving Hénon family $H_{a,-1}$ give a set P' of uniformly hyperbolic parameters such that $P \subset P'$. Namely, the following result proved in [Arai (2007(a))]:

Theorem 16 *If* a *is in one of the following closed intervals, then* $\mathcal{R}(H_{a,-1})$ *is uniformly hyperbolic.*

[4.5383300781250, 4.5385742187500],	[4.5388183593750, 4.5429687500000],
[4.5623779296875, 4.5931396484375],	[4.6188964843750, 4.6457519531250],
[4.6694335937500, 4.6881103515625],	[4.7681884765625, 4.7993164062500],
[4.8530273437500, 4.8603515625000],	[4.9665527343750, 4.9692382812500],
[5.1469726562500, 5.1496582031250],	[5.1904296875000, 5.3366210937500],
[5.5659179687500, 5.6077880859375],	[5.6342773437500, 5.6768798828125],
[5.6821289062500, 5.6857910156250],	[5.6859130859375, 5.6860351562500],
[5.6916503906250, 5.6951904296875],	[5.6999511718750, ∞),

---

[9] i.e., the existence of many regions of hyperbolic parameters in the parameter plane.

[10] The set P is called *"plateau"* because the hyperbolicity of the chain recurrent set $\mathcal{R}(H_{a,b})$ implies the Ω-stability.

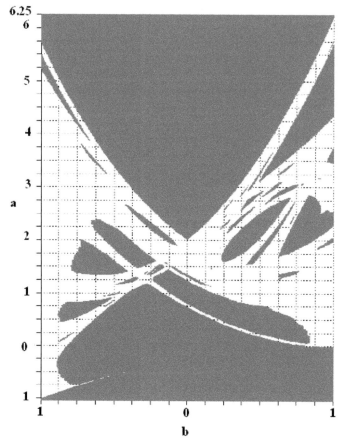

Figure 2.4: Uniformly hyperbolic "plateaus" for the Hénon map $H_{a,b}$. Reused with permission from Arai, Experimental Mathematics. Copyright 2007, Taylor & Francis Publishing.

Thus, it follows that $H_{a,-1}$ has a tangency when a is close to 5.699951171875. Hence, the following theorem was proved in [Arai (2007(a))] using the rigorous computational method developed in [Arai & Mischaikow (2006)]:

Proposition 2.1 *(a) There exists $a \in [5.6993102, 5.6993113]$ such that $H_{a,-1}$ has a homoclinic tangency with respect to the saddle fixed point on the third quadrant.*
 (b) When we decrease $a \in \mathbb{R}$ of the area-preserving Hénon family $H_{a,-1}$, the first tangency occurs in $[5.6993102, 5.699951171875)$.

The generalization of the above results to certain C^2 diffeomorphisms exhibiting hyperbolic invariant sets can be found in [Hoensch (2008)]. These maps are called *Hénon-like diffeomorphisms* and they have the form $H(x, y) = (rx(1 - x) - by, x)$. Hence, the main result proved in [Hoensch (2008)] for Hénon-like maps is given by:

Theorem 17 *There exists a $b_0 > 0$ such that for each $0 < b \leq b_0$ there is a unique $r = r(b)$ with the property that (a) if $r > r(b)$, the non-wandering set Ω (H) is uniformly hyperbolic, (b) if $r = r(b)$, there is a quadratic homo-clinic tangency between the stable and the unstable manifolds of the fixed point $(0,0)$. This homoclinic tangency causes the loss of hyperbolicity of the non-wandering set Ω (H). However, the set Ω (H)$\setminus\mathcal{O}(q)$ is uniformly hyperbolic in the sense that a first-return map on $\Lambda\setminus\mathcal{O}(q)$ is uniformly hyperbolic. ($\mathcal{O}(q)$ is the orbit of the homoclinic tangency). These statements are also true for C^2- perturbations of the Hénon maps, for these perturbations $r > r(b)$ corresponds to the situation of the map being "before the first tangency", and $r = r(b)$ corresponds to the situation of the map being "at the first tangency". These two situations are illustrated by* Fig. 2.5.

Here the sets $l^s_{1,1}$, $l^s_{1,2}$, l^u_1, and l^u_2 in Fig. 2.5 are some special elementary curves defined for the case of Hénon-like maps. Now, it is easy to claim from the above analysis and the theoretical proof given in [Newhouse (1980)] that the Hénon attractor is an example of a quasi-attractor, i.e., the homoclinic tangencies are everywhere dense in the parameter space and the quasi-attractor is a typical limit set for the Hénon map (2.1). From this point, another numerical proof was given to show that the Hénon map given by $h_2(x, y) = (a - x^2 + y, bx)$ is a quasi-attractor [Anishchenko, et al. (1998)]. Indeed, Fig. 2.6 shows the behavior of the stable and the unstable manifolds of the period-1 cycle obtained from the Hénon map h_2 for $a = 1.3$ and $b = 0.3$. We note that this feature is qualitatively the same for manifolds of saddle cycles of other periods. For 2-D mappings, manifolds of saddles are 1-D curves and chaotic attractors are located along the unstable manifolds of saddle cycles repeating their form.

For dissipative 2-D maps, the unstable manifolds of their saddles and their chaotic attractor must *be packed* in some bounded region of the phase plane. Hence, for smooth 2-D maps, the unstable manifolds inevitably undergo a bending in the form of *horseshoe* which leads to *dangerous* tangencies between the stable and unstable manifolds and, respectively,

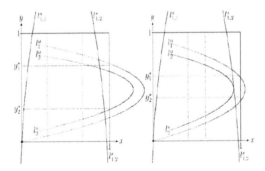

Figure 2.5: Parts of the invariant manifolds before the first tangency when $r > r(b)$ (left) and parts of the invariant manifolds at the first tangency when $r = r(b)$ (right). Reused with permission from U.A. Hoensch , Nonlinearity, (2008). Copyright 2008, IOP Publishing Ltd.

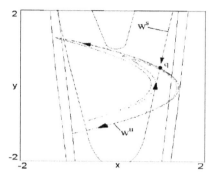

Figure 2.6: Stable and unstable manifolds of saddle points in the Hénon map h_2 for $a = 1.3$ and $b = 0.3$.

to the quasi-attractor. The algorithm described in Section 1.1.8 was used in [Anishchenko, et al. (1998)] to conclude that the attractor governed by the Hénon map f is a quasi-attractor.

In fact, the calculation of the following quantities is a sufficient condition for this purpose:

1. Calculate the angles ϕ between manifolds W^s and W^u of a chaotic orbit for different points of the attractor and the analysis of their statistics.
2. Calculate the distribution of the probabilities of the angle between the manifolds of a chaotic trajectory $P(\phi)$ on the quasi-attractor shown in Fig. 2.7(a).
3. Calculate the probability $P^{\delta\varphi}$ that the angle φ falls within a small neighborhood of zero ($\delta\varphi = 1^0$) as a function of the controlling parameter a. This means that the map f is not hyperbolic. The principle of such an algorithm is $P^{\delta\varphi} = 0$, if at a certain parameter value of a the trajectory does not have the unstable manifold.
4. Calculate the dependencies of Lyapunov exponents of chaotic orbits on the controlling parameters.

Hence, for the case of Hénon map h_2, the results are as follows: (a) Fig. 2.7(a) shows that in the neighborhood of zero, the probability $P(\phi)$ of the angle ϕ is finite and this fact implies the presence of tangency points of manifolds, i.e., a non-robust homoclinic curves of saddle cycles exist along which manifolds of the cycles approach each other tangentially. (b) In Fig. 2.7(b) the dependence of the probability $P^{\delta\varphi}$ in term of a is shown. It is clear that except for a set of a corresponding to $P^{\delta\varphi}(a) \neq 0$ of falling within the neighborhood of zero, there is a denumerable set of the a values for which $P^{\delta\varphi}(a) = 0$ that corresponds to windows of stability of periodic orbits. (c) In Fig. 2.7(c) the dependence of the largest Lyapunov exponent of the Hénon map h_2 on the variable a is displayed. Clearly, there is a set of *jumps* to the regions of negative values, which correspond to periodic windows. This property is typical for quasi-attractors. This last property is explained more carefully in the two following sections: The first is about the compound windows of the Hénon map by Edward Lorenz. The second is about the existence of infinitely many period-doubling bifurcations by E. Sander and J. A. Yorke that gave a general theory about this topic. At this stage, it is very interesting to contrast the works and results on the Lozi map in more detail with respect to the Hénon map or Lorenz-like maps[11]. Indeed, the Lozi map (3.1) was introduced as a simpler non equivalent version of the Hénon family (2.1). Thus, the similarities between the two maps are nonetheless striking. In fact, it is well known that while varying the parameter, the Hénon attractor is obtained via the standard period-doubling bifurcation

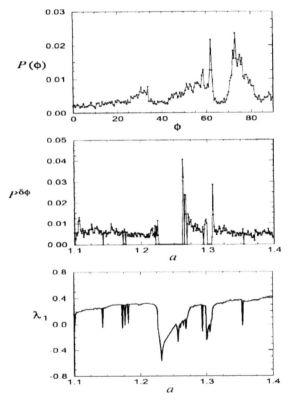

Figure 2.7: Calculation results of the characteristics for the Hénon attractor given by h_2. (a) A distribution of the probabilities of the angle ϕ between stable and unstable manifolds for $a =$ 1.179 and $b = 0.3$, (b) the probability that the angle ϕ falls within the interval $0 < \phi < 1$ ($\delta\varphi =$ 1^0) versus parameter a for $b = 0.3$ and (c) a dependence of the largest Lyapunov exponent on parameter a for $b = 0.3$. Reused with permission from Anishchenko, V. and Strelkova, G, Discrete Dynamics in Nature and Society (1998). Copyright 1998, Hindawi Publishing Corporation.

---

[11] i.e., two-dimensional Lorenz-like systems are the Poincaré return maps to a z =constant section for the classical Lorenz equations, at standard parameter $r = 28$.

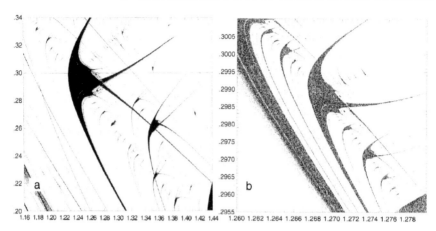

Figure 2.8: (a) Locations (shaded) in parameter space where a solution of equation $x_{n+1} = bx_{n-1} - x_n^2 + a$ (which is equivalent to the Hénon map (2.1)) has a positive probability of not blowing up (i.e., $|x_n|$ not going to infinity), as determined by random searching. Variations in shading indicate variations in probability of a blow-up. Along horizontal line $b = 0.3$ (b) The same as Fig. 2.8(a), but for locations where a solution has a positive probability of being chaotic. Horizontal scales indicate a, vertical scales indicate b. Reused with permission from Lorenz, Physica D, (2008). Copyright 2008, Elsevier Limited.

route to chaos as a typical behavior, unless for Lozi map (3.1), no period doubling route to chaos is allowed, and the attractor goes directly from a *border-collision bifurcation* developed from a stable periodic orbit. Thus, the two chaotic attractors go via different and distinguishable route to chaos as a typical behavior. As claimed in the above sections, the Lozi attractor has some properties close to the Lorenz attractor and it was shown theoretically that the Hénon attractor is an example of a quasi-attractor, while the Lozi map is an example of a quasi-hyperbolic or it is a Lorenz-type attractor. In fact, the formation of chaotic attractors is related to the behavior of manifolds of saddles, i.e., in 2-D maps, these manifolds are 1-D curves and chaotic attractors are located along the unstable manifolds of saddle cycles repeating their form. For dissipative 2-D maps, the unstable manifolds of saddles and the chaotic attractor must be confined in a bounded region of the phase plane. Thus, if the map is smooth, then the unstable manifolds undergo a *bending* in the form of horseshoe. Hence, a *dangerous* tangencies between the stable and unstable manifolds will be created as shown by the various results about the Hénon map presented in Section 2.1. Indeed, Fig. 2.6 shows the behavior of the manifolds of a saddle fixed point (period-1 cycle) for the Hénon map (2.1). Here, it is possible also to show that manifolds of saddle cycles of other periods behave qualitatively the same way. Since homoclinic tangencies are everywhere dense in the parameter space as shown in [Newhouse (1980)], the quasi-attractor is a typical limit set for the Hénon map (2.1). However, piecewise-smooth maps have no dangerous tangencies as shown in [Banerjee, et al. (1998)]. Fig. 1.2 show the behavior of the manifolds of a saddle fixed point for the Lozi map (3.1). This figure show that at the intersection of the stable and unstable manifolds the angle between them is not equal to zero. Thus, the intersection of these manifolds is everywhere transversal and there is no stable periodic orbits originating from the appearance of homoclinic trajectories. As a result, the hyperbolic chaotic set is the only attracting limit set for the Lozi map (3.1), i.e., the quasi-hyperbolic attractor.

2.4 Compound windows of the Hénon-map

In order to see clearly the structure of the chaotic attractor of the Hénonmap, we present in this section the so called *compound windows* observed for several chaotic attractors. This property is absent for some piecewise linear maps including the Lozi. Indeed, it was shown in [Lorenz (2008)] that the shapes and locations of the periodic windows-continua of parameter values for the Hénon map (2.1), for which solutions $x_0, x_1, \ldots$ can be stably periodic, are embedded in larger regions. In these regions, chaotic orbits or orbits of other periods prevail-are found by a random searching procedure and displayed graphically. From some numerical experiments, it was remarked that many windows have a typical shape and are called *compound windows*, i.e., a shape having a central *body* from which four narrow *antennae* extend as show in Fig. 2.8 for the map $x_{n+1} = bx_{n-1} - x^2_n + a$, which is equivalent to the Hénon map (2.1). These windows are often arranged in *bands*, to be called *window streets*. The portion of a window where a fundamental period prevails a stability measure U ($|U| < 1$) was introduced where curves of constant U are found by numerical integration. In fact, there is one line in parameter space where the Hénon map (2.1) reduces to the one-parameter logistic map, and two antennae (bounded by the curves $U = 1$ and $U = -1$) from each compound window intersect this line. But, as either curve with $U = -1$ leaves the line, it diverges from the curve where $U = 1$, crosses the other curve where $U = -1$, and nears the other curve where $U = 1$, forming another antenna. In this case, the region bounded by the numerically determined curves coincides with the subwindow founded by random searching. For given values of m, points in parameter space producing periodic solutions, i.e., $x_0 = x_m = 0$, belong to Cantor sets of curves that closely fit the window streets. While points producing solutions where $x_0 = x_m = 0$ and satisfying a third condition (Approximating the condition that x_n be bounded as $n \to \infty$.), belong to curves called street curves of order m. These curves approximate individual members of the Cantor set and individual window streets. In this structure, compound windows of period $m + m'$ are located near the intersections of street curves of orders m and m' with some exceptions for this general results. See Fig. 2.9.

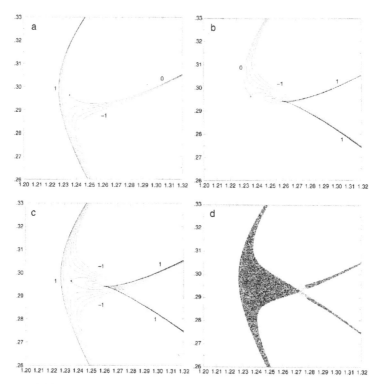

Figure 2.9: (a) Portions of U-curves from $U = 1$ to $U = -1$, at intervals of $1/4$, that extend from 7(21) logistic subwindow. (b) The same as Fig. 2.11(a), but extending from 7(11) logistic subwindow. (c) Superposition of Fig. 2.10(a) and (b). (d) Portion (shaded) of area covered by Fig. 1.11(a)–(c) where period 7 is stable, as determined by random searching. Horizontal scales indicate a, vertical scales indicate b, numbers beside curves indicate U. Reused with permission from Lorenz, Physica D, (2008). Copyright 2008, Elsevier Limited.

2.5 The existence of infinitely many period-doubling bifurcations

In this section, we discuss the existence of infinitely many period-doubling bifurcations for certain type of dynamical systems, namely, generic maps of the form:

$$F : \mathbb{R} \times \mathfrak{M} \to \mathfrak{M} \tag{2.3}$$

where $\mathfrak{M}$ is a smooth locally compact manifold without boundary, typically $\mathbb{R}^N$. The main concepts of this topic are the space of orbits under the Hausdorff metric, the set of flip orbits, the set of nonflip orbits, and cascades.

In [Sander & Yorke (2011)] a general theory[12] of cascades was developed for map (2.3). Indeed, it was observed for some examples that each cascade has infinitely many period-doubling bifurcations, i.e., whenever there are any cascades, there are infinitely many cascades. In particular, it was shown that there is a close connection between the transition through infinitely many cascades and the creation of a horseshoe, i.e., a rigorous explanation for the relation between cascades and chaos. The method of approach follows the general framework for cascades, even for observable systems. These results are valid in the C^∞ approach similar to *Milnor's treatment of Sard's Theorem* [Milnor (1965)] rather than for C^r, for a specific r, i.e., under the following criterion:

Hypothesis 1 (The setting). Let F defined by (2.3) be C^∞- smooth. We refer to it as a parameterized map on $\mathfrak{M}$. The notion of cascades for dynamical systems were first reported in [Myrberg (1962), May (1974)].

Definition 16 *We call a period-k orbit of the parameterized map (2.3) a flip orbit if its Jacobian matrix $D_x F^k (\lambda, x)$ has an odd number of eigenvalues less than -1, and -1 is not an eigenvalue. Otherwise the orbit is non flip.*

Let $PO_{nonflip}(F)$ be the space of nonflip orbits of F in $\mathbb{R} \times \mathfrak{M}$ under the Hausdorff metric. Let A be an open arc, then a cascade is a type of subarc of A defined by:

Definition 17 *Let k denote the smallest period of the orbits in A.*
(a) *A cascade is a half-open subarc that contains orbits with all of the periods k, 2k, 4k, 8k, such that it contains precisely one orbit of period k. We refer to such a cascade as a period-k cascade.*
(b) *A cascade is homeomorphic to $[0, 1)$ where 0 maps to the single period-k orbit.*
(c) *We call an open arc A bounded if there is a compact subset of $\mathbb{R} \times \mathfrak{M}$ that contains all the orbits of A. Otherwise A is unbounded.*
(d) *If a cascade is contained in an unbounded open arc, we call it unbounded cascade, otherwise we call it bounded cascade.*
(e) *(The orbit index): A topological index of a periodic orbit taking on a value in $\{0,-1,+1\}$. A periodic orbit in $PO_{nonflip}(F)$, has an orbit index of either -1 or $+1$.*

In fact, it was shown in [Sander & Yorke (2011)] that if a component is a bounded open arc, then it always contains two disjoint cascades called *paired cascades*. It was proved also that each component has a preferred orientation that can be determined at each hyperbolic orbit by computing its orbit index (see Theorem 18 (M_2). A preferred orientation via this orbit index was used

---

[12] i.e., general criteria for the existence of cascades for parameterized maps.

in Theorem 19 below to show that an open arc in a bounded parameter region contains a cascade. Some examples of dynamical systems illustrate these implications as follows [Yorke & Alligood (1983), Franks (1985)]:

(a) (Maps with Smale horseshoes in dimension 2). Assume that $\mathcal{M}$ is a two-manifold and F is generic. Assume there are parameter values λ_0 and λ_1 for which $F(\lambda_0, .)$ has at most finitely many nonflip orbits, and $F(\lambda_1, .)$ has infinitely many nonflip saddles and at most a finite number of nonflip attractors and repellers. Then F has infinitely many cascades between λ_0 and λ_1.

(b) (Maps with *Geometric Off-On-Off Chaos* in dimension 2). These are maps that first have no chaos, then chaos appears, and then it disappears as is varied. This scenario appears to happen with the time-2π maps of the single-well and double-well Duffing and the forced damped pendulum.

(c) (*Large-scale perturbation of quadratic maps*). Consider the parameterized one-dimensional map:

$$F(\lambda, x) = \lambda - x^2 + g(\lambda, x) \qquad (2.4)$$

where and $x \in \mathbb{R}$, and $g : \mathbb{R} \times \mathbb{R} \to \mathbb{R}$. For each generic, smooth, C1-bounded function g, the map F has exactly the same number of unbounded period-k cascades as occur in the case $g = 0$, and we give a recursive formula for that number. F may have extra cascades, but they are all bounded paired cascades. That is, the number of unbounded cascades is robust under largescale perturbations.

(d) (*Large-scale perturbations of the cubic map*). The results in Corollary. 2.1(c) are not related to whether the base map is quadratic. Similar behavior occurs for the parameterized cubic map (2.4) below.

This parameterized map has infinitely many cascades for a residual set of C^1 functions g that are C^1 bounded. The number of period-k cascades differs from the quadratic case. This is to be expected since the behavior of the cubic map for large λ reflects a shift on three symbols and so is more complex than that of the tent map (3.6).

(e) (*N-dimensional coupled systems*). This is an extension of the result of Corollary 2.1 (c) to the case of N coupled quadratic maps. Such cascades are a collective phenomenon for high-dimensional coupled systems. Corollary 2.1 (a) follows from the fact that all but a finite number of the nonflip orbits in the horseshoe are in components that contain either one or two cascades, and all but a finite number of the distinct nonflip orbits in the horseshoe are in different components.

2.5.1 Components and generic bifurcations in the space of orbits

In this section, we present definitions and properties of both components and generic bifurcations in the space of orbits for the general generic map (2.3). In fact, there is a residual set of (2.3) for which every orbit is either hyperbolic or is a generic bifurcation orbit (Proposition. 2.3 below).

Definition 18 *(a) (Orbits, flip orbits, and nonflip orbits). Write $[x]$ for the orbit (always means periodic orbit) of the periodic point x. By period of an orbit or point, we mean its least period. If x is a periodic point for $F(\lambda, .)$, then we sometimes say $\sigma = (\lambda, x)$ is a periodic point and write $[\sigma]$ or $(\lambda, [x])$ for its orbit. Let $\sigma = (\lambda, x)$ be a periodic point of period p of a smooth map $G = F(\lambda, .)$. We refer to the eigenvalues of σ or $[\sigma]$ as shorthand for the eigenvalues of Jacobian matrix $DG^p(x)$. All the points of an orbit have the same eigenvalues.*

(b) We say that $[\sigma]$ is hyperbolic if none of its eigenvalues have absolute value 1. We say it is a flip orbit if the number of its eigenvalues (adding multiplicities) less than -1 is odd, and -1 is not an eigenvalue.

(c) We call all other orbits nonflip orbits. Define

$$\begin{cases} PO(F) = \{[\sigma] : [\sigma] \text{ is an orbit for } F\} \\ PO_{nonflip}(F) = \{[\sigma] \in PO(F) : [\sigma] \text{ is a nonflip orbit for } F\} \end{cases} \quad (2.5)$$

The set $PO(F)$ (resp. $PO_{nonflip}(F)$) is the space of flip (resp. nonflip) orbits of F in $\mathbb{R} \times \mathscr{M}$ under the Hausdorff metric, i.e., the distance between two orbits in the space $PO(F)$ (resp. $PO_{nonflip}(F)$) is defined using the Hausdorff metric.

Definition 19 *(a) We say that two orbits are close in $\mathscr{M}$ if every point of each orbit is close to some point of the other orbit. The periods of the two orbits need not be the same.*

(b) (Hausdorff metric on sets). For a compact set S of $\mathscr{M}$ and $\epsilon > 0$, let $B(\epsilon, S)$ be the closed neighborhood of S. Let S_1 and S_2 be compact subsets of M. (We are only interested in the case where these sets are orbits). Assume ϵ is chosen as small as possible such that $S_1 \subset B(\epsilon, S_2)$ and $S_2 \subset B(\epsilon, S_1)$.

Then the Hausdorff distance $dist(S_1, S_2)$ between S_1 and S_2 is defined to be ϵ. Let $\sigma_j = (\lambda_j, x_j)$ for $j = 1, 2$ be orbits. We define the distance between $[\sigma_1]$ and $[\sigma_2]$ to be:

$$dist([\sigma_1], [\sigma_2]) = dist([x_1], [x_2]) + |\lambda_1 - \lambda_2| \quad (2.6)$$

For example, we have

$$dist((\lambda, [x(\lambda)]), (\lambda_*, [x_*])) \to 0 \text{ as } \lambda \to \lambda_* \quad (2.7)$$

if $(\lambda, [x(\lambda)])$ is a family of period-$2p$ orbits that bifurcate from the period-p orbit $(\lambda_*, [x_*])$ due to a period-doubling bifurcation. Generally, periodic points are dense in compact chaotic set, but the use of Hausdorff metric changes the geometry. Indeed, let (λ, x) be a saddle fixed point, then every $\mathbb{R} \times \mathcal{M}$ neighborhood of it with a transverse homoclinic intersection has infinitely many periodic points y_m. But the fact that x is a hyperbolic saddle point, implies that some points in the orbit of each y_m are far from x. Hence, the orbits $[y_m]$ do not converge to $[x]$ in the Hausdorff metric.

Definition 20 (a) (*Cascade of period m*). *The term component means a connected component of $PO_{nonflip}(F)$ in the Hausdorff metric. An arc is a set that is homeomorphic to an interval. We call it an open arc if the interval is open or half-open if that describes the interval.*

(b) *A period-doubling cascade[13] of period m is a half-open arc C in $PO_{nonflip}(F)$ with the following properties. Let $h : [0, 1) \to C$ be a homeomorphism.*
(i) *The set of periods of orbits in C is m, 2m, 4m, 8m, ...*
(ii) *m is the minimum period of orbits in the component that contains C.*
(iii) *The set C has no proper connected subset with properties (i) and (ii).*

Note that $h(0)$ will be the only orbit of period m in C and it will be a period-doubling bifurcation orbit. If a component contains a cascade, we refer to the component as a cascade component.

(iv) *For generic F, the following additional property hold ([Sander & Yorke (2011)]): If $\{p_k\}_1^\infty$ is the sequence of periods of the orbits, ordered so that for each k the k +1 orbit lies between (using the ordering induced from $[0, 1)$) the k orbit and the $k + 2$ orbit, then no period will occur more than a finite number of times. That implies $\lim_{k \to \infty} p_k = \infty$.*

As an example of arcs and a cascade, we consider the map:

$$F(\lambda, x) = \lambda - x^2 \tag{2.8}$$

Map (2.8) has exactly one saddle-node bifurcation, and exactly one period-doubling bifurcation for each period 2^k. Indeed, the smallest λ for which there is an orbit is $\frac{-1}{4}$ and that is a saddle-node fixed point $Q = \left(\frac{-1}{4}, \frac{-1}{2}\right)$. There is a unique periodic attractor for each $\lambda \in J = \left(\frac{-1}{4}, \lambda_{Feig}\right)$ where λ_{Feig}

is the end of the first cascade. The number λ_{Feig} is called *Feigenbaum limit parameter* [Sander & Yorke (2011)]. For $\lambda \in J$, the attractors of (2.8) constitute a component C of the attractors in $PO_{nonflip}(F)$. Indeed, for each $\lambda \in J$ there is a unique attracting orbit x in C and for each orbit in C there is a unique λ in J which means that the map on C defined by:

$$(\lambda, [x(\lambda)]) \to \lambda \tag{2.9}$$

---

[13] These cascades can be unbounded because each of them lie in a compact subset of $\mathbb{R} \times \mathcal{M}$.

is a homeomorphism. Thus, C is an open arc since it is homeomorphic to the interval J. In this case, the cascade is a subarc $C_1 \subset C$. Clearly, for $\lambda = \frac{3}{4}$ the orbit in C is the period-doubling fixed point $Q_1 = \left(\frac{3}{4}, \frac{1}{2}\right)$ while for $\lambda > \frac{3}{4}$, the orbits in C have period greater than 1. The cascade C_1 can be defined by:

$$C_1 = \left\{ \text{the orbits in } C \text{ for which } \lambda \geq \frac{3}{4} \right\} \tag{2.10}$$

Hence, C_1 is the smallest subarc of C for which all periods 2^k occur. On the other hand, the arc C is not a maximal arc because there is another arc C_2 of unstable fixed points $\left(\lambda, y(\lambda) = \dfrac{-1+\sqrt{1+4\lambda}}{2}\right)_{\lambda \in \left(-\frac{1}{4}, +\infty\right)}$. The two arcs C and C_2 terminate at the saddle-node fixed point Q. Thus, the component containing C is $R_c = C \cup C_2 \cup \{Q\}$. It is maximal because on one extreme the period goes to ∞ and on the other extreme, $\lambda \to \infty$. Finally, the cascade C_1 is unbounded, since, each point of the arc is a different orbit and the set of λ values in the maximal arc is unbounded. Contrary to the situation of the map (2.8), a cascade is generally quite complicated and does not have such regular behavior and known type of bifurcations. There are three kinds of generic orbit bifurcations as shown in [Robinson (1995)]:

Definition 21 (a) *(Generic orbit bifurcations). Let F satisfy* **Hypothesis.1**. *We say a bifurcation orbit P of F is generic if it is one of the following three types:*
    (i) A generic saddle-node bifurcation (having eigenvalue +1).
    (ii) A generic period-doubling bifurcation (having eigenvalue −1).
    (iii) A generic Hopf bifurcation[14] *with complex conjugate eigenvalues which are not roots of unity.*
  (b) F is generic if each non-hyperbolic orbit is one of the above three types.

As examples illustrating Definition 2.6, the maps with Hopf bifurcations for which the eigenvalues are not complex roots of unity are generic because the roots of unity are countable. While period tripling bifurcations or multiples other than two bifurcations are not generic. Generally, for parameterized maps with more than one unstable dimension, cascades do not contain attractors because the existence of attractors requires the existence of at most one unstable dimension, which implies that there are no Hopf bifurcations.

---

[14] The only bifurcation orbits in the case of a generic Hopf bifurcation are saddle-node bifurcation orbit.

Hypothesis 2 (Generic bifurcations). Assume **Hypothesis 1**. Assume that each orbit of F is either hyperbolic or is a generic bifurcation orbit.

Generally, under **Hypothesis 2**, the period of the orbits in C is locally constant near hyperbolic orbits, saddle-node and Hopf bifurcations. It can change by a factor of two only at period-doubling bifurcations. Hence, we have the following result:

Proposition 2.2 *An arc C in $PO_{nonflip}(F)$ is a cascade if the sequence of periods $\{p_k\}$ of the non-hyperbolic orbits in C (such that orbit $k+1$ is between orbit k and orbit $k+2$) tends to infinity. This case is possible if the sequence is infinite and no period occurs more than a finite number of times.*

The following result was proved in [Sander & Yorke (2011)]:

Proposition 2.3 *(Generic F constitute a residual set). There is a residual set $S \subset C^\infty$ of parameterized maps F satisfying **Hypothesis 1** for which all bifurcation orbits are generic. This residual set S is C^1 dense in the uniform C^1 topology, that is, for each $F \in C^\infty$, there is a sequence $(F_t) \subset S$ such that:*

$$\|F - F_t\|_{C1} \to 0 \text{ as } t \to \infty \tag{2.11}$$

The proof of Proposition 2.3 uses standard transversality arguments with minor changes of the proof given in [Palis & Takens (1987)]. In order to show that components are one-manifolds for the generic map (2.3), we need to define the index orientation as a homeomorphism satisfying some conditions [Sander & Yorke (2011)]:

Definition 22 *(Index Orientation). (a) Assume that a periodic orbit y of period p of a smooth map G is hyperbolic. Define the unstable dimension $\dim_u(y)$ to be the number of real eigenvalues (with multiplicity) having absolute value > 1.*

*(b) Let F satisfy **Hypothesis 2**, and assume that Q is a component that is a one-manifold. Then we know that there is a homeomorphism $h : X \to Q$ where X is either the interval $(-1,+1)$ or a circle, which we will write as $[-1,+1]/\{-1,+1\}$. For each $s \in X$, let $h_\lambda(s)$ denote the projection of $h(s)$ to the corresponding parameter value. Thus h_λ is a map from $\mathbb{R}$ to $\mathbb{R}$, and X and λ both inherit an orientation from the real numbers. Therefore we can describe h as increasing or decreasing at s whenever h_λ is increasing or decreasing at s.*

We say the homeomorphism h is an index orientation for Q if whenever $h(s)$ is a hyperbolic orbit, $h_\lambda(s)$ is locally strictly increasing when $\dim_u(h(s))$ is odd and is locally strictly decreasing when $\dim_u(h(s))$ is even.

The following results were proved in [Sander & Yorke (2011)]:

Theorem 18 *(Components are oriented one-manifolds for generic F). Consider all F as in **Hypothesis 1**. There is a residual set of such F for which each component (of $PO_{nonflip}(F)$)*

(M_1) *is a one-manifold, i.e., is either a simple closed curve or is homeomorphic to an open interval, and*

(M_2) *it has an index orientation.*

The proof of Theorem 18 (M_1) is based on the two following ideas:

1. Show that in a neighborhood of any hyperbolic orbit, its component is an arc.
2. Show the same for each non-hyperbolic orbit and remark that each point in a component has a neighborhood in the component that is an arc.

Proposition 2.4 *(The neighborhood of a generic bifurcation orbit). Assume F satisfies* **Hypothesis** *2. Assume* $P = (\lambda_0, [x_0])$ *is a generic bifurcation orbit in* $PO_{nonflip}(F)$ *and let C be its component. Then P has a neighborhood in C that is an open arc in which it is the only bifurcation orbit.*

The proof of Proposition. 2.4 is achieved by applying *the Center Manifold Theorem* [Guckenheimer & Holmes (1997)] and considering individually the three generic orbit bifurcations given in Definition 21. Then applying the Hopf bifurcation Theorem.

Now, let A be a component of the generic map (2.3) that is an open arc and let m be the minimum period of the orbits in A. Assume that $h(0)$ (see Definition 22) is a hyperbolic orbit[15] with period m. Let $A^- = h((-1, 0))$, $A^+ = h((0, 1))$. Let $Per(u)$ be the period of an orbit $u \in PO(F)$.

Definition 23 *(a) We say that a set of orbits in* $PO_{nonflip}(F)$ *is bounded if the union of its orbits lies in a compact subset of* $\mathbb{R} \times \mathfrak{M}$.
(b) We say an open arc A has a bounded end if either A^- or A^+ is bounded.
(c) We refer to A^- and A^+ as the ends of the component.

Note that the property of boundedness of an end is independent of the manner where the arc A is split. The following result was proved in [Sander & Yorke (2011)]:

Proposition 2.5 *(Bounded Cascades). Assume* **Hypothesis** *2. If a component A is an open arc, and one of its ends is bounded, then that end contains a cascade. If the entire component A is bounded, then A contains two cascades, and these are disjoint.*

The proof of Proposition 2.5 is done for the case where A is bounded because the case of bounded one end uses exactly the same method. The fact that A has no limit points under **Hypothesis 2** induces a property called

---

[15] The point $h(0)$ is not in a cascade because a cascade contains only one orbit of smallest period, it is not hyperbolic.

isolation of generic bifurcation orbits of period $\leq p$, that is, for each period p, the map (2.3) has at most a finite number of non-hyperbolic orbits of period p in each bounded region of the set $\mathbb{R} \times \mathcal{M}$. The orientation for components is discussed in what follows by taking into account that an arc in $PO_{nonflip}(F)$ has two orientations and one of them is consistent with a specific topological invariant called the orbit index defined in [Mallet-Paret & Yorke (1982)], where it is defined for all isolated orbits (for flows) by:

Definition 24 *(Orbit index). Assume that an orbit y of period p of a smooth map G is hyperbolic. Based on the eigenvalues of y, we define $\sigma^+ = \sigma^+(y) = $ the number of real eigenvalues (with multiplicity) in $(1, +\infty)$. $\sigma^- = \sigma^-(y) = $ the number of real eigenvalues (with multiplicity) in $(-\infty, 1)$. The fixed point index of y is defined as*

$$ind(y) = (-1)^{\sigma^+} \tag{2.12}$$

From the definition of fixed point index, it follows that

$$ind(x, G^{pm}) = \begin{cases} (-1)^{\sigma^+} \text{ for } m \text{ odd} \\ (-1)^{\sigma^+ + \sigma^-} \text{ for } m \text{ even} \end{cases} \tag{2.13}$$

Since σ^+ and σ^- are the same for each point of an orbit, we can define the orbit index of a hyperbolic orbit

$$\phi\left([x]\right) = \begin{cases} (-1)^{\sigma^+} + \text{ if } \sigma^- \text{ is even} \\ 0 \text{ if } \sigma^- \text{ is odd} \end{cases} \tag{2.14}$$

From Definition 22 we have the following result:

Proposition 2.6 (a) *If [x] is a nonflip hyperbolic orbit, then we have*

$$\phi([x]) = ind(x) \tag{2.15}$$

(b) *A hyperbolic orbit is a flip orbit if and only if its orbit index is zero. Thus for every $[x] \in PO_{nonflip}(F)$, $\phi([x])$ is ± 1 (never zero).*

The following proposition was proved in [Sander & Yorke (2011)] and it is a stronger version of Theorem 18 (M_2):

Proposition 2.7 *(Each component has an index orientation). Let F satisfy* **Hypothesis 2**, *and let Q be a component. Let $\psi : X \to Q$ be a homeomorphism where X is the circle or interval in Definition 22. Define the homeomorphism $\psi^* : X \to Q$ by $\psi^*(s) = \psi(s)$ for all $s \in X$. Then either ψ or ψ^* is an index orientation.*

Proposition 2.7 states that each component has an index orientation, i.e., the behavior of this index near each generic bifurcation. The restriction of oriented arcs of map (2.3) to a region U with a bounded parameter range is described in the following: **Hypothesis 3** (Orbits near the boundary). Let F satisfy **Hypothesis 2**. Let $\lambda_0 < \lambda_1$, and let $U = [\lambda_0, \lambda_1] \times \mathcal{M}$ and $\partial U = [\{\lambda_0, \lambda_1\}]$

× $\mathfrak{M}$. Assume that all orbits in ∂U are hyperbolic. Assume that all orbits in U are contained in a compact subset of $\mathfrak{M}$.

Definition 25 *Assume* **Hypothesis 3** *and its notation. Let* $p \in PO_{nonflip}(F)$ *be a hyperbolic orbit in* ∂U. *If* p *is oriented in the region* U *by index orientation, then it is called an entry orbit of* U. *That is,* p *is an entry orbit if either* $\lambda = \lambda_0$ *and* ϕ $(p) = +1$, *or* $\lambda = \lambda_1$ *and* ϕ $(p) = -1$. *Otherwise, it is called an exit orbit of* U. *A cascade is said to be essentially in* U *if all but a finite number of its bifurcation orbits are in* U.

The following Theorem was proved in [Sander & Yorke (2011)]:

Theorem 19 *(Cascades from boundaries). Assume* **Hypothesis 3**. *Let IN be the set of entry orbits. Let OUT be the set of exit orbits. Assume that IN contains K elements, and OUT contains J elements. We allow one but not both of the sets to have an infinite number of elements.*

(C_K) *If* $K < J$, *then all but possibly* K *orbits in OUT are contained in distinct components, each of which contains a cascade that is essentially in* U.

Likewise, if $J < K$, *then all but possibly* J *orbits in IN are contained in distinct components, each of which contains a cascade that is essentially in* U.

(C_0) *If* $J = 0$ *or* $K = 0$, *then the non flip orbits of* ∂U *are in one-to-one correspondence with the components that intersect the boundary. Each of these components has one cascade that is essentially in* U.

Corollaries 21(a)-2.1(e) are deduced in [Yorke & Alligood (1983), Franks (1985)]. The comparison with Theorems 18, 19 gives the followings results:

1. In [Yorke & Alligood (1983-1985)], a proof was given for the existence of cascades of attracting periodic points for area contracting maps and for elliptic periodic points for area preserving maps. Especially, Theorem 19 (C_0), was obtained without assuming genericity.
2. Theorems 18, 19 apply to parametrized maps with a large number of unstable dimensions, while Corollaries 2.1(a)-2.1(e) apply only for maps with at most one unstable dimension.
3. Corollaries 2.1(a)-2.1(e) and Theorems 18, 19 require snakes[16] in the generic case and smooth convergence arguments who are no longer apply when there is more than one unstable dimension.

Definition 26 *The Morse index is the number of unstable eigenvalues. If the Morse index is even, the orbit index is either* 0 *or* +1. *If the Morse index is odd, the orbit index is either* 0 *or* −1.

---

[16] A *snake* is a maximal path of orbits that contains no orbits whose orbit index is 0. See [Paret & Yorke (1982)].

It was proved in [Franks (1985)] that there are cascades under some conditions:

Theorem 20 *Let d be an odd integer. Assume that for every non negative integer k, every orbit of the continuous map (2.3) of period $2^k d$ has a Morse index with the same parity (all are odd or all even) at $F(\lambda_0, 0)$, and the opposite parity at $F(\lambda_1, 0)$. Then map (2.3) has cascades.*

In other words, on the boundary $F(\lambda_0, 0) \cup F(\lambda_1, 0)$, all orbits are entry orbits (or alternatively all are exit orbits). The proof of Theorem 20 is based on *the Lefschetz trace formula* [Lefschetz (1926)] which allows the smoothness of F to be relaxed. Here $F(\lambda_0, 0)$ and $F(\lambda_1, 0)$ are smooth maps. Theorem 20 does not give information about the possible type of bifurcations and assure only that the component of $PO(F)$ containing the original hyperbolic orbit of period $2^r d$ (*d* odd) contains flip orbits of period $2^k d$ for all $k \in \mathbb{N}$ on the boundary $F(\lambda_0, 0) \cup F(\lambda_1, 0)$.

2.5.2 Examples of cascades for some classes of functions

In this section, we describe a collection of examples of classes of parametrized maps, each with an infinite number of cascades. The reader can draw the bifurcation diagram since the phenomena described are seen over a wide range of parameter intervals.

2.5.2.1 Parametrized maps with horseshoes

The first example is a parametrized maps with horseshoes in 2-D. A result was proved in [Sander & Yorke (2011)] about this context: (Creating a Smale horseshoe). Let the dimension of $\mathfrak{M}$ be 2. Assume **Hypothesis 2**. Let $W = [0, 1] \times \mathfrak{M}$. Assume the following:

(S_0) $F_0 = F(\lambda_0, 0)$ has at most a finite number of saddle orbits.

(S_1) $F_1 = F(\lambda_1, 0)$ has at most a finite number of orbits that are either attractors or repellers, and all its orbits are hyperbolic.

(S_2) F_1 has infinitely many nonflip saddle orbits.

(S_3) There is a compact subset of W that contains all the orbits in W.

Then there are infinitely many cascades whose components have a bounded end in W. Corollary 2.2 say that the creation of a Smale horseshoe in dimension two implies the existence of infinitely-many cascades and this is a significant generalization of the result proved in [Yorke & Alligood (1983)]. Assumption (S_1) implies that F_1 has no bifurcation orbits, because there are at most countably many bifurcation orbits. Thus, there are at most countably many values of λ at which there are bifurcation orbits. It was conjectured in [Sander & Yorke (2011)] that it is always possible to choose λ_1 to assure that there are only finitely many attracting or repelling orbits. Some additional properties about diffeomorphisms with infinitely many coexisting sinks can be found in [Newhouse (1974), Gorodetski & Kaloshin (2007), Tedeschini-Lalli & Yorke (1985), Nusse & Tedeschini-Lalli (1992)]. There are many examples of dynamical systems that satisfy conditions of Corollary 2.2 for some values of λ_0 and λ_1. The first example of this situation is the *the Ikeda map* :

$$F\,(\lambda, z) = \lambda + 0.9z \exp\left(i\left(\frac{0.4 - 6.0}{1 + |z^2|} \right) \right) \qquad (2.16)$$

defined for $z \in \mathcal{M} = \mathbb{C}$. Map (2.16) models the field of a *laser cavity* [Hammel, et al. (1985)]. For this map a globally attracting fixed point appears at $\lambda = 0$.

For $\lambda = 1.0$, four types of solutions appear: a global chaotic attractor, homoclinic points, an attracting fixed point and no repellers. The second example of this situation is the *Pulsed Rotor map* defined by:

$$F\,(x, y) = (x + y\ (mod\,2\pi),\ 0.5y + \lambda \sin\,(x + y)) \qquad (2.17)$$

where $(x, y) \in \mathcal{M} = \mathbb{S}^1 \times \mathbb{R}$. For $\lambda = 0$, a saddle fixed point and an attracting fixed point appears. The attracting fixed point attracts everything except for the stable manifold of the saddle. A chaotic attractor and a fixed point with a transverse homoclinic point appears for $\lambda = 10$. The third example is the *Geometric Off-On-Off Chaos* defined by:

Definition 27 *Assume the dimension of $\mathcal{M}$ is 2. We call a map $G : \mathbb{R} \times \mathcal{M} \to \mathcal{M}$ a Geometric Off -On-Off -Chaos map if it satisfies the following properties: (D_0) There are values $\Lambda_1 < \Lambda_3$ such that $G_1 = G(\Lambda_1, .)$ and $G_3 = G(\Lambda_3, .)$ each have at most a finite number of orbits, whose total is k.*

(D_1) *There is a $\Lambda_2 \in (\Lambda_1, \Lambda_3)$ for which $G_2 = G(\Lambda_2, .)$ has at most a finite number of orbits that are attractors or repelers, and all of its orbits are hyperbolic.*

(D_2) *G_2 has infinitely many nonflip saddle orbits.*

(D_3) *There is a compact subset of $W = [\Lambda_1, \Lambda_3] \times \mathcal{M}$ that contains all of the orbits in W.*

(D_4) *G satisfies **Hypothesis 2**.*

The following result was proved in [Sander & Yorke (2011)]: Assume G is a Geometric Off-On-Off Chaos map. Then there are infinitely many pairs of cascades in $W = [\Lambda_1, \Lambda_3] \times \mathcal{M}$, where the two cascades of each pair are in the same component of $PO_{nonflip}(G)$. Also there are at most k unbounded cascades that have values only in $[\Lambda_1, \Lambda_3]$. Examples of Definition 27 are geometric

Lorenz models [Guckenheimer (1976)], double-well Duffing equation[17] for $\lambda \in \{1.8, 20, 73, 175, 350\}$ and forced damped pendulum defined by:

$$\frac{d^2\theta}{dt^2} + 0.3\frac{d\theta}{dt} + \sin\theta = \lambda\cos t \qquad (2.18)$$

The method of analysis is based on the investigation of the time-2π map on the set $\mathcal{M} = S^1 \times \mathbb{R}$; that is, the first variable is $(mod2\pi)$ and the second is $\frac{d\theta}{dt} \in \mathbb{R}$. Indeed, at $\lambda = 0$, there are only two periodic orbits, both fixed points, an attractor and a saddle. A global chaotic attractor and homoclinic points appears at $\lambda = 2.5$. For $\lambda \geq 10$, the two fixed points are the only orbits. The friction term $0.3\frac{d\theta}{dt}$, imply that the orbits must lie in a compact subset of $\mathcal{M}$ for $\lambda \in [-10, 10]$ which confirm that the system (2.18) is a Geometric Off-On-Off Chaos map with either $\Lambda_1 = 0$, $\Lambda_2 = 2.5$, and $\Lambda_3 = 10$, or by symmetry, with $\Lambda_3 = 0$, $\Lambda_2 = -2.5$, and $\Lambda_1 = -10$. Finally, system (2.18) has at most $k = 4$ unbounded cascades and an infinite number of bounded pairs of cascades.

2.5.2.2 Large-scale perturbations of a quadratic map

The second example of classes of parametrized maps (each with an infinite number of cascades) is the *Large-scale perturbations* of a quadratic map (2.8) of the form:

$$F(\lambda, x) = \lambda - x^2 + g(\lambda, x) \qquad (2.19)$$

The method for counting cascades of period k for each k for the map (2.19) is based on *the number* $\Gamma(1, k)$ which denote the number of cascades in terms of the *tent map* (2.6). This number is defined by:

Definition 28 (a) *An orbit is nonflip if it has an even number of points in* $(0.5, 1]$.
  (b) *If the number of points is odd, it is a flip orbit. Nonflip period-k orbits are the orbits whose derivative satisfies* $\frac{d(T^k)}{dx}(x) > +1$.
  (c) *The derivative is* < -1 *for flip orbits and no orbits have derivative in* $[-1, 1]$.
  (d) $\Gamma(1, k)$ *is the number of period-k nonflip orbits of the tent map (2.6) and the entry "1" refers to the dimension of x.*

---

[17] It was proved in [Zakrzhevsky (2008)] that this system is chaotic by showing the existence of Smale horseshoe for a restoring force of $u^3 - u$ instead of $u^3 - u$.

The following result was proved in [Sander & Yorke (2011)]: (A large-scale perturbation of the parametrized quadratic map). Assume that $F : \mathbb{R} \times \mathbb{R} \to \mathbb{R}$ has the form (2.19), where $g : \mathbb{R} \times \mathbb{R} \to \mathbb{R}$ is C^∞. Assume that there is $\beta > 0$ such that for all λ and x, we have

$$| g\,(\lambda, 0)\,| < \beta, \text{ and } \left| \frac{\partial g(\lambda, x)}{\partial x} \right| < \beta \qquad (2.20)$$

Then for a residual set of g, for each positive integer k, the number of unbounded period-k cascades is $\Gamma\,(1, k)$, which is the same as for the map (2.8). Corollary 2.4 imply that unbounded cascades cannot be destroyed, i.e., the cascades still exist outside the set $S = \{(\lambda, x) \in [-\gamma, \gamma] \times [-\gamma, \gamma]\}$ for some $\gamma \geq 0$ sufficiently large, if g is chosen so that F is generic. In addition, bounded pairs of cascades can exists for some cases and for $\lambda = \lambda_H$ sufficiently large, the map $F\,(\lambda_H, .)$ is a *two-shift horseshoe map*[18] defined by:

Definition 29 (*Two-shift horseshoe map in dimension one*). *We refer to a C^1 one-dimensional function $G : \mathbb{R} \to \mathbb{R}$ as a two-shift horseshoe map when it has the following properties:*
 (1) there is a closed interval J and two non-empty disjoint intervals $J_1 \subset J$ and $J_2 \subset J$ such that $G(J_1) = G(J_2) = J$.
 (2) $G(x) \in J$ implies $x \in J_1 \cup J_2$.
 (3) $G'(x) < -1$ for $x \in J_1$ and $G'(x) > 1$ for $x \in J_2$.

In other words, there is a one-to-one correspondence between the *unbounded period-M cascades* and the *period-M orbits of nonzero index* in the two-shift horseshoe. The so called *Stem period* describe more precisely this relationship:

Definition 30 (*Stem period*). *Assume* **Hypothesis 2**. *Let C be a cascade in an unbounded component Q. Assume there is a compact subset B of $\mathbb{R} \times \mathfrak{M}$ such that all orbits of C that do not lie in B have the same period. We call that period the stem period of C.*

In the case of large-scale perturbed map (2.3), the stem period is equal to the period of the cascade. If the stem period of a cascade is odd, it is equal to the period of the cascade. A conjecture was formulated in [Sander & Yorke (2011)]:

Conjecture 21 *In general for even k, the stem period need not be equal to the period of the cascade, such as for the classes of non-quadratic parametrized maps*

$$F\,(\lambda, x) = x^3 - \lambda x + g\,(\lambda, x) \qquad (2.21)$$

---

[18] Also, $F\,(\lambda, .)$ is a two-shift horseshoe map for $\geq \lambda_H$.

or the classes of higher-dimensional parametrized maps

$$F_i (\lambda, x_1, ..., x_N) = K_i (\lambda) - x^2_i + g (\lambda, x_1, ..., x_N) \tag{2.22}$$

The following result was proved in [Sander & Yorke (2011)]:

Proposition 2.8 *(Stem Period = Minimum Period). Assume that $F : \mathbb{R} \times \mathbb{R} \to \mathbb{R}$ satisfies **Hypothesis 3** (with dimension $N = 1$), and that for all sufficiently large λ, $F (\lambda_H, .)$ is a two-shift horseshoe map. Let $G = F (\lambda_H, .)$, where λ_H is chosen such that G is a two-shift horseshoe map. Let y be a period-M orbit for G with nonzero orbit index. Let C be the component containing (λ_H, y). Then M is the minimum period of an orbit in C.*

2.5.2.3 Large-scale perturbations of cubic parametrized maps

Before starting, we need to define the so called *generalized n-horseshoe* as follows: Let D be the unit square with side lengths 1 with coordinates (0, 0), (1, 0), (1, 1), (0, 1) in the real plane, or $0, 1, 1 + i$ in the complex plane. Hence the Smale horseshoe map f described in [Smale (1967), Cvitanović, et al. (1988)] consists of the following sequence of operations on the unit square D:

(a) Stretch in the y direction by more than a factor of two.
(b) Compress in the x direction by more than a factor of two.
(c) Fold the resulting rectangle and fit it back onto the square, overlapping at the top and bottom, and not quite reaching the ends to the left and right and with a gap in the middle. Hence the action of f is defined through the composition of the three geometrical transformations defined above.
(d) Repeat the above steps to generate the horseshoe procedure that has a Cantor set structure. Mathematically, the above actions (a) to (d) can be translated as follows [Smale (1967)]:

(1) Contract the square D by a factor of λ in the vertical direction, where $0 < \lambda < \dfrac{1}{2}$, such that D is mapped into the set $[0, 1] \times [0, \lambda]$.
(2) Expand the rectangle obtained by a factor of μ in the horizontal direction, where $2+\varepsilon < \mu$, and map the set $[0, 1] \times [0, \lambda]$ into the $[0, \mu] \times [0, \lambda]$ (the need for this ε factor is explained in step 3).
(3) Steps 1 and 2 produce a rectangle $f (D)$ of dimensions $\mu \times \lambda$. This rectangle crosses the original square D in two sections after it has been bent. The ε in step 2 indicates the extra length needed to create this bend as well as any extra on the other side of the square.
(4) This process is then repeated, only using $f (D)$ rather than the unit square. The nth iteration of this process will be called $f^k (D)$, $k \in \mathbb{N}$.

Now, the generalized n-horseshoe is a horseshoe map formed with the above steps, with the following two exceptions:

1: In Step 2, μ should be greater than $n + \varepsilon$.
2: In Step 3, rather than bending once, the rectangle should be bent n − 1 times, and ε should be large enough to accommodate these bends and still allow the horseshoe to pass through the rectangle n times.

Now, the third example of classes of parametrized maps, each with an infinite number of cascades is the large-scale perturbations of cubic parametrized maps given by:

$$F (\lambda, x) = -\lambda x + x^3 + g (\lambda, x) \tag{2.23}$$

As λ increases, cascades occur for the cubic map $x^3 - \lambda x$ and it forms a 3-horseshoe in the same way as the quadratic map form a 2-horseshoe for λ large. The dynamics of this map can be summarized as follow: (i) For highly negative λ there is one orbit, i.e., the fixed point at zero and it has orbit index 1. (ii) As λ increases, 0 has a period-doubling bifurcation where its period-2 orbit is symmetric about zero. (iii) A non-generic Pitchfork bifurcation occurs, i.e., a symmetry breaking bifurcation in which two new period-2 orbits are created. A generic large-scale perturbations destroys the symmetry and gives a parametrized map of the form (2.23) without a Pitchfork bifurcation. We need to define the *three-shift tent map*:

Definition 31 *(a) Define the three-shift tent map* $H_3 : [0, 1] \to [0, 1]$ *as the piecewise linear map with the absolute value of the slope equal to 3 such that* H_3 *is increasing from 0 to 1 on* $\left(0, \dfrac{1}{3}\right)$ *and* $\left(\dfrac{2}{3}, 1\right)$ *and decreasing from 1 to 0 on* $\left(\dfrac{1}{3}, \dfrac{2}{3}\right)$.

(b) The maximal invariant set of H_3 *is topologically conjugate to the shift map on three symbols.*

In fact, under any sufficiently slowly growing additive large-scale perturbation, the map (2.23) has cascades as shown in [Sander & Yorke (2011)]: (Largescale perturbations of cubic parametrized maps). Let $F : \mathbb{R} \times \mathbb{R} \to \mathbb{R}$ be of the form (2.23). Assume that g is a C^∞ smooth function such that for some $\beta > 0$ such that for all λ and x we have

$$|g (\lambda, 0)| < \beta, \text{ and } \left|\frac{\partial g}{\partial x}\right| < \beta \, |x| \tag{2.24}$$

Then for a residual set of g, for each positive integer $k \neq 2^m$ for $m \geq 0$, the number of unbounded cascades with stem period k is the same as the number of nonflip orbits for the map H_3. For all but possibly one nonflip period-k orbit where $k = 2^m$ ($m \geq 0$), there is an unbounded cascade component with stem period-k through the orbit. It follows from Corollary. 2.5 that the number of cascades of odd prime period p for a large-scale perturbation of the cubic parametrized map is bounded below by $\dfrac{(3^p - 3)}{2p}$.

2.5.2.4 High-dimensional systems

The fourth example of classes of parametrized maps, each with an infinite number of cascades is the system of N quadratic maps plus coupling. The more general Γ function used to describe the nonflip orbits for this case is defined by:

Definition 32 *(Number of periodic points Γ) For positive integer N and $x = (x_1, ..., x_N) \in \mathbb{R}^N$, let $T_N : \mathbb{R}^N \to \mathbb{R}^N$ be the product of N tent maps of the form (2.3), $T_N (x) = (T (x_1) , ..., T (x_N))$: For each k, let $\Gamma (N, k)$ denote the number of nonflip orbits of period k for T_N.*

Note that the values of the number $\Gamma (N, k)$ are related to the number of period-k orbits for the shift map on 2^N symbols. The following result was proved in [Sander & Yorke (2011)]: (Systems of coupled quadratic parametrized maps). Let $F : \mathbb{R} \times \mathbb{R}^N \to \mathbb{R}^N$ and $g : \mathbb{R} \times \mathbb{R}^N \to \mathbb{R}^N$ be smooth, and let each component F_i (for each $i = 1, 2, ...,N$) have the form (2.8) where g is such that for some $\beta > 0$,

$$
G_\beta : \begin{cases}
\|g (\lambda, 0)\| < \beta \\
\|D_x g (\lambda, x)\| < \beta \\
\lim_{\lambda \to \infty} K_i (\lambda) = +\infty \\
\lim_{\lambda \to -\infty} K_i (\lambda) = -\infty
\end{cases}
\tag{2.25}
$$

Then for a residual set of g, for each positive integer k, the number of unbounded cascades with stem period k is $\Gamma (N, k)$.

2.5.2.5 Large perturbations of Hénon families

We note that in a recent work [Sander & Yorke (2009)] it was proved that the Hénon map $H_{\lambda, b} (x, y) = (\lambda + by - x^2, x)$ and its arbitrarily large perturbations have period-doubling cascades. Furthermore, a classification of the period of a cascade in terms of the set of orbits it contains was given. This study is a generalization of the results in [Devaney & Nitecki (1979)] presented in Section 1.5 to a broader class of families. In particular, the Hénon map $H_{\lambda, b}$ forms a horseshoe for any fixed nonzero b, as λ varies from small to large and hence for large positive λ the invariant set in a certain region has dynamics of a Smale horseshoe, i.e., conjugate to the full shift on two symbols and then $H_{a, b}$ has infinitely many cascades. The perturbed Hénon map $F : \mathbb{R} \times \mathbb{R}^2 \to \mathbb{R}^2$ has the form:

$$F(\lambda, x, y) = \begin{pmatrix} \lambda + by - x^2 + g(\lambda, x) + \alpha_1(\lambda, x, y) \\ x + \alpha_2(\lambda, x, y) \end{pmatrix} \quad (2.26)$$

where b is a fixed nonzero constant, and λ is the bifurcation parameter. The functions g and $\alpha = (\alpha_1, \alpha_1)$ satisfy conditions (2.28) below. This class includes for example C^∞ functions that are C^1 bounded. To determine the class of functions allowed for α, let $r > 0$ be any fixed arbitrarily large constant. For sufficiently small $\delta > 0$ depending on r, let

$$\Psi_{(\delta,r)} = \{\alpha : \mathbb{R} \times \mathbb{R}^2 \to \mathbb{R}^2 \in C^\infty : |\alpha(\lambda, x, y)|_1 < \delta \text{ when } |(\lambda, x, y)| > r\} \quad (2.27)$$

where $|.|_1$ denotes the C^1-norm. Note that the set $\Psi_{(\delta,r)}$ has no restrictions other than smoothness in the region where $|(\lambda, x, y)| < r$ and all C^1 functions with compact support belongs to it. In fact, there exists a residual subset (the set depends on g) of the set of allowable functions F (which is open in C^1) in which all periodic orbit bifurcations are generic. Since the function is uniformly bounded, there is a constant $\beta > 0$ such that for all $(\lambda, x, y) \in \mathbb{R}^3$, we have:

$$|g(\lambda, 0)| + |\alpha(\lambda, x, y)| < \beta, \text{ and } \left|\frac{\partial g(\lambda, x)}{\partial x}\right| < \beta. \quad (2.28)$$

The inequalities (2.28) imply that these perturbations are dominated by the standard Hénon terms when λ and x are large. In particular, for sufficiently large $\lambda = \lambda_1$ and sufficiently small $\delta > 0$, the map $F(\lambda_1, .)$ is topologically the same as the Hénon map. In fact, it was shown in [Sander & Yorke (2009)] that for $\lambda = \lambda_1$ (λ_1 is any sufficiently large real number), any non flip orbit in the horseshoe lies in a cascade, and no other orbit for that λ_1 is in the cascade. For a map $F(\lambda_1, .)$ define $MaxInv(\lambda_1)$ to be the union of the trajectories such that all positive and negative iterates are bounded:

Theorem 22 (*Cascades for large perturbations of Hénon families*). *Fix $b \neq 0$, $\beta > 0$, and $r > 0$. Let $g \in G_\beta$ defined by (2.25). For $\delta > 0$, let $\alpha \in \Psi_{(\delta,r)}$ and let F be as in (2.26). Then as long as δ is sufficiently small (depending on r), for every sufficiently large $\lambda = \lambda_1$ depending on β, r, and b, there is a residual set of $\alpha \in \Psi_{(\delta,r)}$ depending on the function g and the constant b for which the following hold:*

1. *$MaxInv(\lambda_1)$ is conjugate under a homeomorphism to a two-shift, and this homeomorphism gives a one-to-one correspondence of the even symbol sequences with the non flip orbits. (Hence for λ_1, we can without confusion refer to a periodic orbit for $F(\lambda_1, .)$ as being even.)*
2. *Each unbounded cascade contains exactly one periodic orbit for $F(\lambda_1, .)$, and it is an even orbit.*
3. *For each even orbit there is a unique unbounded cascade containing that orbit.*

4. *If an even periodic orbit is of period k, and k is odd, the cascade containing it is a period-k cascade. If k is even, then the cascade containing it is a period-j cascade, where $k/j = 2^m$ for some m.*

Theorem 22 says that for every even period-k symbol sequence S, there is exactly one unbounded cascade of F given by (2.26) such that for $\lambda = \lambda_1$. In this case, the cascade contains a unique periodic orbit, and it is the unique period-k orbit of F with the symbol sequence S.

3

Dynamical Properties of the Lozi Mappings

The first part of this chapter is a self-contained introduction to chaos, via the Lozi mappings. The second part presents a rigorous proof of chaos in the Lozi mapping. To maintain uniformity, standard notation is used throughout. Indeed, in Section 3.1 an introduction to chaos via the Lozi maps is presented by means of some quantities defined in Chapter 1. In Section 3.2 we discuss ergodic properties of the Lozi mappings, namely, some general properties, construction of invariant measures and their ergodic properties, the Hausdorff dimension and spectra of singularities. Section 3.3 deals with a relatively new result about grammatical complexity for the Lozi mappings including a short introduction to the concepts, definitions and methods used to calculate the desired transition rules. In Section 3.4, admissibility conditions with some examples for symbolic sequences of the Lozi map are discussed in some detail. In particular, contracting and expanding foliations and their ordering, the pruning front and admissibility conditions are presented.

In the second part of this chapter, we present known results in literature concerning a rigorous proof of chaos in the Lozi mappings. In Section 3.5 we present the method used to show the existence of the strange attractor for $b > 0$. This method is based essentially on the construction of the trapping region of the Lozi mapping and the study of its hyperbolicity in the sense of Definition 12(d). In Section 3.6 we discuss the existence of strange attractors of the Lozi mapping for $b < 0$. Section 3.7 deals with the rigorous proof of chaos in the Lozi map using the theory of transversal heteroclinic cycles based on the study of the properties of periodic solutions and the so called *dual line mapping* along with a study of some particular invariant manifolds. Further properties of the Lozi strange attractor are listed in

Section 3.7.4. The geometric structure of strange attractors in the Lozi map is discussed in Section 3.8 based on several works and conjectures. In Section 3.9 we present the parameter-shifted shadowing property of Lozi maps and in Section 3.10 we clarify the method used to visualize the basin of attraction of Lozi mapping based on the solution of the *first tangency problem* and a geometrical method for constructing this basin. The relations between forward limit sets and the strange attractor of the Lozi mappings is discussed in Section 3.11 based on two fundamental techniques: Estimations of parameter dependence and usefulness and maturity of parameter arcs. Finally, In Section 3.12, we discuss the topological entropy of the Lozi maps. In particular, the most important result in this topic is the monotonicity of this entropy in the vertical direction around some values.

3.1 Introduction to chaos via the Lozi maps

The Lozi map [Lozi (1978)] is given by:

$$L(x, y) = \begin{pmatrix} 1 - a\,|x| + y \\ bx \end{pmatrix} \tag{3.1}$$

or

$$L_1(x, y) = \begin{pmatrix} 1 - a\,|x| + by \\ x \end{pmatrix} \tag{3.2}$$

Formula (3.1) or (3.2) was discovered on June 15 th- 1977 around 11 am, during the presentation of the thesis of Ahmed Intissar and numerically tested just after the talk before 12.30 am. Both formula (3.2) and (3.1) are used in this book according to the original source of a specific result. In fact, these two formulas are equivalent dynamically, which means that their dynamical behaviors are the same. The inverse of the Lozi mapping (3.1) is given by:

$$L^{-1}(x, y) = \left(\frac{y}{b}, x - 1 + \frac{a}{b}\,|y| \right) \tag{3.3}$$

and the inverse of the Lozi mapping (3.2) is given by:

$$L^{-1}_1(x, y) = \left(y, \frac{x + a\,|y| - 1}{b} \right) \tag{3.4}$$

We remark that the inverse mappings (3.3) and (3.4) are also piecewise linear systems. The difference between the Lozi mappings (3.1) or (3.2) and the Hénon mapping (2.1) is that the investigation of the Lozi map is better amenable for analytical treatments as compared to the Hénon map, and for some ranges of bifurcations parameters, the Hénon map is conjugate to some Lozi mappings. In fact, it was proved in [Tresser (1982)] that there exists a region in parameter space such that the two mappings are not topologically conjugate:

Theorem 23 *On the line*

$$b > 0, a = 2 - \frac{b}{2} \tag{3.5}$$

in ab-parameter space, there is a sequence (a_i, b_i) converging to $(2, 0)$ such that no $\mathcal{L}(a_i, b_i)$ is topologically conjugate to a Hénon mapping. Here $\mathcal{L}(a_i, b_i)$ is the Lozi map (3.1) with the parameter (a_i, b_i).

On one hand, the Lozi strange attractors presented by maps (3.1) or (3.2) are characterized by some *angular shapes* that are different from Hénon strange attractors presented by (2.1). In fact, It was shown in [Kiriki (2004)] that there exists an open subset in the region of bifurcation parameters where chaos occurs such that for almost every parameter in this region, the forward limit set of singularity points coincides with the Lozi strange attractor. See Section 3.11 for more details and proofs. On the other hand, the Lozi mappings are not realistic as approximations to smooth flows, but these maps are very helpful tools for developing intuition about the topology of a large class of maps of the *stretch and fold* type. It seems that the Lozi mappings bridge the gap between the Axiom A systems (Recall Definition 14) and more complicated systems like the Hénon map (2.1) called *quasi-attractors* (see Section 1.2.3). We note that when $b = 0$, both Lozi mappings (3.1) and (3.2) are reduced to the tent map given by:

$$T(x) = 1 - a |x| \tag{3.6}$$

Some properties of the map (3.6) are presented in several places of this book. In particular, in Section 3.12.1. When b is sufficiently small the Lozi mappings can be considered as natural 2-D generalization of tent map (3.5). In this case, we have $\mathcal{L}(x, y) = (T(x) + y, bx)$ and $\mathcal{L}_1(x, y) = (T(x) + by, x)$. These two formulas are very helpful since they are used in several places in this book to show some properties of the Lozi mappings when the parameter b is considered small enough. It was shown numerically in [Sprott (2009)], that the Lozi map (3.2) has the Lyapunov exponents $\lambda_1 = 0.47023$ and $\lambda_2 = -1.16338$ for $a = 1.7$ and $b = 0.5$. The Kaplan-Yorke dimension is $D_{KY} = 1.40419$. The Lozi map (3.2) is also maximally chaotic for $a = 2$ and $b = 0$, where the Lyapunov exponents are $\lambda_1 = \ln 2 = 0.69314....$ and $\lambda_2 = -\infty$, with a Kaplan-Yorke dimension of $D_{KY} = 1$. The parameters that maximize the Kaplan-Yorke dimension, are located in the boundary $b = 4 - 2a$ where the solutions become unbounded as shown in [Tél (1982(a-b))]. The greatest dimension occurs for $a = 1.7052$ and $b = 0.5896$, for which the Lyapunov exponents are $\lambda_1 = 0.44836$ and $\lambda_2 = -0.97667$ with a Kaplan-Yorke dimension of $D_{KY} = 1.45907$. Figure 3.1 shows the attractor for this case along with its basin of attraction, whose boundary touches the attractor.

3.2 Ergodic properties of the Lozi mappings

In this section, we present several results showing ergodic properties of the Lozi mapping (3.1). The main source of this section is [Collet & Levy (1984)] where the Bowen-Ruelle measure (Recall Definition 2) for the Lozi mapping was constructed. Since, the Lozi maps are almost everywhere hyperbolic diffeomorphism[1] (Recall Definition 12(d)) as shown in Section 3.5.2, so they have similar properties (such as ergodic properties) to those of Axiom A systems (Recall Definition 14), i.e., there is an invariant measure with the following ergodic properties: absolute continuity with respect to the Lebesgue measure in the unstable direction, ergodicity, K-property, Bowen-Ruelle property, Bernoulli character. This means that the Lozi mapping (3.1) is an intermediate stage between the Axiom A dynamical systems and more complicated systems like the Hénon map (2.1) called *quasi-attractors*. See Section 1.2.3 for more details. Similar results were obtained in [Rychlik (1983(a-b-c))] for the Lozi map using a different proof. Similar results for piecewise C^2 hyperbolic maps were obtained in [Young (1982(a))] as described in Section 3.4.1. In fact, the Lozi map is similar to *Sinai's billiards*[2] studied in [Sinai (1963–1970)]. In particular, the discontinuity of the differential for the Lozi map (3.1) allows the uniform hyperbolicity as in the billiards case.

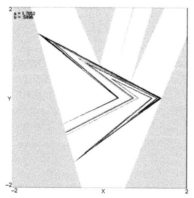

Figure 3.1: The Lozi attractor (3.2) with its basin of attraction obtained for $a = 1.7052$ and $b = 0.5896$.

---

[1] While hyperbolicity occur for the Hénon map (2.1) only on a *Cantor-like sets* of parameters.

[2] The table of the Sinai billiard is a square with a disk removed from its center; the table is flat, having no curvature. This type of maps arises from studying the behavior of two interacting disks bouncing inside a square, reflecting off the boundaries of the square and off each other.

We note that chaos discussed in Sections. 3.5 and 3.6 is proved there in the sense of having a strange attractor with SRB measure. Young in [Young (1998)] discuss this property for Lozi maps and generalized Lozi maps. This work of Young is based on the results given in [Collet & Levy (1984), Young (1985)] to prove the existence of SRB measure with exponential decay of correlations for Lozi-like maps.

3.2.1 General properties of the Lozi mappings

In this section, we present some general properties of the Lozi mappings (3.1) where we use the following notations: M is the Borel σ-algebra of $\mathbb{R}$, if A is a measurable subset of $\mathbb{R}^2$, then M_A is the corresponding factor sub σ-algebra. l and m are the 1-D and 2-D Lebesgue measure, d is the Euclidean distance in $\mathbb{R}^2$. For a partition ξ of the space X_1, $\xi(x)$, $x \in X_1$ is the atom of ξ containing x. Let $S_0^- = Oy$, $S_0^+ = Ox = \mathcal{L}S_0^-$, $S_n^\pm = \mathcal{L}^{\pm n}S_0^\pm$, $n \in \mathbb{N}$, are finite broken lines. From [Misiurewicz (1980)] it follows that $\mathcal{L}^{\pm n}$ is singular on S_n^T , $n \in \mathbb{N}$ and the fields of stable (unstable) directions $E^s(.)$ $(E^u(.))$ are defined outside the set $\bigcup_{n\geq0}S_n^-$ $(\bigcup_{n\geq0}S_n^+)$. Let $\theta^s(x)$ and $\theta^u(x)$ be their angles with respect to the x-axis. Thus, we have the following result:

Lemme 3.1 *(a) The continued fraction expansions for* $\tan \theta^s(x)$ *and* $\tan \theta^u(x)$ *are given by:*

$$
\begin{cases}
\tan \theta^s(x) = a.\varepsilon(x) + \dfrac{b}{\tan \theta^s(\mathcal{L}(x))} \\[3mm]
\tan \theta^u(x) = \dfrac{-b}{a.\varepsilon(\mathcal{L}^{-1}(x)) - \tan \theta^u(\mathcal{L}^{-1}(x))}
\end{cases}
\tag{3.7}
$$

where $\varepsilon(x)$ *is the sign of* x. *(b) If* $\dfrac{b}{a}$ *is small enough, then the continuous fraction expansions (3.6) are convergent.*

If $\lambda(x)$ denotes the expansion factor in the direction $E^u(x)$, i.e., the length of the image by $D\mathcal{L}_x$ of the unit vector in the direction $E^u(x)$, then we have the following result:

Lemme 3.2 *We have*

$$
\lambda(x) = \frac{(a^2 + b^2 + \tan^2(\theta^u(x)) - 2a\varepsilon(x)\tan\theta^u(x))^{\frac{1}{2}}}{(1 + \tan^2\theta^u(x))^{\frac{1}{2}}}
\tag{3.8}
$$

We need the following definitions:

Definition 33 *Let* $W^s(x)$ *(resp.* $W^u(x)$*) be the global stable (resp. unstable) manifold of* x. *The maximal smooth component* $W^s_{loc}(x)$ $(W^u_{loc}(x))$ *of* $W^s(x)$ *(resp.* $W^u(x)$*)*

containing x is called the local stable (local unstable) manifold of x. We say that the point x splits $W^u_{loc}(x)$ into two semi local unstable manifolds.

Let X be the fixed point of the Lozi map $\mathcal{L}$ given by (3.1) with positive coordinates given by (3.117) below. The point Z is the intersection of the positive x-axis with $\overline{W^u_{loc}(X)}$. Let F be the triangle defined by the points Z, $\mathcal{L}(Z)$, $\mathcal{L}^2(Z)$ and H_0 be the triangle XZP, where $P = W^s_{loc}(X) \cap (Z, \mathcal{L}(Z))$. Let $\Lambda_{a,b}$ be the strange attractor of $\mathcal{L}$ which is equal to $\cap^{\infty}_{n=0} \mathcal{L}^n(F)$ as shown by (3.127). Let λ and λ_+ be the infimum and the supremum of $\lambda(x)$ for x in the set $F \setminus \cup_{n \geq 0} S^{\pm}_l; \lambda > 0$.

By using the continued fraction expansion (3.6), it is easy to verify the following result proved in [Collet & Levy (1984)]:

Lemme 3.3 *Let x and y in F be such that for $0 \leq j \leq q$, $\mathcal{L}^j(x)$ and $\mathcal{L}^j(y)$ are on the same side of S_0. Then (i) The angle between $W^s_{loc}(x)$ and $W^s_{loc}(y)$ (if they are defined) is bounded by $2\left(\dfrac{b}{\lambda}\right)^{q-1}$. (ii) If $a < 2$, then the angle between two local stable manifolds is between $\dfrac{2\pi}{5}$ and $\dfrac{-2\pi}{5}$, and the angle between a local stable and a local unstable manifold is greater than $\dfrac{\pi}{5}$. (iii) We have*

$$\left|\frac{\lambda(x)}{\lambda(y)} - 1\right| \leq 2\left(\frac{b}{\lambda}\right)^{q-1} \tag{3.9}$$

It is more convenient to use the set H defined by (3.129). This set has the properties shown in Proposition 2 of [Misiurewicz (1980)], i.e., Proposition. 3.18 in this chapter.

We investigate now the absolute continuity of the unstable foliation. Let W^1 and W^2 be two local unstable manifolds in $\Lambda_{a,b}$. Let $P = P_{W^1 W^2}$ be a map defined from W^1 to W^2 by:

$$x \in W^1 \to P(x) = W^u_{loc}(x) \cap W^2 \text{ if } W^u_{loc}(x) \cap W^2 \neq \varnothing. \tag{3.10}$$

P is defined on the set:

$$\mathfrak{D}(P) = \{x \in W^1 : W^u_{loc}(x) \cap W^2 \neq \varnothing\} \tag{3.11}$$

Thus, we have the following result proved in [Collet & Levy (1984)]:

Proposition 3.1 *Given W^1 and W^2 as above, then there is a constant $L_1 > 0$ which is $\mathfrak{D}(1)$ such that for any Borel subset A of W^1, $A \subset \mathfrak{D}(P)$ we have:*

$$(1 - L_1 (d(W^1, W^2))^{\frac{1}{3}}) l(A) \leq l(P(A)) \leq (1 + L_1 (d(W^1, W^2))^{\frac{1}{3}}) l(A) \tag{3.12}$$

The proof of Proposition 3.1 is based on the results of Section 3.2.2 and Section 3.2.3.

3.2.2 Construction of invariant measures for the Lozi mappings

In this section, we give the method used in [Collet & Levy (1984)] for constructing invariant measures for the Lozi mapping (3.1). We note that the measure μ is unique and its uniqueness can be proved by showing that μ is the Bowen-Ruelle measure. Let $L_0 = W_{Loc}(X)$. Since the strange attractor $\Lambda_{a,b}$ is the closure of the set $\cup_{n\geq 0}\mathcal{L}^n (L_0)$. Then it is natural to obtain an invariant measure for the Lozi mapping (3.1) by iterating the Lebesgue measure supported by the manifold L_0. For this purpose, let $(\mu'_n)_{n\in\mathbb{Z}}$ be a sequence of probability measures on $M_{\Lambda a,b}$ (μ'_n has support in $\mathcal{L}^n (L_0)$) defined by:

$$\mu'_n (A) = \frac{l(\mathcal{L}^n A \cap L_0)}{l(L_0)} \text{ for } A \in M_{\Lambda a,b} \tag{3.13}$$

Hence, the sequence $(\mu_n)_{n\in\mathbb{Z}}$ defined by:

$$\mu_n = \frac{1}{n}\sum_{j=0}^{n-1}\mu'_j \tag{3.14}$$

is a sequence of probability measures on the set $M_{\Lambda a,b}$. The compactness property of $\Lambda_{a,b}$ implies that it is possible to extract a subsequence which vaguely converges to an invariant probability measure μ as shown in [Bourbaki (1969)]. Let $A^{\pm}_\varepsilon$ denote the stripe of width ε around the set $S^{\pm}_0$. The estimate of the μ_n-measures of $A^{\pm}_\varepsilon$ is given in [Collet & Levy (1984)] as follow:

Proposition 3.2 *For any positive number*

$$0 < \tau < \left(1 - \frac{1}{K\log\lambda}\right) \tag{3.15}$$

(K is the integer appearing in Lemma 3.4), there is a positive real number ε_0, such that for $0 < \varepsilon \leq \varepsilon_0$, we have:

$$\mu_n (A^{\pm}_\varepsilon \leq \varepsilon^\tau, \forall n \in \mathbb{N}. \tag{3.16}$$

The proof of Proposition 3.2 was given in [Collet & Levy (1984)] for $A_\varepsilon = A^-_\varepsilon$ by using some geometrical considerations. Indeed, for $n \in \mathbb{N}$, the image $\mathcal{L}^n L_0$ is a segment or a broken line. Let J_n be the set of maximal smooth components contained in $\mathcal{L}^n L_0$. Thus, for $M \in J_n$, the endpoints of M belong to $S_p \cup S_q$ for some p, q, such that $1 \leq p, q \leq n + 2$. Let $k(M) = \inf(p, q)$ and if M belonging to J_n, then $R_p(M)$ be the element of J_{n-p} containing $\mathcal{L}^{-p}M$ for $p \in \mathbb{Z}$. Define recursively a finite sequence of integers $k_i(M)$ by:

$$\begin{cases} k_0(M) = 0 \\ \\ k_1(M) = k(M) \\ \\ k_{i+1}(M) = k_i(M) + k(R_{k_i(M)}(M))) \text{ as long as} \\ \\ R_{k_i(M)}(M) \neq L_0, k_i = k_i(M) \end{cases} \tag{3.17}$$

A lower bound on $k(.)$ for small b was proved in [Collet & Levy (1984)] as follows:

Lemma 3.4 *Assume b is small enough, then there is an integer $K > 4$ and $\theta > 0$ such that if $M \in J_{n'}$, $l(M) < \theta$, and $M \cap S^-_0 \neq \varnothing$, then $k(M) > K$.*

Proof 1 *The fact that b is small enough, and the existence of an integer $K > 2$ such that $0 < p \leq K$ implies $(S^-_0 \cap F) \cap S_p = \varnothing$. Let $\varnothing = \inf d\,(S^-_0 \cap F, S_p \cap F)$, then if $l(M) < 0$, one has $k(M) > K$.*

The basic estimate was also proved recursively for μ'_n, $n \in \mathbb{Z}$ by taking τ be a number verifying (3.14). In this case, the estimate is obvious for $\mu' -1$ since $l(L_0) > 1$, for μ'_n, $n \leq -2$ and ε small enough, $\mu'n\,(A_\varepsilon) = 0$.

Now, let $(\mu_n)_{n \in \mathbb{N}}$ be a subsequence of $(\mu'_n)_{n \in \mathbb{N}}$ which converges weakly to μ. For fixed τ as in Proposition 3.2 we have the following corollary [Collet & Levy (1984)]: We have

(i) For ε small, $n \in \mathbb{N}$, $k \in \mathbb{Z}$,

$$\begin{cases} \mu_n\,(A^\pm_\varepsilon) \leq \varepsilon^\tau \\ \mu\,(\mathcal{L}^k A_\varepsilon) \leq \varepsilon^\tau \end{cases} \tag{3.18}$$

(ii) For $N \in \mathbb{N}$, we define

$$H^N = \left\{ \begin{matrix} x \in \Lambda_{a,b} : \text{one (at least) of the endpoints of} \\ W^u_{loc}(x) \text{ does not lie on } \cup^N_{k=0} S^+_k \end{matrix} \right\} \tag{3.19}$$

There exist $c > 0$ and α, $0 < \alpha < 1$, such that for $n \in \mathbb{N}$,

$$\begin{cases} \mu_n\,(H^N) < cx^N \\ \mu\,(H^N) < c\alpha^N \end{cases} \tag{3.20}$$

(iii) For $\varepsilon \in \mathbb{R}^+_*$, we define

$$H_\varepsilon = \left\{ \begin{matrix} x \in \Lambda_{a,b} : \\ \text{one (at least) of the semi-loc unstable manifolds is shorter than } \varepsilon \end{matrix} \right\} \tag{3.21}$$

For ε small, $\mu_n(H_\varepsilon) < \varepsilon^\tau$, $n \in \mathbb{N}$ and $\mu(H_\varepsilon) < \varepsilon^\tau$.

(iv) For $x \in \Lambda_{a,b}$, $\mu(\{x\}) = 0$.

(v) For $x \in \Lambda_{a,b}$, $\mu(\{W^u(x)\}) = \mu(\{W^s(x)\}) = 0$.

3.2.3 Ergodic properties of the invariant measure

In this section, we discuss ergodic properties of the invariant measure constructed in the previous section. For this purpose, let α and β be two countable partitions which decompose the strange attractor $\Lambda_{a,b}$ of the Lozi map (3.1) and let ζ^+ (ζ^-) be the decomposition of μ.a.a. $\Lambda_{a,b}$ into local unstable (stable) manifolds. The partition ζ^+ is measurable because it is generated by $\cup_{n \geq 0} S^+_n$. The aim of these considerations is the investigation of the properties of the conditional expectations of μ on the unstable foliation by using the sequence $(\mu_n)_{n \in \mathbb{N}}$. Let μ^+ be the restriction of μ to the sub σ-algebra $M^+ \subset M_{\Lambda a,b}$ of the sets ζ^+-saturate. Corollary 6(ii)), imply that for μ.a.a. $W_0 \in \zeta^+$, there are two maximal smooth components I_0 and J_0, contained in $\cup^{N_0}_{k=0} S^+_k$ for some $N_0 \geq 0$ such that the endpoints of W_0 lie on I_0 and J_0. From this result we define the partition α by:

$$\alpha(W_0) = \{W \in \zeta^+ \text{ with endpoints on } I_0 \text{ and } J_0\} \tag{3.22}$$

Let $\zeta^+_N \subset \zeta^+$ be the union of the elements of ζ^+ with endpoints on the set $\cup^N_{k=0} S^+_k$, $N \in \mathbb{N}$. We have that ζ^+_N is a finite union of atoms of α, since, by Corollary 6(ii)), we have:

$$\lim_{N \to \infty} \mu^+(\zeta^+_N) = 1 \tag{3.23}$$

α is a countable partition of ζ^+.

In what follows, we present a partition of $\Lambda_{a,b}$ into parallelograms. Indeed, let P^+_N (P^-_N) be the partitions generated by $\cup^N_{n=0} S^+_n$ ($\cup^N_{n=0} S^-_n$). Thus, we have $\zeta^\pm = \lim_{N \to \infty} P^\pm_N$ and by a simple recursion argument, it is easy to show that the sets $P^\pm_N(x)$, $x \in \Lambda_{a,b}$, are convex sets since S^+_n is a broken line, folded only on $\cup^{N-1}_{i=0} S^+_i$. Let $x \in \Lambda_{a,b}$ such that $W^+ = W^u_{loc}(x)$ and $W^- = W^s_{loc}(x)$ have positive length. Let $N_\pm(x)$ be the smallest integers such that the endpoints of W^- (W^+) lie outside the set $P^+_{N+(x)}(x)$ ($P^-_{N-(x)}(x)$). In this case, $N_\pm$ are finite, $N_\pm$ are measurable functions, so the atom $\beta(x)$ can be defined, for such an element $x \in \Lambda_{a,b}$ by:

$$\beta(x) = N^{-1}_+(N_+(x)) \cap N^{-1}_-(N-(x)) \cap P^+_{N+(x)}(x) \cap P^-_{N-(x)}(x) \tag{3.24}$$

The unstable and stable fibers $\gamma^+(x)$, $\gamma^-(x)$ are usually defined by:

$$\left\{ \begin{array}{l} \gamma^+(x) = \beta(x) \cap W^u_{loc}(x) \\ \gamma^-(x) = \beta(x) \cap W^s_{loc}(x), \text{ i.e., } \gamma^\pm = \beta \vee \zeta^\pm \end{array} \right. \tag{3.25}$$

The convexity of the set $P^\pm_{N\pm(x)}(x)$ and the fact that $N^\pm$ are constant on atoms of $P^\pm_{N\pm(x)}(x) \vee \zeta^\pm$ show that the set $\beta(y)$ is a parallelogram in the following sense:

$$x \in \beta(y) \Rightarrow \begin{cases} \exists! z \in \beta(y), z = \gamma^+(x) \cap \gamma^-(y) \\ \exists! z' \in \beta(y), z' = \gamma^-(x) \cap \gamma^+(y) \end{cases} \quad (3.26)$$

The definition of the atom β shows that it is a countable partition of the strange attractor $\Lambda_{a,b}$ into parallelograms. The application of the usual theorem on *disintegration* [Rohlin (1949)] on the measurable partition ζ^+ show that there is a μ-a.s. unique family $\{\mu_W, W \in \zeta^+\}$ of probability measures on $\Lambda_{a,b}$ such that:

(a) μ_W has μ-a.s. support on W,

(b) for $A \in M_{\Lambda_{a,b}}$ the map $W \to \mu_W(A)$ is in $L^1(\zeta^+, d\mu^+)$, and

$$\mu_W(A) = \mu^+(\mu.(A)) = \int_{\zeta^+} \mu_W(A)\, d\mu + (W) \quad (3.27)$$

For $W \in \zeta^+$, let l_W denote the normalized 1-D Lebesgue measure on W. The following ergodic properties of $\mathcal{L}$ were proved in [Collet & Levy (1984)]:

Proposition 3.3 *(a) The conditional expectations of μ on the local unstable manifolds are the corresponding 1-D Lebesgue probabilities, i.e., $\mu_W = l_W$ for μ.a.e. $W \in \zeta^+$.*

(b) $(\Lambda_{a,b}, \mathcal{L}, \mu)$ is ergodic.

(c) For $n \in \mathbb{N}^*$, $(\Lambda_{a,b}, \mathcal{L}^n, \mu)$ is ergodic.

(d) $(\Lambda_{a,b}, \mathcal{L}, \mu)$ is a K-system.

(e) $(\Lambda_{a,b}, \mathcal{L}, \mu)$ is isomorphic to a Bernoulli shift.

The proof of Proposition 3.3(e) is similar to the one given in [Young (1982(a))] and by applying Proposition 3.3(a) with respect to the set M^+_0 and Lemma 3.5 below and the results in [Young (1982(b))].

The canonical map $(\Lambda_{a,b}, M^+_0 \vee M^-_0) \xrightarrow{P} (\zeta^+ \times \zeta^-, M^+_0 \times M^-_0)$ given by:

$$P(x) = (\zeta^+(x), \zeta^-(x)) \quad (3.28)$$

is almost everywhere defined. Let $v = \mu \circ P^{-1}$ denote the image of μ through P and let $A \in M^+_0 \vee M^-_0$ such that $P(A) M^+_0 \times M^-_0$. By definition we have:

$$\mu^+ \otimes \mu^-(P(A)) =_{w \in \zeta^+} d\mu^+(W) . \mu^-\left(_{x \in A \cap W}\zeta^-(x)\right) \quad (3.29)$$

Hence, the following lemma, proved in [Collet & Levy (1984)]:

Lemme 3.5 *v is absolutely continuous with respect to $\mu^x = \mu^+ \otimes \mu^-$, i.e., $v << \mu^+ \otimes \mu^-$.*

By Radon-Nikodym's theorem [Nikodym (1930), Shilov (1978)], there is a $\mu^{\times}$-integrable function $h : \zeta^{+} \times \zeta^{-} \to \mathbb{R}^{+}$, such that:

$$dv(x) = h(x).d\mu^{\times}(x) \tag{3.30}$$

Definition 34 *For a function f, and for a given $A \in M^{+}_{0} \vee M^{-}_{0}$ such that $v(A) > 0$, the conditional probability $v_{A} = v(. \mid A)$ is given by:*

$$v_{A}(f) = \frac{v(f.\kappa_{A})}{v(\kappa_{A})} = \frac{u^{\times}(f.h\kappa_{A})}{u^{\times}(f.\kappa_{A})} = \frac{u^{\times}(f.h)}{u^{\times}_{A}(h.)} \tag{3.31}$$

or, more briefly:

$$v_{A} = \frac{\mu^{\times}_{A}(h.)}{\mu^{\times}_{A}(h)} \tag{3.32}$$

The above considerations were used in [Collet & Levy (1984)] to prove the *weak Bernoulli property* for the Lozi mappings (3.1) and the unicity of μ, i.e., μ is the Bowen-Ruelle measure:

Proposition 3.4 *(a) $(\Lambda_{a,b}, M, \mu, \mathcal{L})$ is weak Bernoulli, that is μ and $\mu^{\times} \circ P$ coincide on n $(M^{+}_{n} \vee M^{-}_{n})$.*
(b) For $g \in C^{0}(F)$ and m-almost any $x \in F$,

$$s = \lim_{n \to \infty} \frac{1}{n} \sum_{k=0}^{n-1} g\,(\mathcal{L}^{k}x) = \mu\,(g) \tag{3.33}$$

Let

$$\left\{ \begin{array}{l} \mathfrak{A} = \left\{ \begin{array}{l} x \in \cup_{n \geq 0} \mathcal{L}^{n}L_{0} : \exists g \in C^{0}(F) \\ \text{and } s \text{ does not exist or is not equal to } \mu\,(g) \end{array} \right\} \\ \mathfrak{L} = \{x \in F : \forall n, p \ \in \mathbb{N}, W^{s}_{loc}(\mathcal{L}^{n}x) \cap \mathcal{L}^{p}L_{0} \neq \varnothing\} \end{array} \right. \tag{3.34}$$

The following Lemma was proved in [Collet & Levy (1984)]:

Lemme 3.6 *(a) $\forall n \in \mathbb{N}, \mu_{n}\,(\mathfrak{A}) = l\,(\mathfrak{A} \cap L_{0}) = 0$.*
(b) We have

$$m(\mathfrak{L}) = 0. \tag{3.35}$$

The proof of Proposition. 3.4(b) follows from three steps: (1) Consider the points x of $\cup_{n \in \mathbb{N}} \mathcal{L}^{n}L_{0}$ such that, for some $g \in C^{0}(F)$, relation (3.32) does not hold. (2) Prove that the length of the set defined in Step 1 is zero (Lemma. 3.6(a)). (3) Find positive integers p, n such that $W^{s}_{loc}(\mathcal{L}^{n}x)$ crosses $\mathcal{L}^{p}L_{0}$ (Lemma. 3.6(b)) for m-a.e. $x \in F$, i.e., m-almost surely in F, this intersection does not fall in the exceptional set estimated in Lemma. 3.6(a).

The remaining of this section is devoted to the proof of Proposition 3.4(b). First, if $x \in F \setminus \mathcal{L}$, there exists an integer n, and $p \geq n$ such that $W^{s}_{loc}(\mathcal{L}^{n}x) \cap \mathcal{L}^{p}L_{0} \neq \varnothing$. We define a new set $\mathfrak{B}$ by:

$$\mathcal{B} = \{x \in F \setminus \mathcal{L} : \forall n, \in \mathbb{N}, p \geq n, W^s_{loc}(\mathcal{L}^n x) \cap \mathcal{L}^p L_0 \subset \mathfrak{A}\} \tag{3.36}$$

Hence, if $x \in F \setminus (\mathcal{L} \cup \mathcal{B})$, then there is an integer $n \in \mathbb{N}$ and an integer $p \geq n$ such that for some point y of the set $W^s_{loc}(\mathcal{L}^n x) \cap \mathcal{L}^p L_0$ does not belong to $\mathfrak{A}$. Thus, if g belongs to $C^0(\mathbb{R}^2)$, we have:

$$\lim_{m \to \infty} \frac{1}{m} \sum_{j=0}^{m-1} g (\mathcal{L}^j x) = \lim_{m \to \infty} \frac{1}{m} \sum_{j=0}^{m-1} g (\mathcal{L}^j y) = \mu (g) \tag{3.37}$$

The proof is completed if $m(\mathcal{B}) = 0$. For this purpose, let $\mathcal{B}_p$ be the set defined by:

$$\begin{cases} \mathcal{B}_p = \{y \in F \setminus \mathcal{L} : W^s_{loc}(y) \cap \mathcal{L}^p L_0 \subset \mathfrak{A}\} \\ \mathcal{B} = \cap_{p > n \geq 0} \mathcal{L}^{-n} \mathcal{B}_p \end{cases} \tag{3.38}$$

From the second relation of (3.37) it is enough to show that $m(\mathcal{B}_p) = 0$ for every integer p. We have:

$$\mathcal{B}_p \subset \cup_{x \in \mathcal{B} \cap \mathcal{L}^p L_0} (W^s_{loc}(x) \cap F) \tag{3.39}$$

Thus, let W be an unstable segment and consider the sets:

$$\begin{cases} \mathcal{B}_W = \cup_{x \in \mathfrak{A} \cap W} W^s_{loc}(x) \\ \mathfrak{A}_W = \{x \in \mathfrak{A} \cap W : W^s_{loc}(x) \neq \{x\}\} \end{cases} \tag{3.40}$$

and prove that $m(\mathcal{B}_W) = 0$ using the relation $l (\mathfrak{A} \cap W) = 0$, i.e., it is enough to prove:

$$m(\cup_{x \in \mathfrak{A}_W} W^s_{loc}(x)) = 0 \tag{3.41}$$

If the manifolds W^s_{loc}'s are depending smoothly on x, then (3.40) is a consequence of *Fubini's theorem* [Freiling (1986)], else one can use Lemma. 3.3. Indeed, let $\varepsilon > 0$ small enough. The relation $l (\mathfrak{A} \cap W) = 0$, imply that the set $\mathfrak{A} \cap W$ can be covered by a countable union of open disjoint intervals of total length smaller than ε. Let I be one of these intervals and let $\varepsilon' < \varepsilon$ be its length. Let $n \in \mathbb{N}^*$, such that $\cup_0^{n+1} S_{\overline{k}}$ splits I into at most 2^{n+2} segments. If J is such a segment, then by Lemma. 3.3, we have that the dispersion of the angles of $\{W^s_{loc}(x), x \in J\}$ is bounded by $2\left(\dfrac{b}{\lambda}\right)^n$. So, if n is arbitrary and $\lambda > 2b$ we obtain:

$$\begin{cases} m(\cup_{x \in I} W^s_{loc}(x)) \leq 16 \left(\dfrac{2b}{\lambda}\right)^n + l (J) \leq 16 \left(\dfrac{2b}{\lambda}\right)^n + \varepsilon' \\ m(\cup_{x \in \mathfrak{A}_W} W^s_{loc}(x)) \leq 2\varepsilon \end{cases} \tag{3.42}$$

because we have $m(\cup_{x \in 3} W^s_{loc}(x)) \leq 4 \left(\dfrac{b}{\lambda}\right)^n + l (J)$ and $m(\cup_{x \in I} W^s_{loc}(x)) \leq 2\varepsilon'$. Finally, we have $m(\mathcal{B}_W) = 0$. Hence the proof of Proposition. 3.4(b) is completed.

3.2.4 The Hausdorff dimension of the Lozi mappings

The Hausdorff dimension of the Lozi mappings was calculated in several works by several methods. For example it was calculated using the *minimal spanning tree* in [Martínez, et al. (1993)]. In [Hiroaki (1985)] the Lyapunov dimension was calculated for the case of coupled two-dimensional maps. In [Grassberger (1989)] the Lyapunov and dimension spectra of 2-D attractors were calculated, with an application to the Lozi mappings. In [Mcdonough, et al. (1995)] a chaos detector was discovered and it was applied to the Lozi mappings as a criterion for detecting chaos in dynamical systems just like the largest Lyapunov exponent and the different types of dimensions and entropies defined in Chapter 1 for dynamical systems. Some methods for calculating multifractal spectra and multifractal rigidity were presented in [Hiroki, et al. (1987), Barreira, et al. (1997)]. Some recent numerical methods can be found in [Borges & Tirnakli (2004), Szustalewicz (2008)].

In this section, we discuss the calculation of the Hausdorff dimension of the Lozi mappings (3.1) presented theoretically in [Collet & Levy (1984)]. For this aim, let $\chi+$ and $\chi-$ be the characteristic exponents of $(\Lambda_{a,b}, \mathcal{L}, \mu)$, i.e.,

$$\chi\pm = \lim_{n\to\pm\infty} \frac{1}{n} \log \|D_x\mathcal{L}^n\|, \text{ for } \mu.\text{a.e.}x \in \Lambda_{a,b}. \tag{3.43}$$

Let h be the μ-entropy of $\mathcal{L}$. The following result was proved in [Ledrappier & Strelcyn (1982)]:

Theorem 24 *If M is a compact surface and (M, f, m) is an ergodic C^2 dynamical system with characteristic exponents $\chi_1 \geq 0 \geq \chi_2$, then the Hausdorff dimension of m is given by:*

$$HD(m) = h_m(f)\left(\frac{1}{\chi_1} - \frac{1}{\chi_2}\right) \tag{3.44}$$

where $h_m(f)$ is the m-entropy of f. If the limit exists almost everywhere, then $HD(m)$ is defined by:

$$HD(m) = \lim_{\varepsilon\to 0} \frac{\log m(\mathcal{B}(x,\varepsilon))}{\log \varepsilon} \tag{3.45}$$

where

$$\mathcal{B}(x,\varepsilon) = \{y \in M : d(x, y) < \varepsilon\} \tag{3.46}$$

Theorem 23 is still valid for the Lozi mapping (3.1) despite the fact that this map is only almost everywhere of class C^∞. The following results was proved in [Collet & Levy (1984)]:

Proposition 3.5 (a) *For μ-a.e. $x \in \Lambda_{a,b}$ we have*

$$\lim_{\alpha \to 0} \frac{\log \mu (\mathcal{B}(x, \alpha))}{\log \alpha} = h \left(\frac{1}{\chi+} - \frac{1}{\chi-} \right) \tag{3.47}$$

(b) *With the above notations, we have*

$$h = \chi_+ \tag{3.48}$$

The proof of Proposition 3.5(a) is based on the adaptation of the ideas given in [Ledrappier & Strelcyn (1982)] to the Lozi map (3.1) along the two inequalities given by the following lemma:

Lemme 3.7 *We have*

$$
\begin{cases}
\lim_{\alpha \to 0} \inf \frac{\log m(\mathcal{B}(x,\alpha))}{\log \alpha} \geq h. \left(\frac{1}{\chi+} - \frac{1}{\chi-} \right) \mu\text{-}a.e. \\[2mm]
\lim_{\alpha \to 0} \sup \frac{\log \mu(\mathcal{B}(x,\alpha))}{\log \alpha} \leq h. \left(\frac{1}{\chi+} - \frac{1}{\chi-} \right)
\end{cases} \tag{3.49}
$$

Proposition. 3.5(b) is a mere consequence of [Ledrappier & Strelcyn (1982)] and it is easy to check that hypothesis of [Katok & Strelcyn (1980)] are fulfilled, so that the result of [Ledrappier & Strelcyn (1982)] applies. Another result about an upper bound (an explicit estimates) for the Hausdorff dimension of the Lozi map (3.2) was proved in [Ishii (1997b)] as follows:

Theorem 25 *(Hausdorff dimension of the attractors). Assume that*

$$\sqrt{2}a > b + 2, \, b < 4 - 2a, \, b > 0 \text{ is small} \tag{3.50}$$

Then we have

$$1 + \frac{\log (a + \sqrt{a^2 - 4b}) - \log 2}{\log (a + \sqrt{a^2 - 4b}) - \log 2b} \leq \dim_H \mathcal{L}_1 \leq 1 + \frac{\log 2}{\log (a + \sqrt{a^2 - 4b}) - \log 2b} \tag{3.51}$$

Form (3.51) we remark that this estimate becomes sharper when $b > 0$ goes to 0 and the ratio

$$\frac{\text{(The upper bound)} - \text{(The lower bound)}}{\text{(The upper bound)} - 1}$$

tends to $1 - \dfrac{\log a}{\log 2}$ which is close to zero. For example, if $a = 1.7$ and $b = 0.1$ then the estimate gives

$$1.176669.... \leq \dim_H \mathcal{L}_1 \leq 1.247848...,$$

and for the choice $a = 1.7$ and $b = 0.01$ we get

$$1.102712.... \leq \dim_H F \leq 1.135055...$$

The main tool of the proof of Theorem 24 is the so called *Young equality* [Young (1982(b))].

More information about ergodic properties of Lozi mappings can be found in [Levy (2007)]. For *generalized hyperbolic attractors* (see Section 3.4 for more details), one can see [Pesin (1992)]. This class of dynamical systems (including the Lozi attractor) is defined on a Riemannian manifold with singularities. Their attractors have strong hyperbolic behavior. For this class of systems, the existence of a special invariant measure[3] was proved in Section 3.4.3. This measure is an analog of the Bowen-Ruelle-Sinai measure for classical hyperbolic attractors introduced in Section 1.2.1. Some ergodic properties were also studied for this class of systems with respect to this measure. For more details, see Sections 3.4.4 and 3.4.5. In [Chazottes, et al. (2005)] the so called *Devroye inequality*[4] was proved to be true for a large class of non-uniformly hyperbolic dynamical systems introduced in Section 1.2.2. This class includes families of piecewise hyperbolic maps (Lozi-like maps), scattering billiards (e.g., planar Lorentz gas), unimodal and Hénon-like maps. More details about the structure of generalized hyperbolic attractors can be found in [Afraimovich, et al. (1995), Dobrynskii (1999–2005)]. Finally, invariant measures for hyperbolic piecewise smooth mappings of a rectangle were studied in [Jakobson & Newhouse (1996-2000)].

3.2.5 Spectra of singularities of the Lozi mapping

In [Hiroki, et al. (1987)] the spectra of singularities and the generalized dimensions and entropies were explicitly derived for certain type of Lozi strange attractors. The probability measure on these attractors is characterized by the spectrum of singularities $f(a)$ defined by:

Definition 35 *The spectrum of singularities $f(a)$ is a Legendre transformation of the generalized dimensions $D(q)$ $(-\infty < q < +\infty)$, .i.e., $f(a)$ represents the fractal dimension of a set of singularities which have an identical exponent a.*

Generally, for chaotic systems the probability measure has an infinite number of singularities with different scaling exponents a. In [Hiroki, et al. (1987)] relations between the partial dimensions and the expansion rates defined in [Morita (1987), Eckmann & Procaccia (1986)] were established. In [Grassberger (1983)] a relation of the generalized entropies to properties of unstable cycles within chaos was derived. These relations were used in [Hiroki, et al. (1987)] to calculate the spectrum $f(a)$ for two-dimensional

---

[3] i.e., the Gibbs u-measures.
[4] This inequality provides an upper bound for the variance of observables in some special forms.

invertible maps with constant Jacobian, i.e., to obtain $\tau_k(q)$ defined below by the first part of (3.52) explicitly:

Proposition 3.6 [*Gressberger (1985), Halsey, et al. (1986)*] *Let $D_1(q)$ and $D_2(q)$ be the partial dimensions in the stretching and contracting direction on a strange attractor, respectively. The generalized dimensions are given by their sum $D(q) = D_1(q) + D_1(q)$, and various quantities are given by:*

$$\left\{ \begin{array}{l} \tau_k(q) = (q-1)D_k(q)\,, k = 1,2 \\[2mm] \alpha_k(q) = \frac{d\tau_k(q)}{dq}, 1 \\[2mm] f_k(\alpha_k) = \alpha_k q - \tau_k(q) \\[2mm] \tau(q) = \tau_1(q) + \tau_2(q) = (q-1)D(q) \\[2mm] \alpha(q) = \frac{d\tau(q)}{dq} = \alpha_1 + \alpha_2 \\[2mm] f(\alpha) = \alpha q - \tau(q) = f_1(\alpha_1) + f_2(\alpha_2) \end{array} \right. \tag{3.52}$$

The quantities in (3.52) are derived from $\tau_1(q)$ and $\tau_2(q)$. Now, let $\{X_n\}_{n \geq 0}$ be a chaotic orbit, and let $\lambda_k(n)$ be the local expansion rate of nearby orbits in the kth direction at $X_{n'}$, then we have the following result proved in [Morita, et al. (1987)] as follows:

Theorem 26 (*Relation between $\tau_1(q)$ and $\tau_2(q)$*): *The quantities $(\lambda_k(q))_{1 \leq k \leq n}$ defined by the first part of (3.52) are related by the following relation:*

$$\langle \exp\{-n\Lambda_1(n)\tau_1(q) - n\Lambda_1(n)\tau_2(q)\}\rangle \sim 1, n \gg 1 \tag{3.53}$$

where

$$\Lambda_k(n) = \frac{1}{n}\sum_{m=1}^{n}\lambda_k(m) \tag{3.54}$$

and $\langle \,.\,,\,.\,\rangle$ denotes the long-time average with respect to initial point X_0. The quantity $\tau_1(q)$is given by:

$$\tau_1(q) = q - 1 \tag{3.55}$$

If the attractor has no singularity in the stretching direction, then we have that:

$$D_1(q) = 1, \tau_1(q) = f_1(\alpha_1) = 1, \tag{3.56}$$

Thus, the quantity $\tau_2(q)$ can be determined by using (3.53) and (3.54).

Definition 36 *(1) The generalized entropies K (q) for two-dimensional maps with $D_1 (q)=1$ are given by:*

$$\Psi (q) = (q-1) K (q)= -\lim \frac{1}{n} \ln \langle \exp \{-n \Lambda_1 (n) (q-1)\}\rangle \qquad (3.57)$$

(2) ([Morita, et al. (1987)] combination with (3.53)) give the following equality:

$$\Psi (q - \tau_2 (q))= -\tau_2 (q) \ln |J| \qquad (3.58)$$

Where J is the Jacobian.

This means that a Legendre transformation of $\Psi (q)$ leads to the spectrum of dynamical scaling exponents $h (\gamma)$ defined in [Sano, et al. (1986), Eckmann & Procaccia(1986)]. Finally, $\tau_2 (q)$ and hence $f (a)$ can be determined if we know $\Lambda_1 (n)$ and by using (3.57) and (3.58).

Applying the previous results (i.e., Proposition 3.6 and Theorem 25) to the following selected Lozi map given by:

$$\begin{pmatrix} x_{n+1} \\ y_{n+1} \end{pmatrix} = \begin{pmatrix} f (x_n) + by_n \\ x_n - \dfrac{1}{2} \end{pmatrix} \qquad (3.59)$$

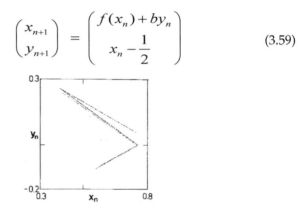

Figure 3.2: The Lozi attractor (3.59)-(3.60) with $\beta = \dfrac{1+\sqrt{5}}{2}$, $b = -0.3$. Reused with permission from Hiroki Hata, Terumitsu Morita, Koji Tomita and Hazime Mori, Hokkaido Mathematical Journal, (1987). Copyright 1987, Hokkaido Mathematical Journal.

$$f (x_n) = \begin{cases} \beta x, \text{ if } 0 \le x < \dfrac{1}{2} \\ \beta (1-x), \text{ if } \dfrac{1}{2} \le x < 1 \end{cases} \qquad (3.60)$$

which is an extension of the tent map (3.6). For $\beta = \dfrac{1+\sqrt{5}}{2}, b = -0.3$, the map (3.59)–(3.60) displays the strange attractor shown in Fig. 3.2.

Thus, it is easy to show the following result:

Proposition 3.7 *(1) The fixed point* **X*** *of the map (3.59)-(3.60), the eigen-values* μ^*_1, μ^*_2 *and eigenvectors* e^*_1, e^*_2 *around* **X*** *are given by:*

$$\begin{cases} \mathbf{X}^* = \begin{pmatrix} x^* \\ y^* \end{pmatrix} = \dfrac{1}{2(\beta + 1 + b)} \begin{pmatrix} 2\beta - b \\ \beta - 1 \end{pmatrix} \\ u^*_1 = -\left(\dfrac{\beta + \sqrt{\beta^2 - 4b}}{2} \right), \ e^*_k = \dfrac{1}{\sqrt{1 + (u^*_k)^2}} \begin{pmatrix} u^*_k \\ 1 \end{pmatrix} \end{cases} \tag{3.61}$$

(b) The local expansion rate λ_1 *(n) in the stretching direction at* $\mathbf{X}_n$ *is given by:*

$$\lambda_1 (n) = \ln | T (n) u_1 (n) | \tag{3.62}$$

where T (n) is the Jacobian matrix at $\mathbf{X}_n$,

$$T (n) = \begin{pmatrix} (-1)^i \beta & b \\ 1 & 0 \end{pmatrix}, \ i = \begin{cases} 0, \ if \ 0 \leq x < \dfrac{1}{2} \\ 1, \ if \ \dfrac{1}{2} \leq x < 1 \end{cases} \tag{3.63}$$

and u_1 *(n) is the unit vector in the stretching direction at* $\mathbf{X}_n$. *The unit vectors* u_1 *(n)'s are obtained iteratively from:*

$$u_1 (n + 1) = \exp (-\lambda_1 (n)) T (n) u_1 (n) \tag{3.64}$$

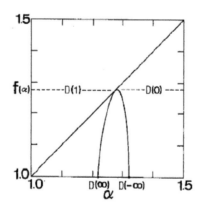

Figure 3.3: The spectrum $f(\alpha)$ for the Lozi attractor (2.59)–(2.60) shown in Fig. 2.2 where D(∞) = 1.220, D(1) = 1.274, D(0) = 1.278, D(−∞) = 1.325. Reused with permission from Hiroki Hata, Terumitsu Morita,Koji Tomitaand Hazime Mori, Hokkaido Mathematical Journal (1987). Copyright 1987, Hokkaido Mathematical Journal.

The initial condition $\mathbf{X}_0$ can be taken on the unstable manifold[5] so that $u_1(0) = e^*_1$. Thus we obtain:

$$\Lambda_1(n) = \frac{1}{n} \sum_{m=1}^{n} \lambda_1(m). \qquad (3.65)$$

and then $\Psi(q)$ from (3.57) and then $\tau_2(q)$ from (3.58). Thus, we obtain the spectrum $f(\alpha)$ which is shown in Fig. 3.3 for $\beta = \dfrac{1+\sqrt{5}}{2}$, $b = -0.3$. Several generalized dimensions and entropies are given in Table (3.66) (Reused with permission from Hiroki Hata, Terumitsu Morita, Koji Tomita and Hazime Mori, Hokkaido Mathematical Journal (1987). Copyright 1987, Hokkaido Mathematical Journal.)

N	$D(\infty)$	$D(5)$	$D(1)$	$D(0)$	$D(-1)$	$D(-5)$	$D(-\infty)$	
	1.220	1.258	1.274	1.281	1.281	1.291	1.325	
10	1.221	1.258	1.273	1.277	1.280	1.291	1.323	
5	1.232	1.260	1.276	1.280	1.283	1.292	1.310	
	$K(\infty)$	$K(5)$	$K(1)$	$K(0)$	$K(-1)$	$K(-5)$	$K(-\infty)$	(3.66)
	0.340	0.409	0.452	0.466	0.476	0.506	0.580	
10	0.340	0.409	0.43	0.464	0.474	0.509	0.579	
5	0.340	0.411	0.459	0.470	0.480	0.502	0.542	

The same work and results have been given also for the Hénon map (2.1).

3.3 Grammatical complexity of the Lozi mappings

In this section, we discuss the notion of *complexity* for the Lozi maps (3.2). This notion was defined and studied in [Wolfram (1984)] and more details can be found in [Hopcroft & Ullman (1979)].

3.3.1 Concepts and definitions

In this subsection, some definitions are presented for more comprehension of the notion of grammatical complexity.

Definition 37 (a) *Grammatical complexity is a quantity measuring the difficulty in describing the topological structure of a map. (b) The complexity of a regular language is the number of states of the minimal finite automaton (see Definition 23(a) below) equivalent to the language.*

It follows from Definition 20 that the grammatical complexity (GC) is a computable object if the map under consideration has hyperbolic structure

---

[5] The reason is the fact that the unstable manifold of the fixed point $\mathbf{X}^*$ is linear near $\mathbf{X}^*$.

(Recall Definition 12(d)) and this uses informations on the existence of Markov partitions. The basic notion here is the *Chomsky hierarchy* [Wolfram (1984)]:

Definition 38 *The Chomsky hierarchy consists of four levels of languages: regular languages, context-free languages, context-sensitive languages and re-cursively enumerable languages, from the bottom.*

Now, some basics of regular languages, finite automata, and the definition of *regular-language complexity* are given in the following definition:

Definition 39 *(a) A language is a set of finite strings denoted by the word "string". (b) An infinite string is denoted by the word "sequence". (c) The string of length 0 is denoted by ε. (d) A regular language is a set of strings expressed by three kinds of operations:*

$$\left\{ \begin{array}{c} concatenation : L_1 L_2 = \{xy : x \in L_1, y \in L_2\} \\ * : L^* = \cup_{j=0}^{\infty} L^j \\ + : L_1 + L_2 = L_1 \cup L_2 \end{array} \right. \tag{3.67}$$

where L^j is the j-fold concatenation of L'.

As an example of Definition 33(d), the set of all strings over the alphabet $\Sigma = \{0, 1\}$ is a regular language expressed as $L' = (0 + 1)^* = \Sigma^*$.

Definition 40 *(a) A finite automaton is a mathematical model $M = (Q, \Sigma, \delta, q_0, F)$, which represents transitions between states with a certain input string:*

$$\left\{ \begin{array}{c} Q = \{q_1, q_2, ..., q_n\} : \ set \ of \ states \\[2mm] \Sigma = \{0, 1\} : \ input \ alphabet \\[2mm] \delta : Q \times \Sigma^* \rightarrow Q : \ transition \ function \\[2mm] q_0 \in Q : \ initial \ state \\[2mm] F \subset Q : \ set \ of \ final \ states \end{array} \right. \tag{3.67}$$

Here $\delta(q, x)$ denotes a function of a state $q \in Q$ and an input string $x \in \Sigma^*$. (b) A regular language is closely related to a finite automaton. (c) We say that a finite automaton is a deterministic finite automaton (DFA) if the transition function $\delta(q, x)$ is uniquely determined by the present state q and an initial input string x, and a non-deterministic finite automaton (NFA) otherwise. (d) For a finite automaton $M = (Q, \Sigma, \delta, q_0, F)$ and a string $x \in \Sigma^*$, if $\delta(q_0, x) \in F$, then we say that x is accepted by M. For a NFA, it suffices that there exists a transition by which the NFA reaches one of the final states among the possible ones with an input. (e) The set of strings

$$L(M) = \{x \in \Sigma^* : \delta\,(q_0,\, x) \in F\} \tag{3.69}$$

is the language accepted by M.

Examples of a DFA and a NFA are shown in Fig. 3.4.

Definition 40(e) is equivalent to the fact that a language L' is regular and L' is accepted by a finite automaton [Hopcroft & Ullman (1979)]. Figures 3.4(a) and 3.4(b) show this situation, i.e., a NFA and a DFA accepting $L' = 1(0 + 01)^*$. Generally, several different automata can represent the same language. The unicity in minimality viewpoint was proved by the *Myhill-Nerode theorem*, so minimal DFAs can be used for comparing different languages [Hopcroft & Ullman (1979)]:

Theorem 27 *(Myhill-Nerode theorem) There exists a unique DFA up to permutation of labeling such that the number of states is minimal among DFAs accepting a given regular language.*

A concrete algorithm minimizing a redundant DFA is presented in [Hopcroft & Ullman (1979)]. Using the notion of minimal DFA, a definition of complexity was proposed in [Wolfram (1984)]:

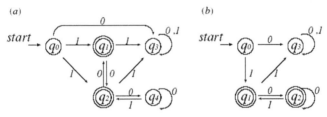

Figure 3.4: (a) Example of a non-deterministic finite automaton. (b) Example of a deterministic finite automaton. In each figure, a node of a graph represents a state. A transition between two states is possible if an input symbol is the same as the one labeled on an arrow connecting the two states. The initial state is q_0, and the final states are denoted by double circles. In (a), if the initial input is 1, the transition from q_0 is allowed to either q_1 or q_2, and if the initial input is 0, the transition is allowed only to q_3. Reused with permission from Ryouichi and Akira, J. Phys. A: Math. Gen, (2004). Copyright 2004, IOP Publishing Limited.

Definition 41 *The complexity of a regular language L' is defined as:*

$$C\,(L') = \log N\,(L') \tag{3.70}$$

where $N\,(L')$ is the number of states of the minimal DFA accepting L'. We call $C\,(L')$ the grammatical complexity (GC) of the language L'.

A series of investigations has been devoted to this subject for some type of dynamical systems as in [Badii & Politi (1997), Grassberger (1986), Auerbach & Procaccia (1990), Hao (1991), Xie (1993),Wang & Xie (1994), Crutchfield & Young (1990), Lakdawala (1996), Xie (1996),Wang, et al. (1999)]. For example in [Grassberger (1986), Wang, et al. (1999)]

the complexity of one-dimensional unimodal maps on an interval was studied. It was shown in [Auerbach & Procaccia (1990), Hao (1991)] that if the kneading sequence of a unimodal map is either periodic or eventually periodic, the language generated from the map is regular in the lowest level of the Chomsky hierarchy [Chomsky (1959)] and the converse is also true [Xie (1993), Wang & Xie (1994)]. In [Crutchfield & Young (1990), Lakdawala (1996), Xie (1996)] it was proved that the language generated at the period-doubling accumulation point of unimodal maps is one in the level located between context-free and context-sensitive. A conjecture was formulated in [Xie (1996), Wang, et al. (1999)] as follow:

Conjecture 28 *Unimodal maps never generate proper context-free lan*guages.

Generally, due to the construction of a proper symbolic dynamics, it is very difficult[6] to carry out an analogous programme in higher dimensional dynamical systems (except for the trivial horseshoe dynamics). However, for the simplest possible examples of higher dimensional problems, concrete algorithm based on the idea of the pruning front [Ishii (1997(a)), de Carvalho & Hall (2002)] was proposed for the Lozi map (3.2) in [Ishii (1997(a))]. An algorithm to generate the pruning front for the Hénon map[7] $H_{a,b}(x, y) = (a + by - x^2, x)$ was presented in [Ryouichi & Akira (2004(b))].

3.3.2 The method of calculation of transition rules

In [Ryouichi & Akira (2004(a))] the grammatical complexity (GC) of the symbol sequences generated from the Hénon map (2.1) and the Lozi map (3.2) was calculated using the methods developed to construct the pruning front (Recall Definition 56(b)). Two major results were founded:

1. The language of symbol sequences is regular in the sense of the Chomsky hierarchy and the corresponding grammatical complexity takes finite values when the map is hyperbolic (recall Definition 12(d)). In this case, the complexity exhibits a self-similar structure[8] as a function of the system parameter.

2. For non-hyperbolic cases, the complexity monotonically increases as the increase of the resolution of the pruning front.

On one hand, the method of calculation is based on the construction of a symbolic dynamics which is conjugate or semi-conjugate to the original dynamics along with a finite automaton accepting the language made out of it. On the other hand, it is known that for a one-dimensional unimodal

---

[6] Because there are no critical points in generic higher dimensional maps.

[7] This algorithm can be applied to the area-preserving case, i.e., $|b| = 1$.

[8] In fact, similarity of the pruning fronts is an origin of this self-similarity property.

map, the critical point divides the interval. Hence, for each subinterval, it is possible to define a symbol, i.e., the itinerary of the critical point controls admissibility of the orbits [Milnor & Thurston (1988)]. The lack of critical points imply the impossibility of an extension of the kneading theory to two-dimensional maps. But, an analogous construction has been explored in [Cvitanovic, et al. (1988), Cvitanovic (1991)]. The main idea is the pruning front with the assumption of two dimensional symbol plane for the horseshoe dynamics. Then observe which orbits become non–admissible (considered pruned from the horseshoe symbol plane) as the system parameter is varied. In this case, the forbidden regions in the symbol plane are called *pruned regions* and they cannot be determined uniquely. Note that the *border* of the primary pruned region is the pruning front which is used to completely specify the admissible orbits in the symbol plane [Cvitanovic, et al. (1988), Cvitanovic (1991)]. In some cases, the pruning front was formulated rigorously [Ishii (1997(a)), de Carvalho & Hall (2002)]. A concrete algorithm to provide the pruning front for a given map is more applicable. A recipe was proposed for the Lozi map (3.2) in [Ishii (1997(a))][9]. For some area-preserving Hénon mappings, an algorithm was proposed in [Ryouichi & Akira (2004(b))] and it is based essentially on the notion of hyperbolicity[10] introduced in Section 1.2.1. Otherwise this procedure will not stop within finitely many steps. In fact, there exist infinitely many intervals to which this algorithm can be applied as shown in Fig. 3.5 for a value of a, say $a \simeq 5.4$, in the so called *longest plateau* of the Hénon map.

In [Ryouichi & Akira (2004(a))] some modifications of this algorithm were made for the non-hyperbolic cases by introducing a certain resolution in the symbol plane, i.e., by limiting the length of forbidden strings and ignore the finer structures of the pruning fronts. The conversion of a transition rule into a finite automaton was given for $a \simeq 5.4$ in the longest plateau. In this case, there is a finite automaton accepting all the admissible strings. Indeed, the forbidden strings of the longest plateau are 0010100 and 0011100, which correspond to the blocks shown in Fig. 3.5. Let L' be the set of all strings which do not contain these forbidden ones. Then, the complement $\bar{L}' = \Sigma^* \setminus L'$ is regular and it is expressed as:

$$\bar{L} = (0 + 1)^* \, (0010100 + 0011100) \, (0 + 1)^*. \tag{3.71}$$

The complement $\bar{L}'$ is accepted by the NFA $M = (Q, \Sigma, \delta, q_0, F)$ shown in Fig. 3.6 which accepts all sequences containing the forbidden strings.

---

9 It is very difficult to find such a simple algorithm for the Hénon mappings.
10 Note that parameter values for which the Hénon map (2.1) has hyperbolic (resp. not hyperbolic) structures has a positive Lebesgue measure. See [Lai, et al. (1993)] for more details.

Thus, we have that if $M = (Q, \Sigma, \delta, q_0, F)$ is a DFA accepting L', then $M^r = (Q, \Sigma, \delta, q_0, Q \setminus F)$ is a DFA accepting $\bar{L'}$. The reason is that if a language L' is regular, then the complement $\bar{L'}$ is also regular as shown in [Hopcroft & Ullman (1979)] and the fact that $x \in \bar{L'}$ and $\delta(q_0, x) \in Q \setminus F$ are equivalent.

Now, the method of construction of a DFA $M' = (Q', \Sigma', \delta', q'_0, F')$ equivalent to M is given by the following steps:

(1) The initial state is $q_0 = \{q_0\}$ and let $\delta'(q_0, 0) = \{q_0, q_1\}$ be one of the elements of Q' (consists of the subsets of Q) since it is possible to move from q0 to either q_0 or q_1 with an input 0.

(2) The NFA can move from q_0 to $\{q_0, q_1\}$ and from q_1 to q_2, respectively, with 0, let the state to which M moves from $\{q_0, q_1\}$ be $\{q_0, q_1, q_2\}$. That is, to complete Q' it suffices to add (the sets of) states which can be reached from q_0 one after another. The set of the final states F' consists of all the elements of Q' including one of the final states of M. If all the nodes have two arcs labelled 0 and 1, then the DFA M', which is shown in Fig. 3.7, is completed.

(3) The calculation of the GC can be done by reducing M' to the minimal DFA. In this case, all the accessible states from $\{q_0, q_1, q_2, q_7\}$ are the final states of M'. The minimal DFA equivalent to M' is shown in Fig. 3.8.

(4) The minimal DFA accepting L' is obtained by exchanging the roles of the final states and the others. If a string containing 0010100 or 0011100 as an input, then the minimal DFA reaches the state $\{q_0, q_1, q_2, q_7\}$ and this string is never accepted. For the longest plateau, we have $C(L') = \log 8$ and the whole procedure is shown in Fig. 3.9.

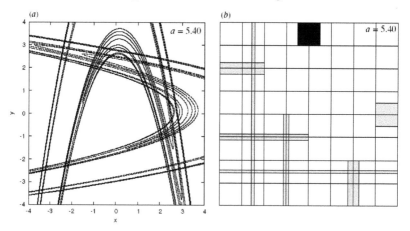

Figure 3.5: The stable and unstable manifolds for the area-preserving Hénon map (2.1) at a = 5.4 and the pruned regions. The primary pruned region is colored black, and its forward and backward images are grey. Reused with permission from Ryouichi and Akira, J. Phys. A: Math. Gen, (2004). Copyright 2004, IOP Publishing Limited.

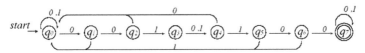

Figure 3.6: A non-deterministic finite automaton accepting $\overline{L'}$. Reused with permission from Ryouichi and Akira, J. Phys. A: Math. Gen, (2004). Copyright 2004, IOP Publishing Limited.

From Fig. 3.10, we conclude that there is a possibility where the GC obtained by the previous procedure can or cannot describe a given dynamical system. The reason is that a primary pruned region can give different automata, i.e., if the forbidden string is 001, it can be expressed by two strings 0010 and 0011 and the lower automaton in Fig. 3.10 does not accept 001, even though this is a forbidden string. In Fig. 3.11, the GC for the area-preserving Hénon map $H_{a,b}$ defined above is plotted as a function of the system parameter a where each figure is a magnification of the upper one. Clearly, above $a = 5.699...$ the *first tangency* between the stable and unstable manifolds occurs [Sterling, et al. (1999)]. Hence the symbolic dynamics forms the binary full shift shown in Fig. 8(b). This shift qualitatively show the same behavior for the Lozi map (3.2). The comparison of the present method with the *continued fractions method* described in [Ishii (1997(a))] shows by Fig. 3.12 that both give almost the same results with a slight difference in the intervals where the GC does not seem to converge.

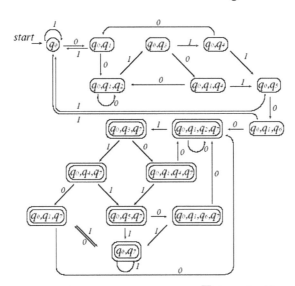

Figure 3.7: A deterministic finite automaton accepting $\overline{L'}$. Reused with permission from Ryouichi and Akira, J. Phys. A: Math. Gen, (2004). Copyright 2004, IOP Publishing Limited.

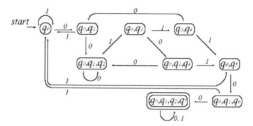

Figure 3.8: The minimal deterministic finite automaton accepting $\overline{L'}$. Exchanging the role of the final states and the others gives the minimal one accepting L'. Reused with permission from Ryouichi and Akira, J. Phys. A: Math. Gen, (2004). Copyright 2004, IOP Publishing Limited.

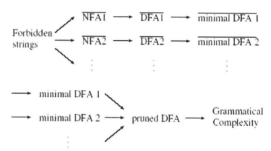

Figure 3.9: Procedure for obtaining the grammatical complexity. The notation $\overline{NFA}$ means a non-deterministic finite automaton accepting $\overline{L'}$, and so on. Reused with permission from Ryouichi and Akira, J. Phys. A: Math. Gen, (2004). Copyright 2004, IOP Publishing Limited.

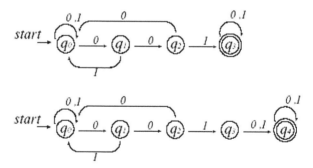

Figure 3.10: Two equivalent non-deterministic finite automata under the condition that infinite sequences are inputed. Reused with permission from Ryouichi and Akira, J. Phys. A: Math. Gen, (2004). Copyright 2004, IOP Publishing Limited.

From Figs. 3.13(a) and 3.13(b) self-similar peaks (simplest possible ones) are clearly visible and it is clear also that the GC gradually increases as the parameter decreases from the first tangency point. In this case, the appearance of such self-similarity can be explained by extracting the pruning fronts for a series of plateaux located at the same positions in each magnified figure. Similar pruning fronts are obtained in [Ryouichi &

Akira (2004b)] for the Hénon map (2.1). However, for the Lozi map (3.2) the largest two pruning fronts in Fig. 3.13(a) are missing and the others are found. The reason is that the shapes of the pruning fronts depend on the ordering of bifurcations. For both cases, there exist ranges (the GC in these regions is zero) of the system parameter where the horseshoe dynamics is realized. But, when the horseshoe structure breaks, the invariant sets of both maps become complicated.[11] In fact, hyperbolicity of the dynamics of both maps makes the corresponding language regular, and then the GC takes a finite value, i.e., the origin of selfsimilar peaks can be attributed to similar-shaped primary pruned regions. The case of divergence for non-hyperbolic parameter intervals will not still hold even if a homoclinic tangency exists. Indeed, at the upper endpoint of the longest plateau of the area-preserving Hénon map (2.1), $a = 5.537...$, two homoclinic points

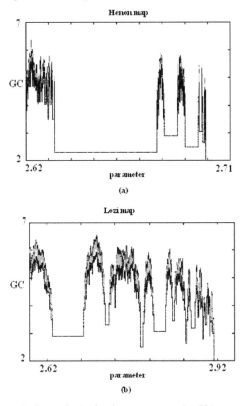

(a)

(b)

Figure 3.11: (a) Grammatical complexity for the area-preserving Hénon map (2.1). Forbidden strings shorter than the specified length are included in the calculations. (b) Grammatical complexity for the area-preserving Lozi map (3.1). Reused with permission from Ryouichi and Akira, J. Phys. A: Math. Gen, (2004). Copyright 2004, IOP Publishing Limited.

---

[11] The pruning fronts may have infinitely many fractal steps.

$(0^\infty 1001 \cdot X 10010^\infty)(X = 0, 1)$ degenerate and they are the last pruned points in the block $001.X100$ as shown in Fig. 3.14. In this case, the corresponding pruned region can be specified only by identifying the missing points $001X$ $1000+0001X1001+0^*1001X10010^*1+10^*1001X10010^*$. Finally, for the Hénon and the Lozi mappings, the shape of the primary pruned region changes in a self-similar manner.

3.4 Admissibility conditions for symbolic sequences of the Lozi map

The symbolic dynamics of the Lozi mappings was the subject of many works. For example in [Zheng (1991)] the dynamics of the Lozi mapping (3.2) was discussed by using symbolic sequences. In this case, two families of symbolic sequences are assigned for two group of lines in the phase plane. The main issue here is that the order of sequences was defined, and the ordering rules were derived. In [Zheng & Liu (1994)] a continuation of the previous work was done for both positive and negative Jacobians. The geometrical structure of foliations associated with attractors was constructed for some typical cases of the Lozi map (3.1) along the determination of critical parameters

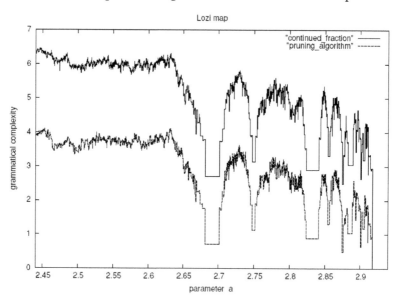

Figure 3.12: Comparison between the result using the continued fractions (Ishii (1997(a))) and the pruning algorithm (Ryouichi & Akira (2004(b))). Although the latter, shifted by two to distinguish the results, tends to be slightly smaller than the former, they coincide at relatively long plateaux. The maximal forbidden strings taken in the calculations is 18. Reused with permission from Ryouichi and Akira, J. Phys. A:Math. Gen, (2004). Copyright 2004, IOP Publishing Limited.

between one- and two-piece attractors. The method of analysis is based essentially on the definition of the ordering rules of symbolic sequences. Some other works on symbolic dynamics for the Lozi maps can be found in [Ishii (1997(b)), Baptista, (2007), Ansari, et al. (2002)].

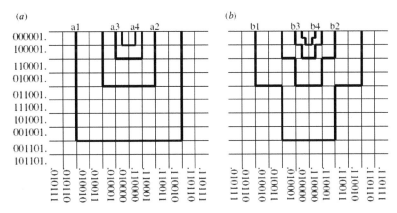

Figure 3.13: Similar pruning fronts observed near the first tangency point. The Hénon map has all the fronts in both (a) and (b), while the Lozi map does not have (a1) and (a2). Reused with permission from Ryouichi and Akira, J. Phys. A: Math. Gen, (2004). Copyright 2004, IOP Publishing Limited.

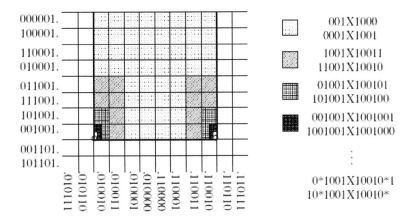

Figure 3.14: Pruning front for the upper endpoint of the longest plateau of the area-preserving Hénon map, $a = 5.537$.... There exists a homoclinic tangency between ($0^*1001.X10010^*$) denoted by circles. However, the primary pruned region, which is the block 001.X100 other than these two points, is specified by a regular expression. Reused with permission from Ryouichi and Akira, J. Phys. A: Math. Gen (2004). Copyright 2004, IOP Publishing Limited.

In [Zheng (1992)] a necessary and a sufficient condition for admissible sequences for the Lozi mappings (3.2) are proposed for $0 < b < 1$. The method of analysis is based on the decomposition of map (3.2) into two coupled onedimensional maps[12] in term of contracting and expanding foliations. In this case, the symbolic sequences assigned to the two classes of foliations are well ordered and the pruning front was constructed from tangencies between them. Indeed, the portion line of the phase plane for the Lozi map (3.2) is the y-axis, thus an L was assigned for $y < 0$, R for $y > 0$ and C for $y = 0$. This means that for a given initial point (x_0, y_0) one can encode its orbit according to the sign of x or y as in [Cvitanovic, et al. (1988), Zheng (1991)] as follows:

$$...s_m...s_1 s_0 \bullet s_0 s_1 s_2...s_n... \tag{3.72}$$

where s_n denotes the x-sign of its nth image, s_m the y-sign of the mth preimage.

Definition 42 *(a) The thick dot $\bullet$ indicates the present position that divides the doubly infinite sequence (3.72) into two semi-infinite sequences, the back-ward sequence $..s_m...s_1 s_0 \bullet$ and the forward sequence $\bullet s_0 s_1 s_2...s_n...$ (b) Points sharing the same forward or backward sequences are called contracting and expanding foliations of the phase space.*

3.4.1 Contracting and expanding foliations and their ordering

In this section, we study the ordering of the contracting and expanding foliations of map (3.2). The analysis is based on symbolic dynamics of onedimensional maps and the so called *forward operator* defined below by (3.85). Let $\varepsilon_n = sgn(x_n)$, and assume that points sharing the same forward sequence $\bullet s_n s_{n+1}...$ fall on the straight line defined by:

$$x - k_n y = \xi_n \tag{3.73}$$

Thus, from Eq.(3.2) we have:

$$x_n - k_n y_n = -(a\varepsilon_n + k_n) x_{n-1} + b y_{n-1} + 1 \tag{3.74}$$

By shifting n to $n + 1$, Eq. (3.74) can be rewritten as:

$$\xi_n = -b^{-1} k_n (\xi_{n+1} - 1) \tag{3.75}$$

where

$$\xi_{n+1} = x_{n+1} - k_{n+1} y_{n+1} \tag{3.76}$$

---

[12] Thus it is possible to extend results about symbolic dynamics for one-dimensional maps to two-dimensional maps.

$$k_n = \frac{b}{a\varepsilon_n + k_{n+1}}$$ (3.77)

The function k_n can be expressed in the form of a continued function as:

$$k_n = \frac{b}{a\varepsilon_n} + \frac{b}{a\varepsilon_{n+1}} + \dots$$ (3.78)

We remark that relation (3.77) defines the map from k_n to k_{n+1} :

$$k_{n+1} = -a\varepsilon_n + \frac{b}{k_n}$$ (3.79)

Then it is easy to prove the following result:

Proposition 3.8 *When*

$$a > 2\sqrt{b}$$ (3.80)

the Lozi map (3.2) has a period-2 orbit that becomes stable nodes or hyperbolic, and the map (3.79) persists the period-2 $\{-k_M, k_M\}$ with

$$k_M = \frac{a}{2} - \sqrt{\left(\frac{a}{2}\right)^2 - b}.$$ (3.81)

In the graph of the mapping function k there are two branches corresponding to $\varepsilon_n = \pm 1$ or $s_n = R$ and L, respectively. In this case, k_n has the same codes as the forward orbit of the point (x_n, y_n). In the interval $[-k_M, k_M]$ the map (3.79) gives a strange repeller and any arbitrary symbolic sequence of R and L corresponds to a point on this repeller, and vice versa. In particular, we have:

$$k (\bullet (RL)^\infty) = k_M, k (\bullet (LR)^\infty) = -k_M$$ (3.82)

From [Hao (1989), Zheng & Hao (1990)] and by using symbolic dynamics of one-dimensional maps, one can conclude that the rules for comparing k of different code sequences are given by:

$$k (\bullet ER\dots) > k (\bullet EL\dots), k (\bullet OR\dots) < k (\bullet OL\dots)$$ (3.83)

where the common strings E and O contains an even and odd total number of R and L.

Hence it easy to verify the following result:

Proposition 3.9 *The convergence of the continued function (3.78) is guaranteed when (3.80) hold and*

$$\begin{cases} k_m = k\,(\bullet R(RL)^\infty) \le k\,(\bullet R...) \le k\,(\bullet\,(RL)^\infty) = k_M \\ -k_m = k\,(\bullet L\,(LR)^\infty) \ge k\,(\bullet L...) \ge k\,(\bullet\,(LR)^\infty) = -k_M \\ k_m = \dfrac{b}{a+k_M} > 0 \end{cases} \tag{3.84}$$

To study equation (3.75), we need to introduce the forward operator **F** defined in [Zheng (1992)] by:

Definition 43 *The forward operator* **F** *is defined by*

$$\mathbf{F}u_n = u_{n+1} \tag{3.85}$$

where $\{u_n\}$ is an arbitrary sequence.

Thus, solution of Eq. (3.75) can be rewritten in terms of **F** as follow:

$$\begin{cases} \xi_n = b^{-1}k_n\,(1 - \mathbf{F}\xi_n) = \left[1 + \dfrac{k_n\mathbf{F}}{b}\right]^{-1}\dfrac{k_n}{b} = \displaystyle\sum_{i=0}^{\infty} q_i \\ q_i = \left(\dfrac{-k_n\mathbf{F}}{b}\right)^i \dfrac{k_n}{b},\ \displaystyle\sum_{i=0}^{\infty} q_i = \dfrac{k_n}{b} - \dfrac{k_n k_{n+1}}{b^2} + \dfrac{k_n k_{n+1} k_{n+2}}{b^3} - \cdots \end{cases} \tag{3.86}$$

Hence, by Eq. (3.86) the line (3.73) is determined with its semi-infinite symbolic sequence. Lines like (3.73) form one class of foliations of the phase space. In a similar way (but the branches R and L are interchanged) the backward symbolic sequence $...s_{n-2}s_{n-1}\bullet$ corresponds to a line:

$$\eta_n = x - h_n y \tag{3.87}$$

and one can verify the following equality:

$$\overline{h}_n = \frac{1}{h_n} = \frac{1}{-a\varepsilon_{n-1} + b\overline{h}_{n-1}} = \frac{1}{-a\varepsilon_{n-1}} + \frac{1}{-a\varepsilon_{n-2}}\ ... \tag{3.88}$$

The map from $\overline{h}_n$ to $\overline{h}_{n-1}$ is given by:

$$\overline{h}_{n-1} = \frac{a\varepsilon_{n-1}}{b} + \frac{1}{b\overline{h}_n} \tag{3.89}$$

which is similar to the map (3.79) for the function k.

Proposition 3.10 *The ordering rules for $\overline{h}$ is given by*

$$\overline{h}\,(...LE\bullet) > \overline{h}\,(...RE\bullet),\ \overline{h}\,(...LO\bullet) < \overline{h}\,(...RO\bullet) \tag{3.90}$$

and

$$\begin{cases} \overline{h}_m \le \overline{h}(...L\bullet) \le \overline{h}_N, -\overline{h}_m \ge \overline{h}(...R\bullet) \ge \overline{h}_N \\ \overline{h}_M = \overline{h}((RL)^\infty \bullet) = \dfrac{k_M}{b}, \ \overline{h}_m = \overline{h}((RL)^\infty L\bullet) = \dfrac{k_m}{b} \end{cases} \tag{3.91}$$

Thus, η_n is given by:

$$\eta_n = 1 - b\overline{h}_{n-1} + b^2\overline{h}_{n-1}\overline{h}_{n-1} - \dots \tag{3.92}$$

Hence, the line (3.92) for backward sequence $\dots s_{n-2}s_{n-1}\bullet$ is determined and lines like (3.92) form the other class of foliations of the phase space. Using the results given in [Greene (1983), Gu (1987)], it is easy to see that the continued function representation of k and $\overline{h}$ is closely related to a *matrix representation*. Thus, if L_n is the Jacobian matrix of the Lozi map (3.1), then Eq. (3.89) can be rewritten as follows:

$$\begin{pmatrix} h_n \\ 1 \end{pmatrix} = \overline{h}_{n-1}\begin{pmatrix} -a\varepsilon_{n-1} & b \\ 1 & 0 \end{pmatrix}\begin{pmatrix} h_{n-1} \\ 1 \end{pmatrix} = \overline{h}_{n-1}L_{n-1}\begin{pmatrix} h_{n-1} \\ 1 \end{pmatrix} \tag{3.93}$$

Assume that

$$a > 1 + b. \tag{3.94}$$

Then for the backward foliation $(s_{n-p}...s_{n-2}s_{n-1})^\infty \bullet$ of a period-p we have that $h_n = h_{n-p}$, thus:

$$\begin{pmatrix} h_n \\ 1 \end{pmatrix} = \overline{h}_{n-1}\overline{h}_{n-2}...\overline{h}_{n-p}L_{n-1}L_{n-2}...L_{n-p}\begin{pmatrix} h_n \\ 1 \end{pmatrix} \tag{3.95}$$

Eq. (3.95) implies that the vector $\begin{pmatrix} h_n \\ 1 \end{pmatrix}$ is an eigenvector of the period. In a similar way, for a periodic forward sequence of period-p we have that

$$\begin{cases} \begin{pmatrix} k_n \\ 1 \end{pmatrix} = k_n\begin{pmatrix} 0 & 1 \\ \dfrac{1}{b} & \dfrac{a\varepsilon_n}{b} \end{pmatrix}\begin{pmatrix} h_{n+1} \\ 1 \end{pmatrix} \\ = k_n L_n^{-1}\begin{pmatrix} k_{n+1} \\ 1 \end{pmatrix} = k_n k_{n+1}...k_{n+p-1}L_n^{-1}L_{n+1}^{-1}...L_{n+p-1}^{-1}\begin{pmatrix} k_n \\ 1 \end{pmatrix} \end{cases} \tag{3.96}$$

Thus, the following result can be proved easily:

Proposition 3.11 (a) *When (3.94) holds, we have* $|\overline{h}| < \overline{h}_M < 1$, *since* $\overline{h}_M < 1$, *for any backward sequence.* (b) *The vector* $\begin{pmatrix} h_n \\ 1 \end{pmatrix}$ *is the unstable direction of the* period, and the line (3.73) is called *backward contracting manifold (BCM)*. (c) The vector $\begin{pmatrix} k_n \\ 1 \end{pmatrix}$ is the stable direction of the period, and the line (3.92) is called *forward contracting manifold (FCM)*.

An example of how to determine the manifolds BCM and FCM is the one given in [Zheng (1991)]. In this case, the ordering of forward sequences rules for the Lozi map (3.2) is given by:

$$\left\{ \begin{array}{l} \bullet ER... > \bullet EL..., \bullet OR... < \bullet OL... \\ ...R\overline{E}\bullet > ...L\overline{E}\bullet, ...R\overline{O}\bullet < ...L\overline{O}\bullet \end{array} \right. \tag{3.97}$$

Hence, the common leading string E contains as even number of R's and $\overline{E}$ an even number of L's, while O and $\overline{O}$ contain an odd number of R's and L's respectively. The manifolds FCM of the greater of two forward sequences is located on the right of the FCM of the smaller. While for two backward sequences the greater BCM is above the smaller, i.e., according to (3.97) the greatest sequences are $\bullet RL^\infty$ and $R^\infty\bullet$, and the smallest are $\bullet L^\infty$ and $R^\infty L\bullet$.

To finish the determination of manifolds BCM and FCM for the Lozi map (3.2), it is convenient to introduce a metric representation for symbolic sequence as in [Cvitanovic, et al. (1988)]. Indeed, for a semi-infinite sequence $\bullet s_1 s_2... ...s_n...$, we assign the number α defined by:

$$\alpha = \sum_{i=1}^{\infty} \mu_i 2^{-i}, \mu_i = \left\{ \begin{array}{l} 0 \\ 1 \end{array} \right., \quad \text{if } (-1)^i \prod_{j=1}^{i} \varepsilon_j = \left\{ \begin{array}{l} 1 \\ -1 \end{array} \right. \tag{3.98}$$

and for a backward sequence $...s_n...s_2 s_1 \bullet$ we assign the number β defined by:

$$\beta = \sum_{i=1}^{\infty} v_i 2^{-i}, \mu_i = \left\{ \begin{array}{l} 0 \\ 1 \end{array} \right., \quad \text{if } \prod_{j=1}^{i} \varepsilon_j = \left\{ \begin{array}{l} -1 \\ 1 \end{array} \right. \tag{3.99}$$

The (α, β)-plane, is called *symbolic plane* and any semi-infinite sequence corresponds to a number between zero and one, and doubly infinite sequence corresponds to a point in the unit square of this plane. Contracting and expanding lines become vertical and horizontal lines, respectively. Thus, using (3.98) and (3.99) we have the following result:

Proposition 3.12 *We have*:

$$\left\{ \begin{array}{l} \alpha\,(\bullet RL^\infty) = \beta\,(R^\infty\bullet) = 1 \\ \alpha\,(\bullet L^\infty) = \beta\,(R^\infty L\bullet) = 0 \\ \alpha\,(\bullet RRL^\infty) = \alpha\,(\bullet LRL^\infty) = \dfrac{1}{2} \\ \beta\,(R^\infty LR\bullet) = \beta\,(R^\infty LL\bullet) = \dfrac{1}{2} \end{array} \right. \tag{3.100}$$

To find the α-value of a forward word (resp. backward word) containing a C, the letter C must be replaced by LRL^∞ or RRL^∞ (resp. by $R^\infty LL$ or $R^\infty LR$). Figure 3.15 shows an example of expanding and contracting foliations (BCMs and FCMs) for the Lozi map (3.2).

In this case, the segment of the unstable manifold passing through the saddle point H_+ in the first quadrant has the greatest sequence $R^\infty \bullet$. The region (the border) of $R^\infty \bullet$ is located above this line since no sequence can be greater than it, while the region (the border) $R^\infty L \bullet$ is located in the lower half of the plane and belongs to the unstable manifold of H_+, and joins with line $R^\infty \bullet$ at the x-axis, i.e., $C\bullet$. In the symbolic plane (α, β) the two border lines $R^\infty \bullet$ and $R^\infty L \bullet$ are mapped to the lines $\beta = 1$ and $\beta = 0$. The shaded area in Fig. 3.15 is bounded by the lines $R^\infty LR \bullet$ and $R^\infty LL \bullet$, which together with the line $C\bullet$ maps (in the symbolic plane) to the single line $\beta = \frac{1}{2}$. Note that there are many regions belonging to single sequences of the type $R^\infty Q \bullet$ or $\bullet PL^\infty$.

3.4.2 The pruning front and admissibility conditions

From Fig. 3.16, it is clear that the BCMs $QR\bullet$ and $QL\bullet$ are tangential with the FCMs $\bullet P$ at the x-axis. In this case, none of BCMs inside the angle spanned by $QR\bullet$ and $QL\bullet$ can intersect with any FCMs right of $\bullet P$. Figure 3.16 shows that in the symbolic plane the tangency rules out of forbidden rectangle, which is the area bounded by the lines $\alpha = 1$, $\alpha = \alpha \ (\bullet P)$, $\beta = \beta \ (QR\bullet)$, $\beta = \beta \ (QL\bullet)$.

Proposition 3.13 *(a) When the line $\bullet RL^\infty$ is tangential with $R^\infty \bullet$ the one saddle is heteroclinic to the other.*

(b) The condition for this tangency is

$$\eta \ (R^\infty \bullet) = \frac{1}{1 + b\bar{h}_r} = \frac{1}{(a + k_1)\left(1 + \dfrac{k_1}{b}\right)} = \xi \ (\bullet RL^\infty) \qquad (3.101)$$

where

$$\bar{h}_r = \bar{h} \ (R^\infty \bullet), \ k_1 = k \ (\bullet L^\infty), \ b\bar{h}_r = k_1 = \frac{a}{2} - \sqrt{\left(\frac{a}{2}\right)^2 + b} \qquad (3.102)$$

(c) We have:

$$a = a_c = 2 - \frac{b}{2} \qquad (3.103)$$

and when $a < ac$, the two lines $R^\infty \bullet$ and $R^\infty L\bullet$ are separated from $\bullet RL^\infty$. Some other line say $\bullet K$ smaller than $\bullet RL^\infty$ is tangential with the line $R^\infty \bullet$.

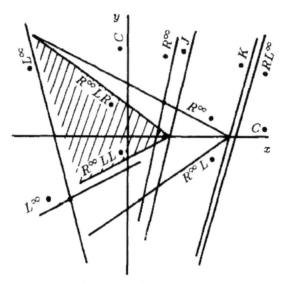

Figure 3.15: Some expanding and contracting lines. Lines $R^\infty\bullet$ and $R^\infty L\bullet$ and the first image of $\bullet K$ form the boundary of the fundamental trapping region. Above the line $R^\infty\bullet$ is the region with a common single sequence $R^\infty\bullet$. The shaded region corresponds to the single line $\beta = \frac{1}{2}$ in the symbolic plane. Reused with permission from Zheng, Chaos, Solitons and Fractals, (1992). Copyright 1992, Elsevier Limited.

Proof 2 *For the proof of (b) we note that Eq. (3.103) can be derived from Eqs. (3.86) and (3.92). For (c) use [Tél (1983(a))] and remark that relation (3.101) leads to* $k_1 = \dfrac{-b}{2}$.

We note that the sequence $\bullet K$ determines the first forbidden rectangle in the symbolic plane in which no strings greater than $\bullet K$ appear in any allowed symbolic sequences. The tangent points are determined one by one when moving from the tangent point of $\bullet K$ and $R^\infty\bullet$ to the right. These points rule out of series of forbidden rectangles, shorter and shorter in the vertical direction. The last tangency (which is tangential with some FCM line denoted by $\bullet J$) is reached at $R^\infty LR\bullet$. Thus, a *fundamental forbidden zone* shown in Fig. 3.16 consists of the whole forbidden area caused by the tangencies located between $\bullet K$ and $\bullet J$.

The two edges $\bullet K$ and $\bullet J$ are called *the pruning front* [Cvitanovic, et al. (1988)]. In the (ξ, η)-plane $\bullet K$ and $\bullet J$ satisfy $\xi = \eta$ and the points inside the fundamental forbidden zone satisfy $\xi > \eta$. The fact that the fundamental forbidden zone and all its images and pre-images form the complete forbidden zones[13] imply that any images and pre-images of the fundamental forbidden zone are also forbidden. On the other hand, points in the symbolic

---

[13] In fact, there is a fundamental forbidden zone which generate all the others and determines the admissibility condition for allowed symbolic sequences.

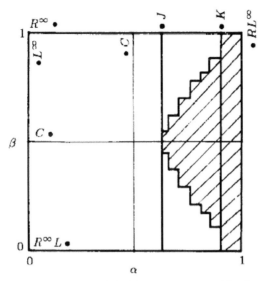

Figure 3.16: Sketch of the fundamental forbidden zone generated by the tangencies on the *x*-axis. Reused with permission from Zheng, Chaos, Solitons and Fractals (1992). Copyright 1992, Elsevier Limited.

plane outside the forbidden zones verifying $\xi < \eta$ and they correspond to allowed sequences. Indeed, let us consider a point $A = (x, y)$ outside of these zones and check the consistency of the signs of x and y with their codes. The values of ξ and η can be calculated from (3.86) and (3.92) and from the definition of ξ and η:

$$\begin{cases} x = \dfrac{h\xi - K\eta}{h - k} \\[2mm] y = \dfrac{\varepsilon - \eta}{h - k} \end{cases} \tag{3.104}$$

We have $|h| > k_M > |k|$, since $|\bar{h}| \le \bar{h}_M$, we have $|h| \ge (\bar{h}_M)^{-1} = \dfrac{b}{k_M}$ and it can be verified that $k^2_M < b$ for $a > 2\sqrt{b}$. From the second part of (3.104) it is clear that the sign of y-coordinate is opposite to the sign of h, which confirm the consistency between y and its code. The consistency for every symbol can be done by shifting the given sequence. To derive admissibility condition for symbolic sequences of the Lozi map (3.2), it is necessary to note that the present dot • must be put somewhere in the doubly asymptotic sequences to divide it into pair of semi-infinite sequences. Thus, an infinite number of points are obtained by shifting the dividing dot leftwards or rightwards. These points are called *shifts of the sequence*. Thus, the admissibility condition was given in [Zheng (1992)] as follows:

Proposition 3.14 *(a) A symbolic sequence is admissible if and only if its shifts never fall inside the fundamental forbidden zone, i.e., A necessary condition for an admissible sequence determined by the first tangency at •K. (b) A sufficient condition given by the last tangency at •J: a symbolic sequence is admissible if its shifts are always smaller than •J. (c) Another sufficient condition is: for a sequence* $...s_{n-2}s_{n-1}s_ns_{n+1}s_{n+2}...$*, if all the shifted back-ward semi-sequences are between* $...s_{n-3}s_{n-2}R•$ *and* $...s_{n-3}s_{n-2}L•$*, and at the same time* $•s_ns_{n+1}s_{n+2}...$ *is the maximum of all the shifted forward sequences, then the given sequence is admissible as long as* $\eta\,(...s_{n-2}s_{n-1}•) \geq \xi\,(•s_ns_{n+1}s_{n+2}...)$.

In the (η, ξ)-plane the admissibility condition can be expressed as follows [Zheng (1992)]:

Proposition 3.15 *A symbolic sequence is admissible if and only if* $\xi \leq \eta$ *is satisfied for all its shifts. The sufficient condition (c) in Proposition 3.14 is then* max (ξ) $\leq$ max (η) .

3.4.3 Examples of admissibility conditions for the Lozi map

In this section, we give an example of admissibility conditions for the Lozi map (3.2) when b approaches zero, in which the Lozi map reduces to the one dimensional tent map (3.6). In this case, expression (3.86) is reduced to the so called λ-*expansion* [Collet & Eckmann (1980)] because $\eta = 1$ is independent of backward sequences by (3.92) and when $b \to 0$ we have $\bar{h} \to -\frac{\varepsilon_{n-1}}{a}$ and $k \to \frac{b\varepsilon_n}{a}$. Hence, the sequence $•K$ is the kneading sequence of the tent map (3.6) and $\xi\,(•K) = 1$. The fundamental forbidden zone is the single rectangle bounded by the lines $\beta = 0$, $\beta = 1$, $\alpha = \alpha\,(•K)$ and $\alpha = 1$. Hence, the symbolic dynamics for the tent map (3.6) is recovered and there is a heteroclinic tangency of two saddles when (3.103) hold. If $b \to 0$, we have $a_c \to 0$, and the fundamental forbidden zone shrinks to the single line $\alpha = 1$. When b increases from zero, we have:

$$\begin{cases} \bar{h}_n = -\dfrac{\varepsilon_{n-1}}{a}, k_n = \dfrac{b\varepsilon_n}{a} \\[2mm] \eta\,(R^{\infty}LL•) = 1 - \dfrac{\dfrac{b}{a}}{1-\dfrac{b}{a}} \\[2mm] \xi\,(•RL^mC) = \dfrac{a(1-a^{-m-2})}{a-1} \end{cases} \qquad (3.105)$$

Eq. (3.105) follows from (3.78), (3.86), (3.93) and (3.92), up to the lowest order of b. If l is some large integer, assume that:

$$2^{-l} < b < 2^{-l+1}, \text{ or } 2 - 2^{-l} < a < 2 - 2^{-l-1}, \tag{3.106}$$

Thus, for $m < l - 3$, we have:

$$\eta \left(R^{\infty}LL\bullet \right) - \xi \left(\bullet RL^{m}C \right) > \frac{2^{-m-2} - 2^{-l+1}}{1 - 2^{-l}} > 0 \tag{3.107}$$

which yields $\bullet J > \bullet RL^{l-3}C$ and this confirms that symbolic sequences consisting of RL^j with $j \leq l - 3$ are always admissible. The existence of period $(RL^m)^{\infty}$ (the discussion for the period $(RL^{m-1}R)^{\infty}$ is similar) requires the intersection of lines $\bullet (RL^m)^{\infty}$ and $(RL^m)^{\infty}\bullet$. If the first condition of (3.106) holds, then we have:

$$\begin{cases} \xi \left(\bullet (RL^m)^{\infty} \right) = \frac{a\left(1 - a^{-m-1}\right)}{(a-1)(1+a^{-m-1})} > \frac{1 - 2^{-m-1}}{1 - 2^{-l-1}} \\ \\ \eta \left(R^{\infty}LL\bullet \right) = \frac{\left[\frac{1 - \left(\frac{b}{a}\right)^m}{1 + \frac{b}{a}} - \left(\frac{-b}{a}\right)^m \right]}{\left[1 + \left(\frac{-b}{a}\right)^{m+1} \right]} < 1 - 2^{-l-2} \end{cases} \tag{3.108}$$

For $m > l$, there is no intersection because $\xi > \eta$. However, the period $(RL^{m-2}RL)^{\infty}$ can still exist in the case where the value of ξ is almost unchanged, while η increases by an amount $\frac{b}{a}$.

Proposition 3.16 *If*

$$1 + b < a < 2 - \frac{b}{2} \tag{3.109}$$

then any periodic orbits become unstable, so the existence of the trapping region implies the existence of a strange attractor.

In this case, i.e., when $a \gg b$, the stable manifolds of the fixed point possesses homoclinic point [Tél (1983(a))]. Its intersection[14] with the unstable manifolds of period-2 points is not necessarily non empty and the strange attractor becomes a two-piece one. The content of Proposition. 3.16 is discussed in detail in Section 3.8. In the symbolic plane, the existence of homoclinic points[15] associated with fixed points is equivalent to the existence of such a string U consisting of R and L that $R^{\infty}UR^{\infty}$ is admissible

---

[14] In the symbolic plane, these tangencies between two different kinds of foliations rule out forbidden regions.

[15] Homoclinic point is defined as a point where a stable and an unstable separatrix, i.e., an invariant manifold from the same fixed point or same family intersect.

in the sense of Proposition. 3.15. We have that the smallest •K for $R^\infty U R^\infty$ is •RLR^∞. If the tangency between Q• and •P is marked by Q o P, then R^∞ o RLR^∞ appears first. The heteroclinic intersection requires the existence of some string V which makes $(RL)^\infty$ V R^∞ admissible and the •K for this sequence can never be smaller than RLR^∞. A tangency of the type $(RL)^\infty$ o $RLRRL...R^\infty$ is a criterion for this intersection, where $RL...R^\infty$ is the greatest forward sequence smaller then the given •K, i.e., •$RLRRL...R^\infty$ is the smallest among the allowed forward sequences starting with $RLRR$. Thus, leading string $RLRR$ makes the forward sequences starting with $(RL)^\infty$ as small as possible, and makes it the greatest among sequences of the type $(RL)^m WR^\infty$, $m = 0, 1, 2, ...$

3.5 Existence of strange attractor for $b > 0$

In this section, we present and discuss the most important results available in the literature about the rigorous proof of chaos in the Lozi mappings. We begin by the first one given in [Misiurewicz (1980)], where the author gave a mathematical proof that the Lozi map (3.1) has a strange attractor $\Lambda_{a,b}$ for the range of parameters defined by:

$$
\mathcal{M}: \begin{cases}
(1)\ 0 < b < 1, a > 0, \\
(2)\ a > b + 1 \\
(3)\ 2a + b < 4 \\
(5)\ b < \dfrac{a^2 - 1}{2a + 1} \\
(6)\ a\sqrt{2} > b + 2.
\end{cases} \qquad (3.110)
$$

and proved that the basin $B\ (\Lambda_{a,b})$ contains a neighborhood of $\Lambda_{a,b}$.

The method of the proof given in [Misiurewicz (1980)] is based essentially on finding of a trapping region[16] of the Lozi map (3.1) and then to prove that the Lozi map (3.1) has a hyperbolic structure (See Definition 12(d)). In fact, its hyperbolic structure must be understood only as the existence of a hyperbolic splitting at the points where it is differentiable. In this case, the stable and unstable manifolds are broken lines and they are not manifolds, these sets are called *(un-) stable manifolds*, rather than (un-) stable sets. This structure cannot be extended continuously on the whole plane, but the (un-) stable manifolds exist at almost all points of the trapping region. With some geometrical properties, Misiurewicz proves also that the intersection of the image of the trapping region is a strange attractor. However, by the *influence of the singularities* in the y-axis, the dynamics of

---

[16] A non-empty set which is mapped with its closure into its interior.

the Lozi maps (3.1) are quite delicate [Ishii (1997(a-b)), Ishii & Sands (1998), Kiriki & Soma (2007)]. See Section 3.11.

The proof of Misiurewicz needs some definitions and it is divided into several propositions and lemmas. Indeed, let g be a homeomorphism of a metric space onto itself that defines a discrete dynamical system; then we have the following definition:

Definition 44 *(a) A local stable manifold at a point A is defined as*

$$\left\{ \begin{array}{l} W^s_{loc}(A) = \{B : \lim_{n \to \infty} \alpha_n = 0, \exists \varepsilon > 0 : \alpha_n < \varepsilon, n \ge 0\} \\ \alpha_n = dist\,(g^n(A), g^n(B)) \end{array} \right. \tag{3.111}$$

(b) A local unstable manifold at a point A is defined as

$$\left\{ \begin{array}{l} W^u_{loc}(A) = \{B : \lim_{n \to \infty} \beta_n = 0, \exists \varepsilon > 0 : \beta_n < \varepsilon, n \ge 0\} \\ \beta_n = dist\,(g^{-n}(A), g^{-n}(B)) \end{array} \right. \tag{3.112}$$

(c) A global stable manifold at a point A is defined as

$$W^s(A) = \cup^\infty_{n=0} g^{-n}(W^s_{loc}(g^n(A))). \tag{3.113}$$

(d) A global unstable manifold at a point A is defined as

$$W^u(A) = \cup^\infty_{n=0} g^n(W^u_{loc}(g^{-n}(A))). \tag{3.114}$$

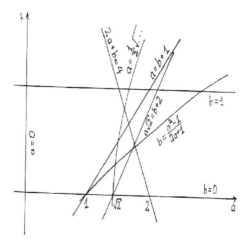

Figure 3.17: Schematic representation of the sets $\mathcal{M}$ and $\mathcal{M}_1$ defined by (3.110) and (3.115) respectively. Reused with permission from Misiurewicz, Ann.N.Y. Acad. Sci (1980). Copyright 1980, Wiley-Blackwell.

Now, let us consider the following set in the ab-plane defined by:

$$\mathcal{M}_1 : a > \frac{1}{2}\sqrt{3b^2 + 4 + \sqrt{(3b^2 + 4)^2 - 32b^3}}$$ (3.115)

It is easy to check that (3.110)-(1) and (3.110)-(5) imply (3.110)-(2), (3.110)-(1) and (3.110)-(6) imply (3.110)-(2), (3.110)-(1) and (3.110)-(6) imply (3.115). Also, the set satisfying (3.110) and (3.115) is open and non-empty as shown in Fig. 3.17. The proof of Misiurewicz is concerned essentially with the range of parameters defined by the range (3.110)-(1).

Proposition 3.17 *Assume (3.110)-(1) and*

$$a + b > 1$$ (3.116)

Then the following properties holds:
(a) *The Lozi map (3.1) is a homeomorphism and it maps linearly the left half-plane onto the lower one and the right one onto the upper one.*
(b) *The map (3.1) has two fixed points given by:*

$$X = \left(\frac{1}{1 + a - b}, \frac{b}{1 + a - b} \right), \text{ if } b < a + 1$$ (3.117)

in the first quadrant and

$$Y = \left(\frac{1}{a + b - 1}, \frac{b}{a + b - 1} \right), \text{ if } b > 1 - a$$ (3.118)

in the third quadrant and they are hyperbolic.

(c) *The eigenvalues of $D\mathcal{L}$*

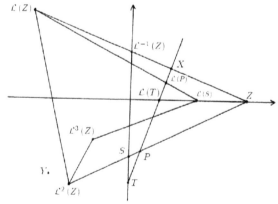

Figure 3.18: Schematic representation of the points and sets used in the proof of Misiurewicz for the Lozi mapping (3.1). Reused with permission from Misiurewicz, Ann.N.Y. Acad. Sci, (1980). Copyright 1980, Wiley-Blackwell.

are $\dfrac{-1}{2}\left(a+\sqrt{a^2+4b}\right)$ and $\dfrac{1}{2}\left(-a+\sqrt{a^2+4b}\right)$ at X, $\dfrac{1}{2}\left(a+\sqrt{a^2+4b}\right)$

and $\dfrac{-1}{2}\left(-a+\sqrt{a^2+4b}\right)$ at Y .

(d) The associated eigenvector to an eigenvalue λ is $\begin{pmatrix}\lambda\\b\end{pmatrix}$.

(e) One half of the stable manifold of X goes to infinity in the first quadrant.

(f) The other half intersects the vertical axis for the first time at the point $T = \left(0,\dfrac{2b-a-\sqrt{a^2+4b}}{2(1+a-b)}\right)$.

(g) The half of the unstable manifold of X, which start to the right intersects the horizontal axis for the first time at $Z = \left(\dfrac{2+a+\sqrt{a^2+4b}}{2(1+a-b)},0\right)$.

(h) A half of the unstable manifold of Y goes to infinity in the first quadrant. The triangle $F = Z\mathcal{L}(Z)\mathcal{L}^2(Z)$ is shown in Fig. 3.18 along with its image under $\mathcal{L}$, and some points which will defined later.

3.5.1 The trapping region of the Lozi mapping

In this section, we present the method on how the trapping region defined by Misiurewicz in [Misiurewicz (1980)] for the Lozi mapping (3.1) was constructed:

Proposition 3.18 *Let a, b satisfy (3.110)-(1), (3.110)-(2) and (3.110)-(3). Then $\mathcal{L}(F) \subset F$.*

Proof 3 For the proof, it is easy to check that: (1) The point Z lies to the right of the origin and $\mathcal{L}(Z)$ lies in the second quadrant, which imply that $\mathcal{L}^2(Z)$ lies in the lower half-plane. (2) The first coordinate of $\mathcal{L}^2(Z)$ is smaller than the first coordinate of Z. Now, the proof is based essentially on the fact that the map (3.1) is a composition of two special mappings. Let S be the point of intersection of the union of segments $\mathcal{L}(Z)\mathcal{L}^2(Z) \cup \mathcal{L}^2(Z)Z$ with the vertical axis. Since S lies below $\mathcal{L}^{-1}(Z)$, then the point $L(S)$ lies to the left of Z. We remark that the map (3.1) is a composition of the affine mapping $(x, y) \rightarrow (1 + y, bx)$ with the mapping $(x, y) \rightarrow (x - \dfrac{a}{b}|y|, y)$ which moves to the left all points except those on the horizontal axis. Hence, if S lies on the line $\mathcal{L}(Z)\mathcal{L}^2(Z)$ or $\mathcal{L}^2(Z)Z$ then $\mathcal{L}(S)$ lies to the right of the line $\mathcal{L}^2(Z)\mathcal{L}^3(Z)$ or $\mathcal{L}^3(Z)Z$ respectively.

If $\mathcal{L}^3(Z) \in F$, then $\mathcal{L}(S) \in F$. We remark that the set $\mathcal{L}(F)$ is a pentagon with the vertices Z, $\mathcal{L}(Z)$, $\mathcal{L}^2(Z)$, $\mathcal{L}^3(Z)$ and $\mathcal{L}(S)$. If $\mathcal{L}^3(Z) \in F$ then they all these vertices belong to F which imply that $\mathcal{L}(F) \in F$. It is easy to see that the set $\mathcal{L}(F)$ lies above the line $\mathcal{L}^2(Z)Z$ and below the line $\mathcal{L}(Z)Z$ since $\mathcal{L}(Z)$ is linear on the left and right half-plane, which confirm that if the point $\mathcal{L}^3(Z)$ does not belong to Z then it lies to the left of the line $\mathcal{L}(Z)\mathcal{L}^2(Z)$.

Now, it is easy to show that the point $\mathcal{L}^3 (Z)$ lies to the right of the line $\mathcal{L}(Z)\,\mathcal{L}^2 (Z)$. Thus, if $\mathcal{L}^2 (Z)$ lies in the left half-plane, then the problem is reduced to the inequality:

$$(4 - 2a - b)\,(1 + a - b)^2 > 0 \tag{3.119}$$

and if $\mathcal{L}^2 (Z)$ lies in the right half-plane then the problem is reduced to the inequalities:

$$\left\{ \begin{array}{l} (4a + 2)\,b^2 + (-2a^3 + a^2 - 2)\,b + (-a^5 + 2a^3 - 3a) < 0 \\ \qquad\qquad \varphi_1 + \varphi_2 > 0 \\ \varphi_1 = -4b^5 + (-a^3 - 3a^2 + 3a - 15)\,b^3 \\ \qquad +(-5a^4 + 2a^3 + 2a^2 - 6a + 7)\,b^2 \\ \varphi_2 = (a^5 + 5a^4 - 4a^3 - 4a^2 + 3a - 1)\,b \\ \qquad +(2a^6 - 4a^4 + 2a^2) + (3a^2 + 13)\,b^4 \end{array} \right. \tag{3.120}$$

Now a trapping region can be defined as follow:

$$\left\{ \begin{array}{l} \overline{\mathcal{L}(Z)\mathcal{L}^2 (Z)} \cup \overline{\mathcal{L}^2(Z)\,Z} \setminus \{\mathcal{L}(Z),\, Z\} \subset Int\,(F) \\ \overline{\mathcal{L}((\mathcal{L}(Z))\mathcal{L}^{-1}(Z))} = \mathcal{L}^2(Z)\,Z \end{array} \right. \tag{3.121}$$

Then the set $\mathcal{L}^{-4} (F)$ is a neighborhood of $\overline{\mathcal{L}^2(Z)\,Z} \cup \overline{\mathcal{L}^2(Z)\,\mathcal{L}\,(Z)} \cup (\mathcal{L}(Z)\mathcal{L}^{-1}(Z))$. The fact that $\overline{\mathcal{L}^{-1}(Z)\,Z}$ is a local unstable manifold of the hyperbolic fixed point X, imply that there exists a rectangle R contained in the first quadrant and its sides are parallel to the eigenvectors of $D\mathcal{L}$ such that:

$$\left\{ \begin{array}{l} \overline{\mathcal{L}^{-1}(Z)\,Z} \setminus \mathcal{L}^{-4}\,(Int\,(F)) \subset Int\,(R) \\ \mathcal{L}(R) \subset Int\,(R) \subset \mathcal{L}^{-4}\,(Int\,(F)). \end{array} \right. \tag{3.122}$$

Thus, let us define the following set:

$$G = R \subset \mathcal{L}^{-4}(F) \tag{3.123}$$

The set G is a compact neighborhood of F and it is a trapping region as will be shown in the following next result proved in [Misiurewicz (1980)]:

Theorem 29 *Let a, b, satisfy (3.110)-(1), (3.110)-(2) and (3.110)-(3). Then $L(G) \subset G$.*

Proof 4 *For the proof, we have $L(R) \subset Int\,(G)$ and $\overline{Z\,\mathcal{L}\,(Z)} \cup \overline{Z\mathcal{L}\,(Z)} \cup \mathcal{L}^4\,(R)$, which imply that $L(F \setminus \mathcal{L}^4\,(R) \subset Int\,(F)$. Thus, we have*

$$L(F \setminus R) = \mathcal{L}^{-4}\,(\mathcal{L}(F \setminus \mathcal{L}^4\,(R))) \subset \mathcal{L}^{-4}\,(Int\,(F)) \subset Int\,(G). \tag{3.124}$$

and it follows that

$$L(G) = L(R) \cup L(G \setminus R) \subset Int\,(G). \tag{3.125}$$

Hence the proof is complete.

The size of the intersection of the image of G is discussed in what follows. First, we need some preliminary results proved in [Misiurewicz (1980)]:

Lemme 3.8 *The sets* $\Lambda_{a,b} = \cap_{n=0}^{\infty} \mathcal{L}^n (G)$ *and* $\widetilde{F} = \cap_{n=0}^{\infty} \mathcal{L}^n (F)$ *are equal.*

Proof 5 *For the proof, we have that the inclusion* $\mathcal{L} (\mathcal{L}^{-4} (F)) \subset \mathcal{L}^{-4} (F)$ *, imply that*

$$\begin{cases} \Lambda_{a,b} = \cap_{n=0}^{\infty} \mathcal{L}^n \left(R \cup \mathcal{L}^{-1} (F) \right) \subset \cup_{k=0}^{\infty} \mathcal{L}^k (W_1 \cup W_2) \\ \\ W_1 = \cap_{n=0}^{\infty} \mathcal{L}^n (R) \setminus \mathcal{L}^{-4} (F) \qquad\qquad (3.126) \\ \\ W_2 = \cap_{n=0}^{\infty} \mathcal{L}^n \left(\mathcal{L}^{-4} (F) \right) \end{cases}$$

$$= \cup_{k=0}^{\infty} \mathcal{L}^k \left(\overline{\mathcal{L}^{-1} (Z) Z} \setminus \mathcal{L}^{-4} (F) \right) \cup \widetilde{F} \subset \widetilde{F}. \qquad (3.127)$$

and the reverse inclusion is a result of the inclusion $F \subset G$.

If (3.115) and (3.110)-(5) are assumed, then (3.115) implies that the point $\mathcal{L}^2 (Z)$ lies to the left of the line XT, and inequality (3.110)-(5) implies that the point of intersection of the line $Z\mathcal{L}^2 (Z)$ with the vertical axis lies above T. Thus, the segments $\overline{XT}$ and $\overline{Z\mathcal{L}^2(Z)}$ intersect with each other in a point P. We note that the point S lies above T and the point $\mathcal{L}(S)$ lies to the right of $\mathcal{L}(T)$ as shown in Fig. 3.18. Let H_0 be the triangle XSP and denote $H = \cup_{n=0}^{\infty} \mathcal{L}^n (H_0)$ and let the symbol Pr means boundary. Thus, we have the following result proved in [Misiurewicz (1980)]:

Proposition 3.19 *Let a, b, satisfy (3.110)-(1), (3.110)-(3), (3.115) and (3.110)-(5). Then (a) $Pr (H) \subset \overline{XT} \cup W^u (X)$. (b) $\mathcal{L}(H) \subset H$. (c) $H \subset F$. (d) The set $\widetilde{H} = \cap_{n=0}^{\infty} \mathcal{L}^n (H)$ is equal to $\widetilde{F}$.*

Proof 6 *For the proof, it is easy to verify that (3.110)-(2) follows from (3.110)- (1) and (3.110)-(5) which imply that the statements (a), (b) and (c) follow immediately from the Definition. To prove the statement (d), let us define by induction the following sets:*

$$H_n = \overline{(\mathcal{L}(H_{n-1}) \setminus H_0)} \qquad\qquad (3.128)$$

Clearly, the set $H1$ = the triangle $X\mathcal{L}(Z)\mathcal{L}(P)$ and by induction for all $n \geq 2$, it follows that the set H_n is either empty or is a triangle with two vertices belonging to $\overline{XP}$, and the third one equal to $\mathcal{L}^n (Z)$ and contained in the set $\mathcal{L}(F) \setminus (H_0 \cup H_1)$. At this end, assume that all sets H_n are non-empty. Thus, we have that all points $\mathcal{L}^n (Z)$, $n \geq 1$, lie in the left half-plane, since the set $\mathcal{L}(F) \setminus (H_0 \cup H_1)$ is contained in the lower half-plane. The facts that the mapping L is linear hyperbolic in the left half-plane and therefore, the points $\mathcal{L}^n (Z)$ converge to the fixed point Y because they stay in a bounded region. Thus, we have $Y \in F \subset Int (G)$, and $W^u (Y) \subset G$ which *makes a contradiction because a half of the manifold $W^u(Y)$ goes to infinity. Hence, there exists a natural number p such that $H_n \neq \emptyset$ for $n \leq p$ and $H_n = \emptyset$ for $n > p$. We have that the polygon H defined above is reduced to the set H defined by:*

$$H = \cup^p_{n=0} \mathcal{L}^n (H_0) \tag{3.129}$$

For $1 \leq n \leq p - 2$, we have

$$\mathcal{L}^n ((\overline{\mathcal{L}(Z)\mathcal{L}^2(Z)})) = \overline{\mathcal{L}^{n+1}(Z)\mathcal{L}^{n+2}(Z)} \tag{3.130}$$

and the set $\mathcal{L}^{p-1}((\overline{\mathcal{L}(Z)\mathcal{L}^2(Z)}))$ is a union of two sets one of them is contained in H_0 and the other one is an interval joining $\mathcal{L}^p(Z)$ with some point of $\overline{XP}$ in the lower half-plane. Thus, the set $\mathcal{L}^p((\overline{\mathcal{L}(Z)\mathcal{L}^2(Z)}))$ is a union of two sets one of them is contained in H_1 and the other one in H_0. Hence, we have $\mathcal{L}^p((\overline{\mathcal{L}(Z)\mathcal{L}^2(Z)})) \subset H$, and

$$\Pr(\mathcal{L}^p(F)) = \mathcal{L}^p(\Pr F) \subset W^u(X) \cup (\overline{\mathcal{L}(Z)\mathcal{L}^2(Z)}) \subset H \tag{3.131}$$

and since H is a simple connected set, we have $\mathcal{L}^p(F) \subset H$. Therefore

$$\widetilde{F} \subset \widetilde{H} \tag{3.132}$$

The proof is hence completed by showing the inverse inclusion that follows from item (c).

The second result is also proved in [Misiurewicz (1980)]:

Theorem 30 *Let a, b, satisfy (3.110)-(1), (3.110)-(3), (3.115) and (3.110)-(5). Then $G = \overline{W^u(X)}$.*

Proof 7 *For the proof, let $A \notin \overline{W^u(X)}$. Then, it is possible to find $\varepsilon > 0$ such that the ball with center A and radius 2ε is disjoint from the manifold $W^u(X)$. Since we have $\overline{XP} \subset W^u(X)$ and by using Proposition. 3.19(a), it follows that for all a sufficiently large, the ball with center A and radius ε is disjoint from the set $\Pr(\mathcal{L}^n(H))$. On one hand, the Lebesgue measure of $\mathcal{L}^n(H)$ converge to 0 as $n \to \infty$ since the absolute value of the Jacobian of $\mathcal{L}$ is equal to $b < 1$. Thus, we have $A \notin \mathcal{L}^n(H)$ for a sufficiently large and therefore $A \notin H$ which imply that $\widetilde{H} \subset \overline{W^u(X)}$. On the other hand, both Lemma. 3.8 and Proposition. 3.19(d), shows that $\widetilde{H} = \Lambda_{a,b}$ and therefore $\Lambda_{a,b} \subset \overline{W^u(X)}$. Finally, we have $\Lambda_{a,b} = \overline{W^u(X)}$, since $X \in Int(G)$, $\mathcal{L}(G) \subset G$ and $\Lambda_{a,b}$ is closed, which imply the inclusion $\overline{W^u(X)} \subset \Lambda_{a,b}$.*

3.5.2 Hyperbolicity of the Lozi mapping

In this section, we give a rigorous proof for the hyperbolicity property for the Lozi mapping (3.1) in the sense of Definition 12(d) with some slight modifications. For this purpose, assume only (3.110)-(1) and (3.110)-(2) holds. Hyperbolicity in this sense means that for every point A there exists a splitting of the tangent space T_A into stable and unstable subspaces. This splitting is continuous and invariant, and for some constants $\alpha_1, \alpha_2, 0 < \lambda_1 < 1 < \lambda_2$ we have the following inequality:

$$|D\mathcal{L}^n v| \leq \alpha_1 \lambda^n_1 \, |v|, \qquad (3.133)$$

for each v from the stable subspace and

$$|D\mathcal{L}^n v| \leq \alpha_2 \lambda^n_2 \, |v|, \qquad (3.134)$$

for each v from the unstable subspace. For the case of Lozi mapping (3.1) the operator $D\mathcal{L}$ is not defined in all points, thus, a *hyperbolic splitting* exists only at the points, for which the derivative exists at the whole trajectory. Also, this splitting cannot be extended to a continuous one due to the same reasons. First, we remark that the points defined by:

$$\left(\frac{1-a-b}{a^2 + (1-b)^2}, \frac{b(1+a-b)}{a^2 + (1-b)^2} \right), \left(\frac{1+a-b}{a^2 + (1-b)^2}, \frac{b(1-a-b)}{a^2 + (1-b)^2} \right) \qquad (3.135)$$

form a periodic orbit of period 2 and the product of the derivatives of $\mathcal{L}$ at these points are equal to:

$$\begin{pmatrix} a & 1 \\ b & 0 \end{pmatrix} \begin{pmatrix} -a & 1 \\ b & 0 \end{pmatrix} = - \begin{pmatrix} a & -1 \\ b & 0 \end{pmatrix}^2 \qquad (3.136)$$

The eigenvalues of the matrix $\begin{pmatrix} a & -1 \\ b & 0 \end{pmatrix}$ are equal to $c = \dfrac{a - \sqrt{a^2 - 4b}}{2}$ and $\dfrac{b}{c}$. Thus, the constants λ_1 and λ_2 cannot be better [17] than c and $\dfrac{b}{c}$ respectively and it is possible to take these constants. The following result was proved in [Misiurewicz (1980)]:

Theorem 31 *Let a, b, satisfy (3.110)-(1) and (3.110)-(2). Then (a) If $A \in \mathbb{R}^2$ is such that $\mathcal{L}$ is differentiable at all points $\mathcal{L}^n (A)$, $n = 0, 1, 2, \dots$ then there exists a one dimensional subspace $E^s_A \subset T_A$ such that*

$$|D\mathcal{L}^n v| \leq C^n \, |v|, \qquad (3.137)$$

for each $v \in E^s_A$ and the constant c comes from Definition 12(d). Moreover $D\mathcal{L}(E^s_A) = E^s_{\mathcal{L}(A)}$. (b) If $A \in \mathbb{R}^2$ is such that $\mathcal{L}^{-1}$ is differentiable at all points $\mathcal{L}^{-n} (A)$, $n = 0, 1, 2, \dots$ then there exists a one dimensional subspace $E^u_A \subset T_A$ such that

$$|D\mathcal{L}^n v| \geq \left(\frac{b}{c} \right)^n |v|, \qquad (3.138)$$

---

[17] i.e., the smallest λ_1 and the largest λ_2.

for each $v \in D\mathcal{L}^{-n}(E^u{}_A)$. Moreover

$$D\mathcal{L}^{-1}(E^u{}_A) = E^u{}_{\mathcal{L}^{-1}(A)}. \tag{3.139}$$

(c) If $A \in \mathbb{R}^2$ is such that $\mathcal{L}$ is differentiable at all points $\mathcal{L}^n(A)$, $n = 0, \pm 1, \pm 2, \ldots$ then

$$T_A = E^s{}_A \oplus E^u{}_A. \tag{3.140}$$

Proof 8 *For the proof, we note that from (3.110)-(1) and (3.110)-(2) it follows that:*

$$0 < c < b < 1 < \frac{b}{c} \tag{3.141}$$

The derivative of $\mathcal{L}$ is given by:

$$D\mathcal{L} = \begin{pmatrix} \pm a & 1 \\ b & 0 \end{pmatrix} \tag{3.142}$$

where the sign depends on a point at which it is taken. Let the vector $\begin{pmatrix} t \\ r \end{pmatrix}$ such that

$$\begin{pmatrix} \pm a & 1 \\ b & 0 \end{pmatrix} \begin{pmatrix} t \\ r \end{pmatrix} = \begin{pmatrix} \bar{t} \\ \bar{r} \end{pmatrix}. \tag{3.143}$$

Hence, we have (1) If $b|\bar{t}| \le c|\bar{r}|$ then $b\,|t| \le c\,|r|$, $|\bar{t}| \le c\,|t|$, $|\bar{r}| \le c\,|r|$. (2) if $|r| \le c\,|t|$ then $|\bar{r}| \le c|\bar{t}|$, $|\bar{r}| \ge \frac{b}{c}\,|r|$, $|\bar{t}| \ge \frac{b}{c}|r|$. Thus, the tangent spaces at different points are identified, then the operators $D\mathcal{L}^{-1}$ (resp. $D\mathcal{L}$) maps the stable sector $\left\{ \begin{pmatrix} \bar{t} \\ r \end{pmatrix} : b|\bar{t}| \le c|\bar{r}| \right\}$ (resp. the unstable sector $\left\{ \begin{pmatrix} t \\ r \end{pmatrix} : |r| \le c|t| \right\}$ into itself and expands all its vectors by a factor $\frac{1}{c}$ (resp. $\frac{b}{c}$) at least. Using inequality (3.141), we confirms that the numbers $\frac{1}{c}$ and $\frac{b}{c}$ are larger than 1. Now, take points A and B such that $\mathcal{L}$ is differentiable at the points $\mathcal{L}^n(A)$ and $\mathcal{L}^{-1}$ at the points $\mathcal{L}^{-n}(B)$, $n = 0, 1, 2, \ldots$ and define the following sets:

$$\begin{cases} E^s{}_A = \cap_{n=0}^{\infty} D\mathcal{L}^{-n} \left\{ \begin{pmatrix} t \\ r \end{pmatrix} \in T_{\mathcal{L}^n(A)} : b|t| \le c|r| \right\} \\ \\ E^u{}_A = \cap_{n=0}^{\infty} D\mathcal{L}^n \left\{ \begin{pmatrix} t \\ r \end{pmatrix} \in T_{\mathcal{L}^{-n}_{a,b}(A)} : |r| \le c|t| \right\} \end{cases} \tag{3.144}$$

It is easy to see that the inequalities (3.137) and (3.138) are results of the properties of stable and unstable sectors defined above and the sets E^s_A and E^u_A are intersections of decreasing sequences of non-empty closed sectors, which confirm that they contain one-dimensional spaces. If C^s_A and C^u_B are vector spaces, then we have:

$$\begin{cases} E^s_A \subset C^s_A = \{v \in T_A : \lim_{n\to\infty} \|D\mathcal{L}^n(v)\| = 0\} \\ E^u_B \subset C^u_B = \{v \in T_B : \lim_{n\to\infty} \|D\mathcal{L}^{-n}(v)\| = 0\} \end{cases} \qquad (3.145)$$

If the sets $T_A \setminus C^s_A$ and $T_B \setminus C^u_B$ are non-empty, then the sets E^s_A and E^u_B are one-dimensional spaces. These sets contain the unstable and stable sectors respectively. The fact that the operators $D\mathcal{L}(A)$ and $D\mathcal{L}^{-1}(B)$ are non-degenerated and the spaces E^s_A, $E^s_{\mathcal{L}(A)}$, E^u_B, $E^u_{\mathcal{L}^{-1}(B)}$ are one-dimensional imply that:

$$\begin{cases} D\mathcal{L}(A)(E^s_A) = E^s_{L(A)} \\ D\mathcal{L}^{-1}(B)(E^u_B) = E^u_{\mathcal{L}^{-1}}(B). \end{cases} \qquad (3.146)$$

and hence the proof of (a) and (b) is completed. Since the intersection of the stable and unstable sectors is $\left\{ \begin{pmatrix} 0 \\ 0 \end{pmatrix} \right\}$, the statement (c) follows immediately.

The linearity property of the map $\mathcal{L}$ in the left and right half-planes and the map $\mathcal{L}^{-1}$ in the lower and upper ones imply the following proposition follows immediately from Theorem 29 and proved in [Misiurewicz (1980)]:

Proposition 3.20 Let a, b, satisfy (3.110)-(1) and (3.110)-(2). Then for a point $A \in \mathbb{R}^2$: (a) If dist ($\mathcal{L}^n(A)$, vertical axis) $\geq ac^n$ for some $\alpha > 0$ and all $n \geq 0$ then the segment $\{A + tv^s_A, |t| < \alpha\}$, (where v^s_A is a unit vector in the stable direction at A) is a local stable manifold at A. (b) If dist ($\mathcal{L}^{-n}(A)$, horizontal axis) $\geq \alpha \left(\dfrac{c}{b}\right)^n$ for some $\alpha > 0$ and all $n \geq 0$ then the segment $\{A + tv^u_A, |t| < \alpha\}$, (where v^u_A is a unit vector in the unstable direction at A) is a local unstable manifold at A.

Again, assume inequalities (3.110)-(1), (3.110)-(2) and (3.110)-(3) and draw a line through the point Y in its stable direction and denote by M the half-plane to the right of it, i.e.,

$$M = \left\{ (x,y) : 2bx + \left(\sqrt{a^2 + 4b} - a\right)y > b\left(\frac{2+\sqrt{a^2+4b}-a}{1-a-b}\right) \right\} \qquad (3.147)$$

Thus, the following lemma was proved in [Misiurewicz (1980)]:

Lemme 3.9 (a) $\mathcal{L}^{-1}(M) \subset M$. (b) There exists a constant $\gamma > 0$ such that for every $\gamma \leq 1$ the intersection of the horizontal line $\mathbb{R} \times \{\gamma\}$ with the set $\mathcal{L}^{-1}(M)$ has length at most γ. (c) $F \subset M$.

Proof 9 *For the proof, the inclusion (a) is a direct result of the formula (3.3). The property (b) follows from the relation (if $y \leq 1$)*

$$
\left\{
\begin{array}{l}
x_1 < x < x_2 \\[2mm]
x_1 = \dfrac{1}{2b}\left(b\left(\dfrac{2+\sqrt{a^2+4b}-a}{1-a-b} \right) - \sqrt{a^2+4b}+a \right) \\[4mm]
x_2 = \dfrac{1}{\sqrt{a^2+4b}-3a}\left(\dfrac{2+\sqrt{a^2+4b}-a}{1-a-b} - 4 \right)
\end{array}
\right\}
\tag{3.148}
$$

for every $(x, y) \in \mathcal{L}^{-1}(M)$. To prove the inclusion (c), we must show that $\mathcal{L}^{-1}(Z) \in M$, which is equivalent to the inequality:

$$
(4 - 2a - b)(a + b - 1)(1 + a - b) > 0,
\tag{3.149}
$$

Inequality (3.149) holds in view of the above assumptions. Now, let $Z \in M$, and $\mathcal{L}^{-2}(Z) \in M$ because for the points A from the lower half-plane the conditions $A \in M$ and $\mathcal{L}^{-1}(A) \in M$ are equivalent. Thus, the vertices of F belong to M, and finally we have $F \subset M$.

Next, let us denote by K_α^s (resp. K_α^u) the set of these points $A \in G$ at which the segment $\{A + tv_{A'}^s, \ |t| < \alpha\}$ (resp. $\{A + tv_{A'}^u, \ |t| < \alpha\}$) is not a local stable (resp. unstable) manifold at A. Let m denote the Lebesgue measure on $\mathbb{R}^2$. The following lemma was proved in [Misiurewicz (1980)]:

Lemme 3.10 *There exists a constant $\delta > 0$ such that $m(K_\alpha^s) \leq \delta\alpha$ and $m(K_\alpha^u) \leq \delta\alpha$ for all $\alpha > 0$.*

Proof 10 *For the proof, by using Proposition. 3.20, we have:*

$$
\left\{
\begin{array}{l}
K_\alpha^s \subset \bigcup_{n=0}^{\infty} \mathcal{L}^{-n}(Q_{\alpha,n}^s) \cap G \\[2mm]
K_\alpha^u \subset \bigcup_{n=0}^{\infty} \mathcal{L}^{n}(Q_{\alpha,n}^s) \cap G
\end{array}
\right.
\tag{3.150}
$$

where

$$
\left\{
\begin{array}{l}
Q_{\alpha,n}^s = \{(x, y) \in \mathbb{R}^2 : \ |x| < \alpha c^n\} \\[2mm]
Q_{\alpha,n}^u = \{(x, y) \in \mathbb{R}^2 : \ |y| < \alpha \left(\dfrac{c}{b}\right)^n\}
\end{array}
\right.
\tag{3.151}
$$

Since the set G is bounded, there exists a constant $\beta > 0$ such that:

$$
m(Q_{\alpha,n}^s \cap G) \leq \alpha\beta c^n
\tag{3.152}
$$

for all α and n. From (3.152), it is easy to show that the absolute value of the Jacobian of $\mathcal{L}^{-1}$ is equal to $\dfrac{1}{b}$ and we have $\mathcal{L}(G) \subset G$. Therefore, we have:

$$m(L^{-n}(Q^s_{a,n}) \cap G) \le m(\mathcal{L}^{-n}(Q^s_{a,n} \cap G)) \le \alpha\beta\left(\frac{c}{b}\right)^n \qquad (3.153)$$

Since $0 < c < b$, we have:

$$m(Q^s_{a,n}) \le \alpha\beta\frac{1}{1-\dfrac{c}{b}} \qquad (3.154)$$

and it is possible to show that the whole first quadrant is contained in M. Thus, using Lemmas. 3.9(a) and (c), we have:

$$\mathcal{L}(G) \subset G = R \cup \mathcal{L}^{-4}(F) \subset M \cup \mathcal{L}^{-4}(M) = M \qquad (3.155)$$

Hence, $G \subset \mathcal{L}^{-1}(M)$, and again Lemma. 3.9(a) imply that:

$$\mathcal{L}^{-n}(G) \subset \mathcal{L}^{-1}(M), n \ge 0. \qquad (3.156)$$

The use of the fact that the absolute value of the Jacobian of $\mathcal{L}$ is equal to b and Lemma. 3.9(b) give the following inequality:

$$m(\mathcal{L}^n(Q^u_{a,n}) \cap G) = b^n m(Q^u_{a,n} \cap \mathcal{L}^{-n}(G)) \le \alpha\gamma c^n \qquad (3.157)$$

for all $n \ge 0$ and $0 < \alpha \le 1$. Hence, we have:

$$m(Q^u_{a,n}) \le \alpha\gamma\frac{1}{1-c}, \qquad (3.158)$$

for all $0 < \alpha \le 1$. Now, let us consider the constant δ defined by:

$$\delta = \max\left(\frac{\beta}{1-\dfrac{c}{b}}, \frac{\gamma}{1-c}, m(G)\right) \qquad (3.159)$$

and we get:

$$\begin{cases} m(Q^s_{a,n}) \le \delta\alpha \\ m(Q^u_{a,n}) \le \delta\alpha \end{cases} \qquad (3.160)$$

for all $n \ge 0$ and $\alpha > 0$. Hence the proof is completed.

The next result proved in [Misiurewicz (1980)] is about the existence of manifolds for the Lozi mapping (3.1):

Theorem 32 *Let a, b, satisfy (3.110)-(1), (3.110)-(2) and (3.110)-(3). Then at almost all $A \in G$, there exists linear local stable and unstable manifolds and broken linear global stable and unstable manifolds (of infinite length).*

Proof 11 *For the proof, the definition of the sets Q^s_α and $Q^u_{\alpha'}$ imply that the linear local stable and unstable manifolds exists for all $\hat{A} \in G \backslash \cap_{\alpha>0} Q^s_\alpha \cup \cap_{\alpha>0} Q^u_{\alpha'}$ and by Lemma. 3.10, we have:*

$$m(\cap_{a>0}Q^s_{\alpha} \cup \cap_{a>0}Q^u_{\alpha}) = 0, \tag{3.161}$$

and these manifolds exist at almost all $A \in G$. Obviously, a broken linear stable (resp. unstable) manifold at $A \in G$ of length at least α can be obtained by taking a local linear stable (resp. unstable) manifold of length αc^k (resp. $\alpha \left(\dfrac{c}{b}\right)^k$) at $\mathcal{L}^k(A)$ (resp. $\mathcal{L}^{-k}(A)$) and then its image under $\mathcal{L}^{-k}$ (resp. $\mathcal{L}^k$). This process cannot be done only for $A \in \mathcal{L}^{-k}(Q^s_{\alpha c^k})$(resp. $A \in \mathcal{L}^k Qu \left(Qu_{\alpha \cdot \left(\frac{c}{b}\right)^k} \right)$). By using Lemma. 3.10, we conclude that the measures of these bed sets converge to 0 as $k \to \infty$. In this case, a broken linear stable (resp. unstable) manifold of length α exists at almost every points of G, which implies that broken linear stable and unstable manifolds of infinite lengths exist at almost every points of G. Hence the proof is complete.

Now, assume additionally inequality (3.110)-(6) which is equivalent to $\dfrac{c}{b} > \sqrt{2}$. Then the following lemma was proved in [Misiurewicz (1980)]:

Lemme 3.11 *Let $I \subset G$ be a segment contained in some unstable manifold. Then there exists $n \geq 0$ and a segment $I_1 \subset \mathcal{L}^n(I)$ such that I_1 intersects both coordinates axis.*

The proof is by contradiction, assume that such n and I_1 do not exist. The set $\mathcal{L}^k$ (I) is a broken line and the mapping $\mathcal{L}$ is linear in the left and right half-planes. Hence, if a segment $J \subset \mathcal{L}^k(I)$ intersects the y- axis then its image $\mathcal{L}(J)$ is an union of at most two segments, each of them intersecting the x-axis. Then the image $\mathcal{L}(J)$ does not intersect the y-axis, and therefore $\mathcal{L}^2(J)$ consists of at most two segments. If the interval J does not intersect the y-axis, then its image $\mathcal{L}(J)$ is a segment and $\mathcal{L}^2(J)$ consists of at most two segments. Therefore, in both cases the image $\mathcal{L}^2(J)$ consists of at most two segments. Hence, the image $\mathcal{L}^{2k}(J)$ consists of at most 2^k segments for $k = 0, 1, 2, ...$ Since $I \subset G$, we get $\mathcal{L}^{2k}(I) \subset G$ and hence the length of $\mathcal{L}^{2k}(I)$ is at most $2^k.diam(G)$. On the other hand, the interval I is a segment of an unstable manifold and thus the length of $\mathcal{L}^{2k}(I)$ is at least $\left(\dfrac{c}{b}\right)^k$ times the length of I. Finally, if k is large enough, we have $\left(\dfrac{c}{b}\right)^k \leq 2$, which is a contradiction to the above assumptions. Hence the proof is complete.

Assumption (3.110)-(5) and Fig. 3.18 and the fact that the point T lies below S and $\mathcal{L}(T)$ to the left of $\mathcal{L}(S)$, imply immediately the following result proved in [Misiurewicz (1980)]:

Lemme 3.12 *Each segment contained in $\mathcal{L}(F)$ and intersecting both coordi- nate axes, intersects $\overline{XT}$.*

The strategy used in [Misiurewicz (1980)] to find a strange attractor for the Lozi mapping (3.1) is based on the ideas that a strange attractor must be equal to the intersection of images of some its neighborhood, and the

Lozi mapping restricted to this set is topologically transitive.[18] Clearly, the set $\Lambda_{a,b}$ (with G is its neighborhood) is the natural candidate for an attractor. The main issue here is to prove that the Lozi mapping restricted to this set is topologically transitive while possibly even some stronger property holds as shown in [Misiurewicz (1980)]:

Theorem 33 *Let a, b, satisfy (3.110)-(1),(3.110)-(3),(3.110)-(5) and (3.110)- (6). Then $\mathcal{L}\,|\,\Lambda_{a,b}$ is topologically mixing, i.e., for every open subsets U, V of $\mathbb{R}^2$ such that $U \cap \Lambda_{a,b}$ and $V \cap \Lambda_{a,b}$ are non-empty, there exists n_0 such that for every $n \ge n_0$ the set $\mathcal{L}^n (U) \cap V \cap \Lambda_{a,b}$ is non-empty.*

For the proof let U, $V \subset \mathbb{R}^2$. Since (3.115) follows from (3.110)-(1) and (3.110)- (5), Theorem 30 implies that the sets $W^u (X) \cap U$ and $W^u (X) \cap V$ are non- empty. Let $A \in W^u (X) \cap V$. Because the set $\mathcal{L}^{-k} (A)$ converges to I as $k \to \infty$, there exists $k_0 \ge 0$ such that the segments $\overline{L^{-n}(A)X}$ is contained in $W^u (X)$ and in the first quadrant. The linear hyperbolicity in the first quadrant of the mappings $\mathcal{L}^{-1}$ defined by (3.3) imply that there exists $k_1 \ge k_0$ and a neighborhood V_1 of A, such that $\mathcal{L}^{-k_1} (V_1)$ intersects the set $\mathcal{L}(F)$ in the same way as $\overline{XT}$. This means that each segment contained $\mathcal{L}(F)$ and intersecting both x-axes and y-axes, intersects also $\mathcal{L}^{-k_1} (V_1)$. In this case, for $k \ge k_1$, every $\mathcal{L}^{-k} (V_1)$ has the same property. The fact that the set $W^u (X) \cap U$ is non-empty, implies that there exists a segment $I \subset W^u (X) \cap U$, and by using Lemma. 3.11, we concludes that there exists $n_1 \ge 0$ such that some subsegment of $\mathcal{L}^{n_1} (I)$ intersects both x-axes and y-axes. Thus, it intersects also the set $\mathcal{L}^{-k} (V_1)$ for all $k \ge k_1$, which imply that $\varnothing \ne \mathcal{L}^{n-n_1} (\mathcal{L}^n (I) \cap \mathcal{L}^{n_1-n} (V_1)) \subset callL^n (U) \cap V \cap W^u (X)$ for all $n \ge n_1 + k_1$. Hence the proof is complete.

This proof shows that the set $\Lambda_{a,b}$ is infinite and it is not a whole manifold, and it was called *strange attractor* in [Misiurewicz (1980)]. Some additional informations (density) about the structure of the mapping calL are listed in the following result proved in [Misiurewicz (1980)]:

Theorem 34 *Let a, b, satisfy (3.110)-(1),(3.110)-(3),(3.110)-(5) and (3.110)- (6). Then $W^u (X) \cap G$ is dense in G and $W^u (X) \cap \Lambda_{a,b}$ is dense in $\Lambda_{a,b}$.*

Proof 14 *Theorem 30 and Lemmas. 3.11 and 3.12 show immediately the density of $W^u (X) \cap \Lambda_{a,b}$ in $\Lambda_{a,b}$. Theorem 32 and Lemmas. 3.11 and 3.12, imply the density of $W^u (X) \cap G$ in G, by showing that there is no segment of an unstable manifold with all images contained in R. The method of analysis is based on tow ideas, namely the map $calL\,|_R$ is linear hyperbolic and the set R is bounded.*

For other values of a and b when (a, b) leaves the regions defined by (3.110)-(1) to (3.110)-(6), Misiurewicz in [Misiurewicz (1980)] claims the results summarized in the following proposition:

---

[18] i.e., For every two open non-empty sets, some images of the first one intersects the second one.

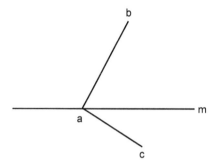

Figure 3.19: The sets *l* and *m* intersect weak transversally.

Proposition 3.21 *(1) If (a, b) crosses the line $a = b + 1$, then the periodic orbit (3.135) becomes an attractive one and $calL(F) \subset F$. The same result was obtained numerically when (a, b) crosses the line $2a = b+3$. (2) If (a, b) crosses the line $2a+ b = 4$, then if $2a- b > 4$, then there is no trapping region[19] containing X. (3) If (a, b) crosses the line $a = \frac{1}{2}\sqrt{3b^2 + 4 + \sqrt{(3b^2 + 4)^2 - 32b^3}}$, then maybe some sets other than H with similar properties can exists. (4) If (a, b) crosses the line $b = \frac{a^2 - 1}{2a + 1}$, then previous results hold. (5) If (a, b) crosses the line $a\sqrt{2} = b+2$, then results analogous to Lemmas. 3.11 and 3.12 are possible. (6) If $\frac{b}{c} > \sqrt[3]{\sqrt{2}}$ then it is possible to analyze some longer piece of $W^u(X)$. (7) For the values $a = 1.7$ and $b = 0.5$,[20] all previous theorems hold.*

(1) The proof is by direct calculations. (2) We have that $calL(F)$ is no longer contained in F. Thus, $W^u(X)$ intersects $W^u(Y)$ and therefore there are points arbitrarily close to X images of which eventually escape to infinity along $W^u(X)$ which implies that there is no trapping region containing X or maybe there is no trapping region at all. But it is possible to prove it only when $2a - b > 4$. In this case, the expanding coefficient $\frac{b}{c}$ is larger than 2 and hence if I is a segment which has a direction from the expanding sector then $calL^n(I)$ contains a segment of length at least $\left(\frac{b}{2c}\right)^n$ times the length of I. Thus, the diameters of the images of any non-empty open set grow to infinity. (3) Remark that $\overline{XT} \cap \overline{L^2(Z)Z} = \emptyset$. (4) Remark that the condition (3.110)-(5) is sufficient, but not necessary for T to lie below S and these points change their relative position much later. (5) Remark that the expanding coefficient $\frac{b}{c}$ becomes smaller than $\sqrt{2}$. (6) Remark that there is no periodic point of period 2 close to the axes, and hence the images of every segment of $W^u(X)$ contain a long segment.

---

[19] Maybe there is no trapping region at all.
[20] Studied numerically in [Lozi (1978)].

Finally, we note that the region of parameters in which the strange attractor of the Lozi map (3.1) appears is vastly larger in numerical results given in [Tèl (1983(a-b))] than that proven theoretically in [Misiurewicz (1980)] and described in the above sections. This phenomenon also occurs in the parameter space of a dynamical system where the results given by numerical methods are better than one obtained analytically when using a specific definition of chaos, i.e., Li-Yorke's definition [Li (1975)], the Smale horseshoe transformation [Smale (1967)] and transversal homoclinic points [Guckenheimer & Holmes (1983)],...etc. This fact implies that the structure of the strange attractor obtained numerically is richer than that given by one of the above definitions. Indeed, for the Hénon map (2.1) this phenomenon was studied in [Hénon (1976), Helleman (1980), Zhang (1984), Yan & Qian (1985)]. Finally, we note that in [Misiurewicz (1980)] the unicity of the Lozi attractor is not discussed and its existence is not also checked for $b < 0$.

3.6 Existence of strange attractor for $b < 0$

In this section, we present a proof of the existence of strange attractor of the Lozi map (3.1) for $b < 0$. Indeed, it was proved in [Cao & Liu (1998(a))] that the Lozi map (3.1) admits an open set E in the parameter space when $b < 0$ such that if $(a, b) \in E$, the map (3.1) displays a strange attractor $\Lambda_{a,b}$ and its basin $B(\Lambda_{a,b})$ contains a neighborhood of itself, i.e., the union of the transversal homoclinic points and weak transversal homoclinic points are dense in the set $\Lambda_{a,b}$. The concept of weak transversal homoclinic points comes from the fact that the Lozi map (3.1) is not a diffeomorphism. So, the notion of transversality cannot be introduced for this case.

Definition 45 *Let l be a set consists of the segment ab and ac and m be a line. If l and m intersect at point a as shown in Fig. 3.19, it is said that l and m intersect weak transversally, otherwise non-transversally intersect.*

First, it is easy to check that when $b < 0$, the Lozi map (3.1) is topologically conjugate to the map $\mathcal{L}'(x, y)$ defined by:

$$\mathcal{L}'(x, y) = \left(1 - a|x| + \sqrt{|b|}\, y, -\sqrt{|b|}\, x\right) \qquad (3.162)$$

Let us consider the map:

$$\widehat{\mathcal{L}}'(x, y) = (1 - a|x|, 0) \qquad (3.163)$$

Thus, map (3.163) can be regarded as a perturbation of the map (3.162) in a bounded region. The construction of the trapping region for the Lozi map (3.1) can be done via the following method: Let q_{ab} be the hyperbolic fixed point of the Lozi map $\mathcal{L}'$ given by (3.162) in the second quadrant. It is clear that the right branch of the unstable manifold $W^u(q_{ab})$ of q_{ab} intersects

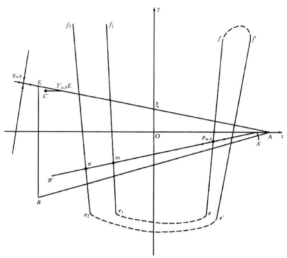

Figure 3.20: The trapping region is the triangle *ABE* where $p_{a,b}$ and $q_{a,b}$ are the fixed points of the Lozi map (3.1).

the *y*-axis at point $\xi = (0, y_1)$, let $A (x_a, 0) = \mathcal{L}' (\xi)$. We have that segment $\xi A \subset W^u (q_{ab})$, $AB = \mathcal{L}' (\xi A)$ and $B (x_{b'}, y_b) = \mathcal{L}' (A)$ when (a, b) is located near the value $(2, 0)$. We also have, $|y_1| \leq c\ |b|$ and $(x_{b'}, y_b)$ is near the point $(1,-1)$ as shown in Fig. 3.20. The proof given in [Cao & Liu (1998(a))] is based on several results summarized in the following proposition:

Proposition 3.22 *(a) For the tent map (3.6), if $a_0 < a_1 < 2$ and a_0 is near 2, then $T^3 (0) - T^2 (0) \geq d$, for $a \in [a_0, a_1]$, where d is a constant. (b) When $a < 2$ and near 2, if $T^n (0) < 0, n \geq 2$, then $\dfrac{dT^a (0)}{da} < 0$ and $\left|\dfrac{dT^a (0)}{da}\right| \geq a^{n-1}$. (c) For $a \in [a_0, a_1]$, there exists $b_0 > 0$ and small. When $|b| < b_0$, there exists a region G_1 and N_1 such that $\mathcal{L}' (G_1) \subset G_1$ and $(\mathcal{L}')^{N_1} (G_1) \subset Int (G_1)$ (the interior set of G_1). (d) There exist a_0, a_1, $a_0 < a_1 < 2$, and a_0 is near 2, $N > 0$, such that $T^N (0) \in I \subset Int [p_a, p'_a]$ for $a \in [a_0, a_1]$, where p_a is the reversed fixed point of T, such that $T^2 (p'_a) = p_a$, and I is a closed interval in $Int [p_a, p'_a]$ for $a \in [a_0, 2]$. (e) There exists $b_0 > 0$ and small, for $a \in [a_0, a_1]$ as in Proposition. 3.22(d). If $|b| < b0$, then there is a $\delta > 0$ such that*

$$(\mathcal{L}')^N \left[(-2\delta, 2\delta) \times \left[\dfrac{-1}{2}, \dfrac{1}{2}\right] \right] \subset Int I \times \left[\dfrac{-1}{2}, \dfrac{1}{2}\right] \tag{3.164}$$

and $Int I \times \left[\dfrac{-1}{2}, \dfrac{1}{2}\right]$ is contained in a region whose boundary is contained in W^u $(p_{a,b}) \cup W^s (p_{a,b})$. (f) There exists $b'_0 > 0$, small such that if $(x, y) \in G_1$, then $\omega (x, y) \subset W^u (p_{a,b})$ for $a \in [a_0, a_1]$, $|b| < b'_0$. (g) The stable sector

$$C_s = \left\{ (t,r) : \sqrt{|b|} \, |t| \le |c| \, \|r\| \right\}, \text{ and the unstable sector } C_u = \left\{ (t,r) : \sqrt{|b|} \, \|r\| \le |c| \, |t| \right\}$$

(where $c = \dfrac{a - \sqrt{a^2 - 4b}}{2}$ *) have the following properties:*

$$\begin{cases} (D\mathcal{L}')^{-1} C_s \subset C_s \\ (D\mathcal{L}')^{-1} C_u \subset C_u \\ \left| (D\mathcal{L}')^{-1} v \right| \ge \dfrac{1}{|c|} |v|, v \in C_s \\[2mm] |D\mathcal{L}'v| \ge \left| \dfrac{b}{c} \right| |v|, v \in C_u \end{cases} \qquad (3.165)$$

(h) If a segment $F_1 \subset G_1$ satisfies that both $\mathcal{L}'F_1$ and $(\mathcal{L}')^2 F_1$ intersects the y-axis, and $T'_{a,b}F_1$ is a segment, then there exists a segment $L \subset F_1$ such that $(\mathcal{L}')^2 L$ intersects $W^s (p_{a,b})$ transversally. (l) If $(a, b) \in [a_0, a_1] \times (-b'_0, 0)$, then $\overline{W^u (p_{a,b})}$ is topologically transitive.

Proof 16 *(a) It is easy to see that $T^3 (0) - T^2 (0) = a (2 - a) \ge a_0 (2 - a_1) = d > 0$.*

*(b) For the tent map (3.6), we have $T^2 (0) = 1 - a^2 + a < 0$ and $\dfrac{dT^2 (0)}{da} = -2a + 1$
< 0 and $\dfrac{dT^2 (0)}{da} = 2a - 1 > a$. Thus, the following inequalities holds:*

$$\begin{cases} \dfrac{dT^a (0)}{da} = \dfrac{\partial T(a,x)}{\partial x} \cdot \dfrac{dT^{a-1} (0)}{da} + \dfrac{\partial T(a, T^{a-1} (0))}{\partial a} = a \cdot \dfrac{dT^{a-1} (0)}{da} + T^{a-1} (0) < 0 \\[2mm] \dfrac{dT^a (0)}{da} \le -a.a^{n-2} = -a^{n-1} \Rightarrow \left| \dfrac{dT^a (0)}{da} \right| \ge a^{n-1} \end{cases} \qquad (3.166)$$

(c) Now let $a \in [a_0, a_1]$ such that $a_0 < a_1 < 2$ and a_0 is located near 2 and let us consider the following application:

$$\pi_1 : \mathbb{R}^2 \to \mathbb{R}, \pi_1 (x, y) = x \qquad (3.167)$$

let $E = E (x_b, y_b)$ be the point in the segment $q_{ab}A \subset W^u (q_{ab})$ such that $\pi_1 E = \pi_1 B$. Form Proposition. 3.22(a), it follow that:

$$\pi_1 \, \hat{\mathcal{L}}^{\prime 3} (0, y_1) - \pi_1 \, \hat{\mathcal{L}}^{\prime 2} (0, y_1) \ge d > 0 \qquad (3.168)$$

thus we have:

$$\pi_1 \, (\mathcal{L}')^3 (0, y_1) - \pi_1 \, (\mathcal{L}')^3 (0, y_1) \ge \dfrac{d}{2} > 0 \qquad (3.169)$$

for small values of the parameter b. Also, we have:

$$\pi_1 \, (\mathcal{L}' (E)) = 1 - a \, |x| + y_e > 1 - a \, |x| + y_b \qquad (3.170)$$

Inequality (3.170) implies that the point $\mathcal{L}' (E)$ lies to the right of point $\mathcal{L}'$ (B) = C and the image of vertical segment EB under the action of map $\mathcal{L}'$ is a *horizontal segment ($\mathcal{L}'$ (E)) C which lies above the x-axis and to the right of the vertical segment BE. Let G_1 be the region bounded by segments of EA and BA which are contained in W^u (q_{ab}) and EB is a vertical segment. Hence, we have the following inclusions:*

$$\begin{cases} (\mathcal{L}'\, E)\, C \subset G_1 \\ (\mathcal{L}'\, E)\, C - T'_{a,b}\, E \subset G_1 \\ \mathcal{L}'\, EA \subset EA \cup BA \\ (\mathcal{L}')^3\, BA \subset IntG_1. \end{cases} \qquad (3.171)$$

From the expansion of unstable manifold W^u (q_{ab}), we conclude that there exists $N_1 > 0$ such that $(\mathcal{L}')^{N_2}$ (E) $\in$ BA which implies that:

$$\begin{cases} \mathcal{L}'\, G_1 \subset G_1 \\ (\mathcal{L}')^{N_1}\, G_1 \subset IntG_1. \end{cases} \qquad (3.172)$$

(d) The first inclusion of (3.172) implies that there is a fixed point p_{ab} for the map (3.162) in G_1 and the second inclusion implies that $p_{ab} \in IntG_1$ and it is a hyperbolic fixed point because (a, b) is located in the vicinity of (2, 0) which implies that W^u (p_{ab}) $\subset IntG_1$. Now, if a is near 2, then we have T^2 (0) = −1 and T^n (0) < 0, then $\dfrac{dT^a(0)}{da} \geq a^{n-1}$, which imply that there exists $\bar{a}$ and n and interval I such that $\bar{a}$ is near 2 and T^n (0) = 0 and $I \subset [p_a, p'_a]$ for a $\in [\bar{a}, 2]$. So, we have:

$$[p_a, p'_a] \subset [-1, 1] = T^{n+1} ([\bar{a}, 2]\,, 0). \qquad (3.173)$$

Now, if N = n + 1, we have a_0, $a_1 \in [\bar{a}, 2]$ such that T^N (0) $\in IntI \subset [p_a, p'_a]$. (e) By using Proposition. 3.22(d), it follows that T^N (0) $\in IntI \subset [p_a, p'_a]$ for a $\in [a_0, a_1]$, this imply that there exists a neighborhood of 0, say U = (−2δ, 2δ) such that T^N (U) $\subset IntI$, for a $\in [a_0, a_1]$. Thus, we have:

$$(\hat{\mathcal{L}}')^N \left[U \times \left[\frac{-1}{2}, \frac{1}{2}\right] \right] \subset IntI \times \{0\}, \qquad (3.174)$$

for a $\in [a_0, a_1]$, this implies that there exists small $b_1 > 0$ such that:

$$(\mathcal{L}')^N \left[U \times \left[\frac{-1}{2}, \frac{1}{2}\right] \right] \subset IntI \times \{0\}, a \in [a_0, a_1],\ |b| < b_1 \qquad (3.175)$$

Now, let $\overline{p_1} < 0$, $\overline{p_2} < 0$ such that $T_{}(\overline{p_2}) = p_a$, $(T)^2 (\overline{p_2}) = p'_a$, and define the stable set of $p_a \times \{0\}$ relative to the map $\mathcal{L}'$ as follow:

$$W^s (p_a \times \{0\}) = \{x \in \mathbb{R} : T^n (x) = p_a,\ n \geq 0\} \qquad (3.176)$$

The unions of the vertical line and $p^a \times \mathbb{R}$, $p^d \times \mathbb{R}$, $\overline{p^a} \times \mathbb{R}$, and $\overline{p^2} \times \mathbb{R}$x are contained in W^s ($pa \times \{0\}$). The fact that the compact part of the stable manifold depends on the map, implies that there exists $b_2 > 0$ such that if $|b| < b_2$, then we have that the manifold W^s ($p_{a,b}$) contains the segment $l = ef$, $l' = e'f'$, $l_1 = e_1f_1$, $l_2 = e_2f_2$, that are C^1 near the segments $p_a \times [-1, 1]$, $p'_a \times [-1, 1]$, $\overline{p_1} \times [-1, 1]$, $\overline{p_2} \times [-1, 1]$ respectively, where $p_{a,b}$ is the continuation of the fixed point $p_a \times \{0\}$ of the map (3.163) for $b = 0$. Thus, we have that the segments l, l'', l_1, l_2 are almost vertical segments. Because the manifold W^s ($p_{a,b}$) is connected and it is the image of W^s_{loc} ($p_{a,b}$) under the action of map $(\mathcal{L}')^{-n}$, $n \geq 0$, there exists curves $\widehat{ff'}$, $\widehat{e_1e}$, $\widehat{e_2e'}$ such that l connects l' by $\widehat{ff'}$, l_1 connects l by $\widehat{e_1e}$, l_2 connects l' by $\widehat{e_2e'}$. Let m, n be the intersection points of the segments l_1 and l_2 with the segment $A'B' \subset W^u$ ($p_{a,b}$) respectively. Now, it is possible to construct a region G' whose boundary is the union of the segments $\overline{me}_1$, $\widehat{e_1e}$, $\widehat{e_1f}$, $\overline{ef'}$, $\widehat{ff'}$, $\widehat{fe'}$, $\overline{e'e}_2$, $\overline{e_2n}$, which are contained in the manifold W^s ($p_{a,b}$) and of the segment $\overline{mn}$ which is contained in the manifold W^u ($p_{a,b}$). If $b_0 = \min \{b_1, b_2\}$, then we have the following inclusions:

$$(\mathcal{L}')^N \left[(-2\delta, 2\delta) \times \left[\frac{-1}{2}, \frac{1}{2} \right] \right] \subset IntI \times \left[\frac{-1}{2}, \frac{1}{2} \right] \subset G', a \in [a_0, a_1], |b| < b_0 \quad (3.177)$$

Thus, it follows that:

$$(\mathcal{L}')^N \left(\left[(-2\delta, 2\delta) \times \left[\frac{-1}{2}, \frac{1}{2} \right] \right] \cap G_1 \right) \subset G', a \in [a_0, a_1], |b| < b_0. \quad (3.178)$$

(f) If $C_{a,0}$ is a Cantor set in the x-axis that is a hyperbolic set of the map (3.6), then it is easy to check that the set of the map (3.163) whose forward orbit never enters $[-\delta, \delta] \times [-1, 1]$ is contained in the manifolds W^s ($C_{a,0}$) which is the union of vertical segments. The fact that hyperbolic sets are persistent under perturbations of the system [Shub (1993)] implies that there exists $b_3 > 0$ such that if $|b| < b_3$ then the Cantor set $C_{a,b}$ of the map (3.162) whose full orbits remain outside of $[-\delta, \delta] \times [-1, 1]$ is hyperbolic. Hence, the corresponding set for the map (3.163) is contained in the stable set of $C_{a,b}$ which depends smoothly on the map, hence, we have that the manifold W^s_{loc} (x, y) for $(x, y) \in C_{a,b}$ is an almost vertical segment connecting the upper boundary and downward boundary of the region G_1. Now, if $(x, y) \in G_1$ and $(\mathcal{L}')^n (x, y) \notin U_1 = [(-2\delta, 2\delta) \times \left[\frac{-1}{2}, \frac{1}{2} \right]] \cap G_1$ $(n \geq 0)$, then there exists a point $(x_0, y_0) \in C_{a,b}$ such that $(x, y) \in W^s_{loc} (x_0, y_0)$. We have that $W^s_{loc} (x_0, y_0)$ intersects $A'B' \subset W^u$ (p_a) at point (x', y') because $(x_0, y_0) \notin \mathcal{L}'U_1 \cup (\mathcal{L}')^2 U_1$ and $W^s_{loc} (x_0, y_0)$ is an almost vertical segment. Thus, we have:

$$\lim_{n \to +\infty} |(\mathcal{L}')^n (x, y) - (\mathcal{L}')^n (x', y')| = 0 \quad (3.179)$$

Relation (3.179) imply that $\omega\,(x,\,y) \subset \overline{W^u}\,(p_{a,b})$. Also, we have that if $(x,\,y)$ $\in G_1$ and $(\mathcal{L}')^{n_0}\,(x,\,y) \in U_1$ for some $n_0 \geq 0$, then $(\mathcal{L}')^{n_0+N}\,(x,\,y) \in G'$. Now, the fact that the boundary of G' is contained in $W^u\,(p_{a,b}) \cup W^s\,(p_{a,b})$ and the map $\mathcal{L}'$ is dissipative implies that $\omega\,(x_0,\,y_0) \subset \overline{W^u}\,(p_{a,b})$ for $(x_0,\,y_0) \in G'$ which confirm the inclusion $\omega\,(x,\,y) \subset \overline{W^u}\,(p_{a,b})$. If $b'_0 = \min\,\{b_0,\,b_3\}$, then we have:

$$\omega\,(x,\,y) \subset \overline{W^u}\,(p_{a,b}),\,(x,\,y) \in G_1,\,a \in [a_0,\,a_1],\,|b| < b_0. \tag{3.180}$$

(g) To show that $\overline{W^u}\,(X)$ is topologically transitive, we must give an estimate of eigenvalues of the map $\mathcal{L}'$ for $(a,\,b) \in [a_0,\,a_1] \times (-b'_0,\,0)$. For this purpose, it is sufficient to give the proof for the stable sector:

Let

$$D\mathcal{L}' = \begin{pmatrix} \pm a & \sqrt{|b|} \\ \sqrt{|b|} & 0 \end{pmatrix} \tag{3.181}$$

$$\begin{pmatrix} t_1 \\ r_1 \end{pmatrix} = D\mathcal{L}'\begin{pmatrix} t \\ r \end{pmatrix} \tag{3.182}$$

if $(t_1,\,r_1) \in C_s$, that is

$$\sqrt{|b|}\,|t_1| \leq |c|\,|r_1|, \tag{3.183}$$

then

$$\begin{cases} \sqrt{|b|}\,|t| \leq |c|\,|r|, \\ |t| = \dfrac{|r_1|}{\sqrt{|b|}} \geq \dfrac{|t_1|}{|c|}, \\ |r_1| = \sqrt{|b|}\,|t| \leq |c|\,|r| \end{cases} \tag{3.184}$$

Thus, we have

$$\begin{cases} (D\mathcal{L}')^{-1}\,C_s \subset C_s \\ \dfrac{1}{|c|}\,|(t_1,r_1)| \leq |(t,r)| \end{cases} \tag{3.185}$$

As a remark, we have that $\left|\dfrac{b}{c}\right| > \sqrt{2}$ for $(a,\,b) \in [a_0,\,a_1] \times (-b'_0,\,0)$ since $\left|\dfrac{a}{c}\right| = \dfrac{a - \sqrt{a^2 - 4b}}{2}$.

(h) Let $(0,\,y)$ be the point of intersection of $\mathcal{L}'\,(F_1)$ with y-axis. It has $|y| \leq 2b$ and $\pi_1\,(\mathcal{L}'\,(0,\,y)) = 1 - a\,|x| + y = 1 + y$. Also, $(\mathcal{L}')^2\,(F_1)$ intersects the y-axis at point $(0,\,y_1)$ and $\mathcal{L}'\,(F_1)$ is a segment or the union of two segments that intersect the x-axis at point $(1 + y,\,0)$. Thus, we have that $(\mathcal{L}')^2\,(F_1)$ contains a segment

connecting the point $(0, y_1)$ and the point $(1 + y, 0)$ must intersect the manifold W^s $(p_{a,b})$ transversally. This fact implies that there exists a segment $L \subset F_1$ such that the segment $(\mathcal{L}')^2$ (L) intersects the manifold W^s $(p_{a,b})$ transversally.

(l) If U and V are open sets in $\overline{W^u}$ $(p_{a,b})$, then, we have that $U = U_1 \cap \overline{W^u}$ $(p_{a,b})$ and $V = V_1 \cap \overline{W^u}$ $(p_{a,b})$, where U_1 and V_1 are open sets in $\mathbb{R}^2$ and U_1 contains a segment $L_0 \subset W^u$ $(p_{a,b})$. First, if neither L_0 nor $\mathcal{L}'$ (L_0) intersects the y-axis, we have:

$$| (\mathcal{L}')^2 \, (L_0) | \geq \left| \frac{b}{c} \right|^2 \, | L_0 | > 2 \, | L_0 | \qquad (3.186)$$

Second, if both L0 and L' (L_0) intersect the y-axis, then (see the proof of Lemma. 3.22(h)), there exists a segment $\overline{L_0} \subset L_0$ such that the segment $\mathcal{L}'$ $(\overline{L_0})$ intersects W^s $(p_{a,b})$ transversely. Third, if either L_0 or $\mathcal{L}'$ (L_0) intersects the y-axis, then the set $(\mathcal{L}')^2$ (L_0) is the union of two segments, and there exists a segment $L_2 \in (\mathcal{L}')^2$ (L_0) such that:

$$| L_2 | > \frac{1}{2} \left| \frac{b}{c} \right|^2 \, | L_0 | > | L_0 | . \qquad (3.187)$$

A repetition of the above arguments gives two cases: (1) There exists n and a segment $I_0 \subset L_0$ such that both $(\mathcal{L}')^n$ (I_0) and $(\mathcal{L}')^{n+1}$ (I_0) intersects the y-axis. (2) There exists a segment $L_{2n} \subset (\mathcal{L}')^{2n}$ (L_0) such that:

$$| L_{2n} | > \left(\frac{1}{2} \left| \frac{b}{c} \right|^2 \right)^n \, | L_0 | \qquad (3.188)$$

The fact that W^u $(p_{a,b}) \in G_1$ and the length of the segment in G_1 is finite implies that there must be n_0 such that case (1) holds. This means the existence of a segment $I \subset I_0$ such that segment $(\mathcal{L}')^{n+1}$ (I_0) intersects W^u $(p_{a,b})$ transversely. By using the λ-Lemma, there exists $N > 0$ such that $(\mathcal{L}')^m$ $(I) \cap V_1 \neq \varnothing$, $(m \geq N > 0)$, i.e.,

$$(\mathcal{L}')^{m+n+2} \, (U) \cap V \neq \varnothing, \, m \geq N > 0. \qquad (3.189)$$

Hence, the proof is complete.

By using Proposition. 3.22 the following result was proved in [Cao & Liu (1998(a))]:

Theorem 35 (a) ω $(x, y) \subset \overline{W^u}$ $(p_{a,b})$ for $(x, y) \in G_1$, and $(a, b) \in [a_0, a_1] \times (-b'_0, 0)$. (b) The unions of the transversal and weak transversal homoclinic points are dense in W^u $(p_{a,b})$ as in Theorem 34(a).

For the proof (a) Proposition. 3.22(l), implies that $\overline{W^u}$ $(p_{a,b})$ is topologically transitive and Proposition. 3.22(f) says that for $(x, y) \in G_1$, we have $\omega(x, y) \subset \overline{W^u}$ $(p_{a,b})$. Therefore, the manifold W^u $(p_{a,b})$ attracts a neighborhood of itself and it is indecomposable which means that it is a strange attractor.

(b) Since $\Lambda_{a,b} = \overline{W^u(p_{a,b})}$, it is sufficient to prove that the unions of the transversal and weak transversal homoclinic points are dense in $W^u(p_{a,b})$. Indeed, let a segment $I \subset W^u(p_{a,b})$ and repeat the arguments as in Proposition. 3.22(l), it follows that there exists n such that $(\mathcal{L}')^n(I)$ contains a segment I_0 which intersects $W^u(p_{a,b})$ transversally. Thus, the unions of the transversal and weak transversal homoclinic points are dense in $W^u(p_{a,b})$.

Finally, an important point might be to highlight the differences between Sections 3.5 and 3.6. We remark that the features of the Lozi map for $b > 0$ are distinct from $b < 0$, and there are two different approaches required to study the chaotic properties in these parameter regimes, i.e., $b > 0$ versus $b < 0$?. Indeed, in Section 3.5 we give a rigorous proof for the hyperbolicity property for the Lozi mapping (3.1) in the sense of Definition 12(d) with a slight modifications for $b > 0$. The strategy used in [Misiurewicz (1980)] to find a strange attractor for the Lozi mapping (3.1) is based on the ideas that a strange attractor must be equal to the intersection of images of some its neighborhood, and the Lozi mapping restricted to this set is topologically transitive. See the last paragraph before Section 3.6. In Section 3.6, we present a proof of the existence of strange attractor of the Lozi map (3.1) for $b < 0$. Indeed, it was proved in [Cao & Liu (1998(a))] that the Lozi map (3.1) admit an open set E in the parameter space when $b < 0$ such that if $(a, b) \in E$, the map (3.1) displays a strange attractor $\Lambda_{a,b}$ and its basin $B(\Lambda_{a,b})$ contains a neighborhood of itself, i.e., the union of the transversal homoclinic points and weak transversal homoclinic points are dense in the set $\Lambda_{a,b}$. The concept of weak transversal homoclinic points comes from the fact that the Lozi map (3.1) is not a diffeomorphism. So, the notion of transversality cannot be introduced for this case.

3.7 Rigorous proof of chaos in the Lozi map using the theory of transversal heteroclinic cycles

We note that the notations used in this section are completely independent from those used in the previous one. The reason is that this section gives a study for the Lozi map $\mathcal{L}_1$ given by (3.2), but we note that all properties discussed here also hold for Lozi map (3.1) since it is equivalent to map (3.2). Now, in [Liu, et al. (1992(a))] the existence of a strange attractor of the Lozi mapping $\mathcal{L}_1$ given by (3.2) in the parameter region

$$0 < b < 1, 1 + b < a < 2 - \frac{b}{2} \qquad (3.190)$$

is rigorously proved using the theory of transversal heteroclinic cycles. The structure of this strange attractor consists entirely of unstable manifolds of infinite hyperbolic and periodic points. Indeed, it was proven that all periodic points are of the saddle type in the parameter region:

$$0 < b < 1, a > 1 + b \qquad (3.191)$$

When $a = \dfrac{5}{8}, b = \dfrac{9}{25}$, the structure of the strange attractor and the relation of the invariant manifolds was investigated by using the so called *the method of dual line mapping*.

Let A (X for the Lozi map (3.1)) be the fixed saddle in the first quadrant of the Lozi map (3.2). Let u_A be the first segment of the unstable manifold of A. Let q be the intersection point between u_A and the x-axis and Rq, 'the image of q, LRq' the image of Rq. Let Δ be a trigon with vertices q, Rq, LRq (this is similar to the triangle F defined in Section 3.5.1). The following result was proved in [Tél (1982(a–b)-1983(a))]:

Theorem 36 *If the image of RLq is settled into Δ, i.e., condition (3.190) holds, then Δ is the trapping region of the Lozi mapping.*

3.7.1 Properties of periodic solutions of the Lozi map

The Lozi mapping $\mathcal{L}_1$ given by (3.2) can be written equivalently as:

$$\begin{cases} R : (x, y) \to (1 - ax + by, x) \text{ for } x \geq 0 \\ L : (x, y) \to (1 + ax + by, x) \text{ for } x \leq 0. \end{cases} \qquad (3.192)$$

In this case, its arbitrary iterative sequence is composed of R and L. Assume that the subscript 1 denotes the case where $y \geq 0$ and 2 for $y \leq 0$. Thus, any point in phase space (x, y) can be expressed as one of $_1L, _2L, _1R, _2R$, i.e., any R (resp., L) in an iterative sequence is accompanied by the subscript 1 (resp., 2).

Definition 46 *All sequences satisfying this matched condition are called probable sequences or $P - K$ solutions of the Lozi mapping.*

For example the sequences $_1R_1L_2R_1L, ..._{,2} L_2R_1LR...$ are probable sequences, but $_2R_2L_1R_1L...$ is not. For Lozi mapping $\mathcal{L}_1$ given by (3.2) there are two *probable fixed points* $_1R$ (resp., $_2L$) in the first quadrant (resp., the third quadrant). The existence of a $P - K$ solution of (3.2) can be reduced to solving a system of linear algebraic equations which give a unique solution by *Cramer's rule* and in this case, the number of probable periodic solutions of (3.2) is infinite and depend on the (a, b) values:

Proposition 3.23 (a) *The numbers of the probable $P - K$ solutions are:*

$$P(k) = \frac{2^k - \sum_{i \in \rho} ip(i) - 2}{k} \quad (k \geq 2) \qquad (3.193)$$

where $\rho = \{j : j$ is a divisor of $k, 1 < j < k\}$.

(b) When $k = 2, 3, 4$ the sequences of the probable $P - K$ solutions are given by:

$$\left\{ \begin{array}{c} P-2: \quad {}_1L_2R \\ P-3: \quad {}_1R_1L_2R;_1 \; L_2L_2R \\ P-4: \quad {}_1R_2R_1L_2R;_1 \; R_2L_2R;_1 \; L_2L_2L_2R \end{array} \right. \tag{3.194}$$

(a) The proof is completed by reduction.

(b) The proof is direct.

Proposition. 3.23(a) implies that $P(2) = 1$, $P(3) = 2$, $P(4) = 3$, $P(5) = 6...$, $P(9) = 59$, $P(10) = 99...$etc.

Proposition 3.24 *Assume that $a > 0$, $b > 0$, then we have*

(a) The fixed point ${}_1R[1/ (1 + a - b), 1/ (1 + a - b)] = A$ exists for $1+a > b$, and it is a stable node for $a \in (0, 1 - b)$ or a saddle point for $a \in (1 - b, \infty)$.

(b) The fixed point ${}_2L [-1/(a - (1 - b)),-1/(a - (1 - b))] = B$ exists for $a > 1 - b$, and it is a saddle point.

(c) The P–2 point ${}_1L_2R \left[\dfrac{(-(a-(1-b))}{(a^2+(1-b)^2)}, \dfrac{(a+(1-b))}{(a^2+(1-b)^2))}, \dfrac{((a+(1-b))}{(a^2+(l-b)^2)}, \dfrac{-(a-(1-b))}{(a^2+(1-b)^2))} \right]$

exists for $a > 1 - b$. It is stable for $a \in (1 - b, 1 + b)$ or of the saddle type for $a \in (1 + b, \infty)$.

(d) All probable $P - K$ points are only of either the stable or the saddle type for $0 < b < 1$.

The proof is by using linear analysis and algebraic calculations.

In the following result proved in [Liu, et al. (1992(a))] we discuss the conditions under which the stability of all probable $P - K$ points disappear.

Theorem 37 *When*

$$0 < b < 1, a > 1 + b \tag{3.195}$$

all probable periodic points and fixed points of the Lozi mapping (3.2) are of the saddle type.

It suffices to analyze the properties of matrices obtained from k-products of the Jacobian matrices $\begin{pmatrix} -a & b \\ 1 & 0 \end{pmatrix}$ and $\begin{pmatrix} a & b \\ 1 & 0 \end{pmatrix}$ in the regions R and L respectively. We have:

$$\left\{ \begin{array}{l} M^{-1}\begin{pmatrix} -a & b \\ 1 & 0 \end{pmatrix} M = \begin{pmatrix} \lambda_1 & 0 \\ 0 & \lambda_2 \end{pmatrix} = J, N^{-1}\begin{pmatrix} a & b \\ 1 & 0 \end{pmatrix} N = \begin{pmatrix} \lambda_3 & 0 \\ 0 & \lambda_4 \end{pmatrix}, \\[2mm] M = \begin{pmatrix} \lambda_1 & \lambda_2 \\ 1 & 1 \end{pmatrix}, N = \begin{pmatrix} \lambda_3 & \lambda_4 \\ 1 & 1 \end{pmatrix}, \lambda_{1,2} = \frac{-a \pm \sqrt{a^2 + 4b}}{2}, \lambda_{3,4} = -\lambda_{1,2} \\[2mm] H = M^{-1}, N = N^{-1}, M = \frac{1}{\sqrt{a^2 + 4b}}\begin{bmatrix} -a & 2\lambda_2 \\ 2\lambda_3 & a \end{bmatrix}, H^2 = I_2, \end{array} \right. \tag{3.196}$$

Thus, matrices $\begin{pmatrix} -a & b \\ 1 & 0 \end{pmatrix}$ *and* $\begin{pmatrix} a & b \\ 1 & 0 \end{pmatrix}$ *can be diagonal and finite products of them can be rewritten as* $(-1)^j \phi$, *where j is the number of copies of* $\begin{pmatrix} a & b \\ 1 & 0 \end{pmatrix}$ *contained in the multiplication,* ϕ *is a matrix obtained from finite products of H and J which can be used to proving that all probable P − K points are of the saddle type because the factor* $(-1)^j$ *has no effect for qualitative properties. Now, Assume that condition (3.191) holds, thus we have* $\lambda_1 < -1, 0 < \lambda_2 < 1$. *Let* w_1 *and* w_2 *be the eigenvectors corresponding to the eigenvalue* λ_1 *and* λ_2 *respectively. It is easy to verify that the eigenvalues of JH and their eigenvectors are given by:*

$$\left\{ \begin{array}{l} \mu_1 = \frac{a + \sqrt{a^2 - 4b}}{2} > 1, v_1 = \left(4b, a^2 - \sqrt{a^4 - 16b^2} \right) \\[2mm] 0 < \mu_2 = \frac{a - \sqrt{a^2 - 4b}}{2} < 1, v_2 = \left(a^2 - \sqrt{a^4 - 16b^2}, 4b \right) \end{array} \right. \tag{3.197}$$

(1) w_1 *belongs to the expansive region of JH, i.e.,*

$$JH\begin{pmatrix} 1 \\ 0 \end{pmatrix} = \frac{1}{\sqrt{a^2 + 4b}}\begin{pmatrix} -\lambda_1 & a \\ 2 & b \end{pmatrix} \tag{3.198}$$

Since we have:

$$\left\{ \begin{array}{l} \lambda_1 a^2 - (a^2 + 4b) > 2b\sqrt{a^2 + 4b} \ (a-1) > 0 \\[2mm] \| JHw_1 \| > \| w_1 \| . \end{array} \right. \tag{3.199}$$

(2) v_1 *belongs to the expansive region of J, i.e.,*

$$Jv_1 = \left(\begin{array}{c} 4\lambda_1 b \\ \lambda_2 \left(a - \sqrt{a^4 - 16b^2} \right) \end{array} \right) \tag{3.200}$$

Thus, we have:

$$|Jv_1| > |v_1| \tag{3.201}$$

Inequality (3.201) can be proved using the following formulas:

$$
\begin{cases}
\|Jv_1\|^2 - \|v_1\|^2 = c_0 = c_1 > c_2 > 0 \\[2mm]
c_0 = 16b^2\left(\lambda_1^2 - 1\right) - \left(1 - \lambda_2^2\right)\left(a^2 - \sqrt{a^4 - 16b^2}\right)^2 \\[2mm]
c_1 = 16b^2\left(\left(\lambda_1^2 - 1\right) - \frac{4b^2\left(1-\lambda_2^2\right)}{(\lambda_3 - \mu_2)^4}\right) \\[2mm]
c_2 = 16b^2\left(\left(\lambda_1^2 - 1\right) - \frac{4b^2\left(1+\lambda_2\right)}{(\lambda_3 - \mu_2)^3}\right)
\end{cases}
\tag{3.202}
$$

(3) Jv_1 belongs to the expansive region of JH. Indeed, it is sufficient to prove that $J\begin{pmatrix}1\\1\end{pmatrix}$ belongs to the expansive region of JH because the angle between v_1 and w_1 is less than $\dfrac{\pi}{4}$, the angle between $\begin{pmatrix}1\\1\end{pmatrix}$ and w_1. We have

$$
\begin{cases}
JH\left(J\begin{pmatrix}1\\1\end{pmatrix}\right) = \frac{1}{\sqrt{a^2+4b}}\begin{pmatrix} -a\lambda_1^2 & -2b \\ 2b\lambda_1 & +a\lambda_2^2 \end{pmatrix} \\[3mm]
\left\|JH\left(J\begin{pmatrix}1\\1\end{pmatrix}\right)\right\|^2 = \lambda_1^2 + \lambda_2^2.
\end{cases}
\tag{3.203}
$$

Since we have:

$$
\begin{cases}
\Xi_1 = \Xi_2 = \Xi_3 > 0 \\[2mm]
\Xi_1 = \left(a\lambda_1^2 + 2b\lambda_2\right)^2 + \left(2b\lambda_1 + a\lambda_2^2\right)^2 - \left(a^2 + 4b\right)\left(\lambda_1^2 + \lambda_2^2\right) \\[2mm]
\Xi_2 = a^2\left(\lambda_1^4 + \lambda_2^4\right) + \left(4b^2 - a^2 - 4b\right)\left(\lambda_1^2 + \lambda_2^2\right) + 4ab\lambda_1\lambda_2\left(\lambda_1 + \lambda_2\right) \\[2mm]
\Xi_3 = a^2\left[a^4 - \left(\lambda_1^2 + \lambda_2^2\right)\right] + 4b\left(\lambda_1^2 + \lambda_2^2\right)\left(a^2 - 1\right) + 2b^2\left[2\left(\lambda_1^2 + \lambda_2^2\right) - a^2\right] \\[2mm]
\left\|JH\left(J\begin{pmatrix}1\\1\end{pmatrix}\right)\right\| > \left\|J\begin{pmatrix}1\\1\end{pmatrix}\right\|.
\end{cases}
\tag{3.204}
$$

Let S be the region which takes v_1 and Jv_2 as its boundaries. By using Propositions 3.24(a), (b) and (c), it is easy to show that any vector in S becomes larger under the action of J and JH and still belongs to S which confirm that for any string of J and JH (also L and R), there is an eigenvector in S where the absolute value of whose eigenvalue is greater than 1. By considering the relation $\det(L) = \det(R) = -b$, the proof is hence complete.

3.7.2 Dual line mapping of Lozi mapping

The notion of *dual line mapping* was used in [Liu, et al. (1992(a))] to find the invariant manifolds of the fixed point A. Indeed, it was observed numerically that the unstable manifold of the hyperbolic fixed point resembles in its shape the strange attractor of the Lozi mapping (3.2). So, the dual line mapping of the Lozi map (3.2) was used to construct the stable and unstable manifolds of the hyperbolic fixed point A. Now, it is easy to show using the inverse mapping L_1^{-1} given by (3.4) that the portion of any straight line in the right half-plane is mapped into a half-line in the upper half-plane. Assume that the line coordinates are (w_1, w_2, w_3) which satisfies $w_1 x + w_2 y + w_3 = 0$ in the usual convention and let $\overrightarrow{W} = (w_1, w_2, w_3)^T$; then the right mapping R for W is $\overrightarrow{W} \to R\overrightarrow{W}$, where

$$R = \begin{pmatrix} 0 & \frac{1}{b} & 0 \\ 1 & \frac{a}{b} & 0 \\ 0 & \frac{-1}{b} & 1 \end{pmatrix} \text{ and } R^{-1} = \begin{pmatrix} -a & 1 & 0 \\ b & 0 & 0 \\ 1 & 0 & 1 \end{pmatrix} \tag{3.205}$$

The left mapping L for W is $\overrightarrow{W} \to L\overrightarrow{W}$, where

$$L = \begin{pmatrix} 0 & \frac{1}{b} & 0 \\ 1 & \frac{-a}{b} & 0 \\ 0 & \frac{-1}{b} & 1 \end{pmatrix} \text{ and } L^{-1} = \begin{pmatrix} a & 1 & 0 \\ b & 0 & 0 \\ 1 & 0 & 1 \end{pmatrix} \tag{3.206}$$

The eigenvalues of R (L) are λ_A^{-1}, 1 and Λ_A^{-1} (λ_B^{-1}, 1 and Λ_B^{-1}) such that:

$$|\lambda_A^{-1}| < 1 < |\Lambda_A^{-1}|, \ |\lambda_B^{-1}| < 1 < |\Lambda_B^{-1}| \tag{3.207}$$

The eigenvector for λ_A^{-1} is the unstable manifold segment u_A of the fixed saddle A in the first quadrant. The eigenvector corresponding to Λ_A^{-1} is just the stable manifold segment S_A of A. The three fixed lines for the mapping L are: the line at infinity (eigenvalue = 1), the stable segment S_B and the unstable segment u_B of the fixed saddle B in the third quadrant corresponding to the eigenvalues λ_B^{-1} and Λ_B^{-1} respectively. In this case, it is possible to show that every line segment tend to S_A as limit, with the exception of u_A under repeated inverse mappings because, any vector, other than the line at infinity and not collinear with S_A, tend to becoming collinear with u_A in the limit when applied repeatedly by the left multiplication of R. Similar conclusions can be given for the mapping L as follows: every line segment in the finite plane except S_B will tend to u_B under repeated L mappings. Also every line segment in the finite plane except u_B will tend to S_B under repeated inverse mappings of L. In the intercept notation,[21] it is easy to see that every straight line in the finite plane (except the x- and y-axes) can be represented by the pair of its intercepts (ξ, η). Indeed, the Lozi mapping (3.2) can be rewritten as follows:

$$\begin{cases} R : (\xi, \eta) \rightarrow \left(1 + b\eta, \dfrac{\xi(1+b\eta)}{b\eta + a\xi}\right), \\[4mm] L : (\xi, \eta) \rightarrow \left(1 + b\eta, \dfrac{\xi(1+b\eta)}{b\eta - a\xi}\right). \end{cases} \tag{3.208}$$

Expressions (3.203) are more convenient for numerical computation, and their inverses are gven by:

$$\begin{cases} R^{-1} : (\xi, \eta) \rightarrow \left(\dfrac{\eta(\xi-1)}{\xi - a\xi}, \dfrac{\xi-1}{b}\right), \\[4mm] L^{-1} : (\xi, \eta) \rightarrow \left(\dfrac{\eta(\xi-1)}{\xi + a\xi}, \dfrac{\xi-1}{b}\right). \end{cases} \tag{3.209}$$

In this case, the image of a line v, under the mapping R is denoted by vR. For example vRL is the image of the line vR under the mapping L. The above properties can be summarized as follows:

(a) Fixity property

$$\begin{cases} u_A R = u_{A'}\ u_A R^{-1} = u_{A'}\ u_B L = u_{B'}\ u_B L^{-1} = u_{B'} \\ S_A R = S_{A'}\ S_A R^{-1} = S_{A'}\ S_B L = S_{B'}\ S_B L^{-1} = S_B \end{cases} \tag{3.210}$$

(b) Asymptoticity property

$$\begin{cases} vR^{\infty} = u_A \text{ for } v \neq S_{A'}\ vR^{-\infty} = S_A \text{ for } v \neq u_{A'} \\ vR^{\infty} = u_B \text{ for } v \neq S_{B'}\ vR^{-\infty} = S_B \text{ for } v \neq u_{B'} \end{cases} \tag{3.211}$$

More properties of the periodic orbits for the Lozi maps (3.1) or (3.2) can be found in [Auerbach (1990)] where a scaling function for the eigenvalues of the unstable periodic orbits of strange sets embedded in two dimensions was introduced. This scaling function was obtained analytically for the Lozi map (3.2) as a convergent series in b. Note that this function converges only for uniformly hyperbolic systems (Recall Definition 14). This fact implies the inadequacy of this description for other systems.

---

[21] The conventional representation of a straight line is done by its x- and y-intercepts ξ and η.

3.7.3 Invariant manifolds of the fixed point A

The exact expressions of different segments of the invariant manifolds of A are found in [Liu, et al. (1992(a))] by using the following method:

(1) Start from $u_{A'}$ when the lower right meets the axis, replace it by the left image of u_A, i.e., its adjacent segment is $u_A L$.

(2) At the same time, replace the upper left of u_A by the right image of $u_A L$, i.e., the adjacent segment is $u_A LR$ which proceeds downward to the right until it meets the x-axis.

(3) The segment $u_A LR$ has two parts: the left portion and the right portion, because the junction point of $u_A LR$ with u_A is to the left of the y-axis. Thus, $u_A LRL$ and then $u_A LR^2$ are the line segments following $u_A L$.

(4) The segment $u_A L^2$ (the left image of $u_A L$) is the next following $u_A LR$. In this case, there is no left image of this $u_A L^2$ in the unstable manifold $u_{A'}$ and no sequences of symbols which begin with $u_A L^3$. Such a sequence is quite different for different values of a and/or b.

In the intercept notation, the values of these segments were computed in [Liu, et al. (1992(a))] and they were used to determine which combinations of the L's and R's do not appear in the sequence. In this case, the unstable manifold U_A (resp. the unstable manifold segment S_A) was expressed uniquely as a sequence of symbols of L's and R's.

Proposition 3.25 (a) For $a = \dfrac{8}{5}$ and $b = \dfrac{9}{15}$, the symbol for U_A is given by:

U_A : starting from the lower right of A,

$u_{A'}$, $u_A L$, $u_A LRL^2$, $u_A LRL^2$, $u_A L^2 R$, $u_A LRL^2 R$, $u_A LRLR^2$,
$u_A LRLRL$, $u_A LR^3 L$, $u_A LR^4$, $u_A LR^2 LR$, $u_A L^2 R^3$, $u_A LRL^2 R^3$,
$u_A LRL^2 RLR^3$, $u_A LRL^2 R^2 LR$, ...,

or starting from the upper left,

$u_{A'}$, $u_A LR$, $u_A L^2$, $u_A LRL^2$, $u_A LRLR$, $u_A LR^3$, $u_A LR^2 L$,
$u_A L^2 R^2$, $u_A LRL^2 R^2$, $u_A LRL^2 RL$, $u_A LRLR^2 L$,
$u_A LRLR^3$, $u_A LRLRLR$, ...

(b) For $a = \dfrac{8}{5}$ and $b = \dfrac{9}{15}$, the symbol S_A is given by: $S_{A'}$, $S_A L^{-1}$, $S_A L^{-2}$,
$S_A L^{-1} R^{-1}$, $S_A L^{-2} R^{-1}$, $S_A L^{-3}$, $S_A L^{-1} R^{-1} L^{-1}$,
$S_A L^{-1} R^{-2}$, $S_A L^{-2} R^{-2}$, $S_A L^{-2} R^{-1} L^{-1}$, $S_A L^{-4}$, $S_A L^{-3} R^{-1}$,
$S_A L^{-1} R^{-1} L^{-1} R^{-1}$, $S_A L^{-1} R^{-3}$, $S_A L^{-2} R^{-3}$, $S_A L^{-2} R^{-1} L^{-1} R^{-1}$,
$S_A L^{-4} R^{-1}$, $S_A L^{-3} R^{-1} L^{-1}$, $S_A L^{-3} R^{-2}$, $S_A L^{-1} R^{-1} L^{-1} R^{-1}$,
$S_A L^{-1} R^{-1} L^{-1} R^{-1} L^{-1}$, $S_A L^{-1} R^{-3} L^{-1}$, $S_A L^{-1} R^{-4}$, ...

For the structures of U_A and S_A: The points A and B are fixed saddles, C and D are $P-2$ saddles, h, k and g are homoclinic points, e and f are heteroclinic points. It is easy to see that U_A resembles the strange attractor of the Lozi map (3.2). Generally, the set $\underline{U}_A$ is the strange attractor.

As examples of some characteristics of the symbol sequences of U_A and S_A we give the following results:

Proposition 3.26 *(a) There exists a subsequence of S_A:*

$S_{A'}$ $S_A L^{-1}$, $S_A L^{-2}$, $S_A L^{-3}$, $S_A L^{-4}$, $S_A L^{-5}$, ..., *in which the limit of this subsequence exists and we have*

$$S_A L^{-\infty} = \lim_{n \to \infty} S_A L^{-n} = S_{B}. \tag{3.212}$$

The whole stable manifold S_B is in the limiting set of S_A and it includes S_B itself and inverse right image of S_B : S_B : $S_{B'}$, $S_B R^{-1}$, where

$$S_B R^{-1} = S_A L^{-\infty} R^{-1} = \lim_{n \to \infty} S_A L^{-n} R^{-1}. \tag{3.213}$$

This means that there exists a subsequence $\{S_A L^{-n} R^{-1}\}$ in the sequence of S_B.

(b) The unstable manifold U_A itself is in the limiting set of U_B and its sequence of segments is u_{B}, $u_B R$, $u_B R^2$, $u_B RL$, $u_B R^2 L$, $u_B R^3$, ..., with a sub-sequence $u_{B'}$, $u_B R$, $u_B R^2$, $u_B R^4$, ..., which converges to $u_B R^\infty = u_A$. In fact, any segment of $U_{A'}$ say $U_A LRL$, can be expressed as

$$u_A LRL = u_B R^\infty LRL = \lim_{n \to \infty} u_B R^n LRL. \tag{3.214}$$

(c) In the sequence of $U_{A'}$ there exist two subsequences:
(1) $u_{A'}$ u_A (LR), u_A (LR)^2 , u_A (LR)^3, ...
and
(2) $u_A L$, $u_A L$ (RL) , $u_A L$ (RL)^2, $u_A L$ (RL)^3 , ...,

where each of them tends to a limit corresponding to the first eigenvector. In the intercept notation, we have:

$$\begin{cases} u_A (LR)^\infty = \left(\dfrac{195 + 105\sqrt{7}}{696}, \dfrac{25 + 125\sqrt{7}}{696} \right) = \text{the starting segments } u_C \\[2ex] u_A L (LR)^\infty = \left(\dfrac{235 + 25\sqrt{7}}{232}, -\dfrac{4170 - 875\sqrt{7}}{2088} \right) = \text{the starting segments } u_D \end{cases} \tag{3.215}$$

The starting segments in (3.115) are those of the two $P-2$ saddle points C and D.

(d) Intersections of a stable manifold and an unstable manifold from two saddles of two different periods give heteroclinic points, e and g.

Proof 21 *(b) Remark that the unstable manifold U_B starts with the segment u_B and will remain in the third quadrant since its left portion is in the third quadrant. Also, its upper right portion meet the y-axis, thus behaving somewhat like u_A.*

(c) Remark that the limits of these sequences are governed by the eigenvalues of matrices LR and RL, given by: $-\dfrac{575+200\sqrt{7}}{81} = -13.6,$ *and* $\dfrac{575-200\sqrt{7}}{81} = 0.566.$

Also, the eigenvalues at the two periodic saddles C and D are just the reciprocals of the nontrivial eigenvalues of RL and LR, which gives the following equalities:

$$u_A (LR)^\infty = u_C, \; u_A L (RL)^\infty = u_D \qquad (3.216)$$

It is possible to show that upon iteration, the whole unstable manifolds U_C and U_D are located within the limiting set of U_A. Thus, the P − 2 points do not lie on the unstable manifold U_A, but on the closure of U_A. The reason is that the slopes of the segments of U_A are rational numbers while those of U_C and U_D are not. Similar method can be used to find other periodic subsequences in the U_A and conjecture that the set $\overline{U}_A$ contains the unstable manifolds of other P − K points.

3.7.4 Further properties of the Lozi strange attractor

In this section, we present further properties of the Lozi strange attractor displayed by the map (3.2). Indeed, Theorem 38 below was proved in [Tèl (1983(a-b))] and guarantees the existence of the trapping region. Also, the following result proved in [Liu, et al. (1992(a))] guarantees the existence of the strange attractor:

Theorem 38 *When (3.190) holds, then the Lozi mapping (3.2) has a strange attractor.*

*For the proof, if condition (3.190) hold, then Theorem 36 implies that all probable periodic points of the Lozi mapping (3.2) are of the saddle type. This fact implies that there are no stationary solutions and stable periodic solutions in the set Δ. Also, the first condition of (3.190) confirms the non existence of a **KAM** invariant curves in Δ. Thus, a strange attractor, must exist in Δ for the same parameters when it is observed numerically.*

In a second step, we will present some results related to the existence of transversal heteroclinic cycles:

Definition 47 *(a) Let f be a mapping. Supposing $p_1, \ldots, p_n$ are hyperbolic fixed points of f. If $W^u (p_i)$ and $W^s (p_{i+1})$ $(i = 1, 2, \ldots, n, p_{n+1} = p_1)$ intersect transversally, we say f possesses a transversal heteroclinic n-cycle.*
(b) Supposing $p_1, \ldots, p_n$ are hyperbolic k-periodic points of f. If $W^u (p_i)$ and $W^s (p_{i+1})$ $(i = 1, 2, \ldots, n, p_{n+1} = p_1)$ intersect transversally, we say f possesses a k-periodic transversal heteroclinic n-cycle.

(c) *Supposing $p_1, ..., p_n$ are hyperbolic periodic points of whose periods are different. If $W^u (p_i)$ and $W^s (p_{i+1})$ $(i = 1, 2, ..., n, p_{n+1} = p_1)$ intersect transversally, we say f possesses a different periodic transversal heteroclinic n-cycle.*

Thus, the following result was proved in [Liu, et al. (1992(a))] about transversal heteroclinic cycles for the Lozi mapping (3.2):

Theorem 39 *(a) Chaos in the sense of the Smale horseshoes can appear in any kind of transversal heteroclinic cycle.*

  (b) The closures of the unstable manifolds (stable manifolds) of those hyperbolic periodic points (or fixed points) that form various kinds of transversal heteroclinic cycle are identical.

Now, it is easy to show the following result:

Proposition 3.27 *(1) When condition (3.190) holds , the fixed point A, P −2 points $_2R_1L$, and P − 4 points $_1R_1R_1L_2R$ are interior points in Δ.*

  (2) These points are saddles.

  (3) The unstable manifolds of these points are also contained in the trapping region.

  (4) There are at least six invariant sets of the Lozi mapping in Δ.

  (5) The number of invariant sets of the above type contained in Δ is infinite.

  (6) The strange attractor is a whole structure containing the unstable manifolds of almost all the hyperbolic periodic points in Δ by way of transversal heteroclinic cycles.

For the proof, (1) and (2) can be obtained by tedious algebraic computations. (3) The result follows immediately from the definition of unstable manifold. (4) This is a result of (3). (5) See [Xie & Chen (No date)]. (6) This is a conjecture formulated from the fact that the Lozi map (3.2) has a one stable strange attractor in the trapping region obtained by numerical computation. Theorem 37(b) illustrates that this conclusion is correct. The same conclusion was also shown for the Hénon mapping (2.1) in [Liu & Cao (1991)].

With b being fixed, the Lozi map (3.2) displays some phenomena as follows:

(1) When $a > 1 + b$, some pieces of chaos appear in the region Δ .

(2) These pieces are enlarged until degeneracy occurs as a increases gradually.

(3) The number of pieces decreases and finally when a is large enough the chaotic region becomes one piece.

This phenomenon can be explained by using the theory of transversal heteroclinic cycle, i.e., the strange attractor consists of two pieces of chaos. First, S_A and U_A do not form any kind of transversal heteroclinic cycle with the invariant manifolds of any periodic points contained in the two pieces of chaos. Second, as soon as S_A and U_A can form any kind of transversal heteroclinic cycle with the invariant manifolds of some periodic points contained in the two pieces of chaos, the two pieces of chaos merge into one and the degeneracy phenomenon occurs. Various phenomena for the Lozi mapping (3.2) are as follows:

(a) W^u (B) does not belong to the strange attractor. In fact, W^u (B) and S_A intersect transversally, but W^s (B) does not intersect with the unstable manifolds of any periodic point contained in the strange attractor. This means that there are no transversal heteroclinic cycles including B, i.e.,

$$\overline{W^u (B)} \supset W^u (A), \overline{W^u (B)} \not\subset W^u (B). \qquad (3.217)$$

(b) Theorem 37(b) implies that the unstable manifolds of other $P - K$ points cannot be contained in $\overline{U}_A$ because various kinds of transversal heteroclinic cycle exists. (c) Let e be an intersection point between W^u (A) and W^s (D). It is easy to show that a different periodic 2-cycle exists, i.e., $\overline{W^u (A)} = W^u (D)$, that is an intersection points between S_A and W^u (D) exists. Numerically, the existence of points e and f is more important than that of points k and h. The reason is that they play an important role in the creation of a transversal heteroclinic cycle. (d) Sufficient condition for the existence of transversal homoclinic point of the fixed point A was given in [Tèl (1983(a-b))]. This condition implies that the existence of transversal homoclinic point does not guarantee the existence of the strange attractor in the sense of numerical computations. Hence, the following conjecture was formulated in [Liu, et al. (1992(a))]:

Conjecture 40 *The existence of a single Smale horseshoe does not imply the existence of a strange attractor.*

3.8 Geometric structure of strange attractors in the Lozi map

In [Cao & Liu (1998(a))] the structure of the strange attractors in the Lozi map (3.1) was further studied and all results showed in [Liu, et al. (1993), Liu & Cao (1991), Zhou & Liu (1988), Liu, et al. (1992(b))] were confirmed to be correct. Indeed, according to [Liu & Cao (1991), Liu, et al. (1993)], the strange attractors of a kind of two dimensional map may be W^u (p), (W^u (X) for the Lozi map (3.1)) where p is a hyperbolic fixed point of the map. A conjecture about several properties in the set W^u (p) was formulated:

Conjecture 41 *(1) The set of the periodic points forms a dense set in $\overline{W^u (p)}$.*
(2) All periodic points are hyperbolic.
(3) The transversal monoclinic points and heteroclinic points are dense in $\overline{W^u (p)}$.
(4) Any two hyperbolic periodic points form a transversal heteroclinic cycle.

(5) $\overline{W^u(p)}$ is the closure of the basin of the strange attractors.

The following result was proved in [Cao & Liu (1998(a))]:

Proposition 3.28 *If the parameters (a, b) satisfy Misiurewicz conditions (3.110) then the strange attractor possesses the following properties:*

(1) *The union of the transversal homoclinic points and weak transversal homoclinic points are dense in $\Lambda_{a,b}$.*
(2) *All periodic points are hyperbolic.*
(3) *The set of periodic points forms a dense set in $\Lambda_{a,b}$.*
(4) *Any two hyperbolic points forms a transversal heteroclinic cycles or a weak transversal heteroclinic cycle.*

Proof 24 *(1) If a segment contained in $\mathcal{L}(F)$ intersects x-axis and y-axis, then it intersects $\overline{XT} \subset W^s_{loc}(X)$ transversally. The fact that $\Lambda_{a,b} = \overline{W^u(X)}$ and $W^u(X)$ $=_{n\geq 0} \mathcal{L}^n\left(\overline{Z\mathcal{L}(Z)}\right)$, implies that it is sufficient to prove that a given segment $I \subset$ $Z\mathcal{L}(Z)$ contains a transversal homoclinic point or weak transversal homoclinic point. Assume for example that the segment I does not intersect y-axis. From Section 3.5.2, it follows that if I is contained in an unstable manifold of some point, then there exists n and $I_1 \subset \mathcal{L}^n(I)$ such that I_1 intersects x-axis and y-axis. This means that I_1 intersects $W^s_{loc}(X)$ and therefore I_1 contains a transversal homoclinic point. Thus, the segment I contains a transversal homoclinic point or a weak transversal homoclinic point.*

(2) *The proof was given in [Liu, et al. (1992)] as shown in Section 3.6.*
(3) *The proof is by using (1) and the construction of Smale horseshoes when the transversal homoclinic point appears, which implies the density of periodic points in $\Lambda_{a,b}$.*
(4) *The proof is equivalent to the fact that if $p_1, p_2 \in Per(\mathcal{L}) \cap \Lambda_{a,b}$, then $W^s_{loc}(p_j)$ intersects $W^s_{loc}(p_i)$, $i, j = 1, 2$ transversally or weak transversally. From [Liu & Cao (1991)], it is sufficient to prove that $\forall p_i \in Per(\mathcal{L}) \cap \Lambda_{a,b}$, $W^s(p_j)$ intersects $W^u(X)$ and $W^u(p_j)$ intersects $W^s(X)$ transversally or weak transversally. From (1) $W^u(p_j)$ intersects $W^s(X)$ transversally or weak transversally. So, the proof is complete if we show that $W^s(p_j)$ intersect $W^u(X)$. Indeed, assume $p_i \in$ the triangle SZX because the orbit of p_i inter-sects the triangle SZX. Now, given $I \subset W^s(p_j)$ and assume I does not intersect x-axis. Hence, $\mathcal{L}^{-1}(I)$ is a segment. Thus $\mathcal{L}^{-2}(I)$ contains one segment or two segments. If $\mathcal{L}^{-1}(I)$ does not intersect x-axis, then $\mathcal{L}^{-2}(I)$ is a segment or it consists of two segments and the common point is in y-axis and the length of one of them $\geq \lambda I$, where $\lambda = \frac{1}{2}\left(\frac{1}{c}\right)^2 > 1$. By iteration, we have a segment $I_n \subset \mathcal{L}^{-2n}(I)$ such that $|I_n| \geq \lambda^n I$. Thus, for sufficient large n, the segment I_n contains the point in the interior of SZX and the point outside of it. We have also $I_n \subset W^s(p_j)$, so I_n can not intersect segment SX. This implies that I_n intersects segment $SZ \subset W^u(X)$ or segment $ZX \subset W^u(X)$ transversally or weak transversally.*

For the case where $b < 0$, we know from Section 3.7 that the Lozi map (3.1) is conjugate to the map (3.162) and the previous steps of the proof can be reconsidered for $(a, b) \in [a_0, a_1] \times (-b'_0, 0)$. The following result was proved in [Cao & Liu (1998(a))]:

Proposition 3.29 *If $b < 0$ and the parameter (a, b) satisfy condition in [Cao & Liu (1998(a))], i.e., $(a, b) \in [a_0, a_1] \times (-b'_0, 0)$, then the strange attractor $\Lambda_{a,b}$ has the same properties as in Proposition. 3.28.*

Proof 25 (1) First, if a segment $F \subset G$ satisfies that both $\mathcal{L}'(F)$ and $(\mathcal{L}')^2(F)$ intersect y-axis and $\mathcal{L}'(F)$ is a segment, then there exists a segment $L' \subset F$ such that $(\mathcal{L}')^2(L')$ intersects $W^s(p_{a,b})^{22}$ transversally. Let $(0, y)$ be the intersection point of $\mathcal{L}'(F)$ with y-axis, it has $|y| \leq 28\,\pi_1(\mathcal{L}'(0, y)) = 1 - a\,|x| + y = 1 + y$. The application π_1 is defined in Section 3.5.1. The fact that $|b|$ is small and $(\mathcal{L}')^2$ (F) intersects y-axis at point $(0, y_1)$ and $\mathcal{L}'(F)$ is a segment or the union of two segments that intersect x-axis at point $(1 + y, 0)$ implies that $(1 + y, 0)$ is near $(1, 0)$, which means that the segment connecting the point $(0, y_1)$ and $(1 + y, 0)$ must intersect $W^s(p_{ab})$ transversally. Thus, there exists a segment $L' \subset F$ such that the segment $(\mathcal{L}')^2(L')$ intersects $W^s(p_{ab})$ transversally.

(2) The remaining of the proof follow that of Proposition 3.28.

For the basin of $\Lambda_{a,b}$, the following result was proved in [Cao & Liu (1998(a))]:

Proposition 3.30 *If the parameters (a, b) satisfy conditions (3.110) or [Cao & Liu (1998(a))], i.e., $(a, b) \in [a_0, a_1] \times (-b'_0, 0)$ as in Proposition. 3.22(e), then the basin $B(\Lambda_{a,b})$ of $\Lambda_{a,b}$ have the following properties:*

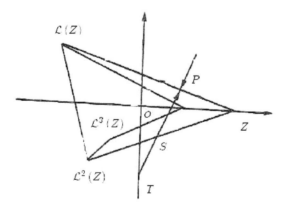

Figure 3.21: Schematic representation of the triangle F with some points of the Lozi map (3.1). Reused with permission from Cao and Liu, Communications in Nonlinear Sciences and Numerical Simulations. (1998). Copyright 1998, Elsevier Limited.

---

[22] The stable manifold of fixed point $p_{a,b}$ of the map (3.162).

(1) $\overline{W^s(X)} = \overline{B(\Lambda_{a,b})}$, *that is, the closure of the stable manifold of the hyperbolic fixed points X is the closure of the basin $B(\Lambda_{a,b})$ of $\Lambda_{a,b}$.*

(2) *When (a,b) satisfies the conditions (3.110) (as in [Misiurewicz (1980)]) the basin of $\Lambda_{a,b}$, is given by $B(\Lambda_{a,b}) = \bigcup_{n \geq 0} \mathcal{L}^n(D_1)$ where the domain D_1 is the triangle ZXS.*

(3) *When (a, b) satisfies the conditions in [Cao & Liu (1998(a))], i.e., $(a, b) \in [a_0, a_1] \times (-b'_0, 0)$ as in Proposition. 3.22(e), the basin of $\Lambda_{a,b}$, is given by $B(\Lambda_{a,b}) = \bigcup_{n \geq 0} \mathcal{L}^n(D_2)$ where the domain D_2 is the triangle ATS.*

Proof 26 *The proof of (2) and (3) are based on several geometric techniques.*

Specific structures of strange attractors for the Lozi mappings and other systems can be found in several works. For example, in [Auerbach, et al. (1988)] it was shown that the calculation of the $f(\alpha)$ function can be reduced to counting periodic orbits for multifractal and hyperbolic strange attractors. In [Morita, et al. (1988)] the dynamics of scaling properties of chaos was studied for invertible 2-D maps. The main issue here is the relation between the partial dimensions of strange attractors in the expanding and contracting directions. This relation was derived in terms of the *local expansion rates* of nearby orbits and making a criterion for studying chaos. An extension of Smale's transverse homoclinic theorem and chaotic phenomena were given for the Lozi maps in [Dong & Lin (1996)]. In [Cleveland (1999)] the rotation for attractors in the Lozi mapping were studied in a very detailed manner. For general properties of chaos in 2-D mappings the reader can see [Abu-Saris, et al. (2003), Anishchenko, et al. (1998-2000(a–b)-2002-2004), Belykh (1995)].

3.9 Parameter-shifted shadowing property of Lozi maps

In [Kiriki & Soma (2007)] certain shadowing properties were studied for the Lozi map (3.1) with the y-axis as their singularity set and strange attractors. In fact, it was proved that there exists a nonempty open subset O of the Misiurewicz domain $\mathcal{M}$ defined by (3.110) such that, for any $(a, b) \in O$, the Lozi map (3.1) with strange attractor $\Lambda_{a,b}$ defined by (3.127) has the *parameter-shifted shadowing property* taking into account that this map does not satisfy the hyperbolicity condition (Recall Definition 12(d)) on *the singularity set*, i.e., the set of nondifferentiable points. The problem of parameter-fixed shadowability is still open even for the Lozi map (3.1) and the Hénon map (2.1). It was known from [Coven, et al. (1988)] that one parameter family of maps with singularity sets, do not have the parameter-fixed shadowing property for many parameter values. An example of this situation for 1-D maps is the tent map (3.6) studied in [Coven, et al. (1988)]. It was proved that this map has the parameter-fixed shadowing property for almost every slope

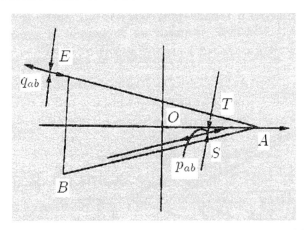

Figure 3.22: Schematic representation of the domains $D_{1,2}$ with some points of the Lozi map (3.1). Reused with permission from Cao and Liu, Communications in Nonlinear Sciences and Numerical Simulations (1998). Copyright 1998, Elsevier Limited.

a in the open interval $I = \left(\sqrt{2}, 2\right)$. But does not have this property for any a in a certain dense subset of I. It was also proved that, for any $a \in I$, the tent map (3.6) has the parameter-shifted shadowing property. As an example of 2-D maps that share the same property is the one studied in [Nusse & Yorke (1988)] by using the kneading theory. Since the Lozi map (3.1) is a natural two-dimensional generalization of tent maps (3.6) when b is sufficiently small, we presume that the Lozi map (3.1) does not have the parameter-fixed shadowing property for many parameter values. The shadowing property associated with parameter-shifting for one-parameter families of two-dimensional maps is defined in [Kiriki & Soma (2007)] by:

Definition 48 *Let $\{f_a\}_{a \in J}$ be a set of maps on $\mathbb{R}^2$, where J is an open interval in $\mathbb{R}$. For $\delta > 0$, the sequence $\{x_n\}_{n \geq 0}$ is called a δ-pseudo-orbit of f_a if*

$$\|f_a(x_n) - x_{n+1}\| \leq \delta \tag{3.218}$$

for any integer $n \geq 0$. For $a \in J$, we say that f_a has the parameter-shifted shadowing property (for short PSSP) if, for any $\varepsilon > 0$, there exist $\delta = \delta(a, \varepsilon)$ such that any δ-pseudo-orbit $\{x_n\}_{n \geq 0}$ of f_a can be ε-shadowed by an actual orbit of $f_{\tilde{a}}$, that is, there exists a $y \in \mathbb{R}^2$ such that:

$$\|f^n_{\tilde{a}}(x_n) - y\| \leq \varepsilon \tag{3.219}$$

for any $n \geq 0$.

If $\tilde{a} = a$, i.e., the shift of parameter value is not allowed, then the definition is the same as that of the original (parameter-fixed) shadowing property. Using Definition 48, the following result was proved in [Kiriki & Soma (2007)]:

Theorem 42 *There exists a nonempty open set O of the Misiurewicz domain $\mathcal{M}$ defined by (3.110) such that, for any $(a, b) \in O$, the Lozi map $\mathcal{L}$ given by (3.1) has the parameter-shifted shadowing property in a one-parameter family $(\mathcal{L}(\tilde{a}, b))_{\tilde{a} \in J}$ fixing b, where J is a small open interval containing a. Here $\mathcal{L}(\tilde{a}, b)$ is the Lozi map (3.1) with the bifurcation parameter $(\tilde{a}, b)$.*

On one hand, conditions of Theorem 42 about the open set O in $\mathcal{M}$ are more stringent than necessary. Indeed, the same result holds in a larger open subset of $\mathcal{M}$ and the set O can be represented concretely. The arguments (pairs of pseudo-orbit entries x_n and short segments l_n) used in the proof of Theorem 42 are not similar to those used to prove the parameter-fixed or shifted shadowing property for various maps. These arguments are suitable only for the Lozi map (3.1). On the other hand, suitable modifications of the proof makes its applicable to almost generalized piecewise hyperbolic maps given in [Hu & Young (1995)] and Section 3.4.1. The geometric Lorenz attractors studied in [Kiriki & Soma (2005)] share the same property proved by using such entrysegment pairs. The parameter-shifted stochastic stability (See Definition 50 below) for the Lozi map (3.1) was studied also in [Kiriki & Soma (2007)] for $(a, b) \in O$ (defined in Theorem 42) with respect to the Lebesgue transition probabilities defined by:

Definition 49 *For any $\delta > 0$, set*

$$D_\delta = \{y \in \mathbb{R}^2 : |y| \leq \delta\} \tag{3.220}$$

Then, the Lebesgue transition probability $p_{\delta,a,b}(. \mid x)$ is defined by:

$$p_{\delta,a,b}(A \mid x) = \frac{m^{(2)}\{y \in D_\delta, \mathcal{L}(x) + y \in A\}}{\pi \delta^2} \tag{3.221}$$

for any Borel set A in $\mathbb{R}^2$, where $m^{(2)}$ is the 2-dimensional Lebesgue measure.

Let G be the compact subset of $\mathbb{R}^2$ with $\mathcal{L}(G) \subset Int\ G$ as in Section 3.5.1, and let $\Lambda_{a,b}$ be the strange attractor of map (3.1) defined by (3.127). The following result was proved in [Kiriki & Soma (2007)]:

Theorem 43 *For any $(a, b) \in O$ and any strictly monotone increasing sequence $\{a_n\}_{n \geq 0}$ converging to a, let $\mu_{\delta_n, a_n, b}$ be $\mu_{\delta_n, a_n, b}$-invariant probability measures on G with $\delta_n = \dfrac{a - a_n}{500}$. Then any limit of $\{\mu_{\delta_n, a_n, b}\}$ is a Sinai-Ruelle-Bowen measure (Recall Definition 2) on $\Lambda_{a,b}$.*

At this end, the definition of *the parameter-shifted stochastic stability* is given in [Kiriki and Soma (2007)] by:

Definition 50 *The property of converging to the SRB-measure with parameter-shifting defined in Theorem 41 is called a parameter-shifted stochastic stability.*

For the subject of stochastic stability of Lozi mappings $\mathcal{L}_1$ given by (3.2), the reader can see [Stojanovska (1989)] where it was proved that this map acting on a compact trapping region in $\mathbb{R}^2$ and its Sinai-Bowen-Ruelle measure is stable under small random perturbations. This is a generalization to the piecewise hyperbolic maps case of the results given in [Kifer (1974–1986–1988), Young (1986)] for Axiom A attractors. Indeed, for $\delta > 0$, let Ω_δ be the set defined by:

$$\Omega_\delta = \{h : \mathbb{R}^2 \to \mathbb{R}^2 : h = \mathcal{L}_1 + \omega, \, |\omega| \leq \delta\} \tag{3.222}$$

Then the main result in [Stojanovska (1989)] is as follows:

Theorem 44 *Let R' be the trapping region of Lozi mappings $\mathcal{L}_1$ given by (3.2) and let $\Lambda_{a,b}$ be its attractor defined in (3.127). Let $\delta > 0$ be sufficiently small. For each $\varepsilon > 0$, let v_ε be the Borel probability measure on Ω_δ defined by (3.222) and let p_ε be invariant measure for the process defined by:*

$$p_\varepsilon(A\backslash X) = v_\varepsilon \, (g : gX \in A) \tag{3.223}$$

Then p_ε tends to the unique SBR measure on $\Lambda_{a,b}$ as $\varepsilon \to 0$.

A proof of Theorem 44 is based on the results in [Misiurewicz (1980), Young (1986), Collet & Levy (1984)], i.e., the main results in Section 3.5.

3.10 The basin of attraction of Lozi mapping

In [Baptista, et al. (2009)] the basin of attraction B of the Lozi map (3.1) was determined in the region of parameters bifurcations given by (3.110) where the strange attractor $\Lambda_{a,b}$ exists, i.e., it was proved that B is a region of the plane whose contraction relies on two points: The first is X_1 given by (3.224) below, which is the intersection of the stable manifold of the third quadrant fixed point with the y-axis and the second point is the first tangency point T_1 defined by (3.230) below. Note that the point Z defined in Proposition 3.17(g) has some importance when studying the set $\Lambda_{a,b}$ but it does not contribute to the description of the subset of points (of the plane) whose iterates go to $\Lambda_{a,b}$. Let X_1 be the intersection of the stable manifold of the fixed point Y given by (3.118) with the y-axis:

$$X_1 = \left(0, \frac{b\left(2 - a + \sqrt{a^2 + 4b}\right)}{\left(a - \sqrt{a^2 + 4b}\right)(a + b - 1)}\right) \tag{3.224}$$

We have the following definition:

Definition 51 *Given a point $P_1 \in \Lambda_{a,b}$ we call its itinerary the symbolic sequence $\pi(P_1) = \ldots \varepsilon_{-2}\varepsilon_{-1} * \varepsilon_0\varepsilon_1\varepsilon_2\ldots$, where the symbols ε_n are defined as follow:*

$$\varepsilon_n = \begin{cases} -1, & \text{if } \mathcal{L}^n \ (P_1)_x < 0 \\ *, & \text{if } \mathcal{L}^n \ (P_1)_x = 0 \\ +1, & \text{if } \mathcal{L}^n \ (P_1)_x > 0 \end{cases} \tag{3.225}$$

For $n \in \mathbb{Z}$, where * plays here the role of joker, i.e., * is both −1 and +1 and $\mathcal{L}^n$ $(P_1)_x$ denotes the first component of the point $\mathcal{L}^n \ (P_1)$.

3.10.1 The first tangency problem

The first tangency (a problem solved in [Ischii (1997(a))]) of the Lozi map (3.1) can be obtained using the point X_1 given by (3.224). See Fig. 2.5 for the geometric description of the first tangency (problem). We notice that the first tangency parameters give the boundary of the full shift region and therefore largest topological entropy (log2) region. Indeed, under certain conditions on the parameters, the tangencies are given by a certain point T_1 on the x-axis and its preimages. From the expression of the inverse map (3.3), we have that $\mathcal{L}^{-1} \ (P_1)$ is a point on the y-axis, therefore $\varepsilon_{-1} \ (T_1)$ is the joker * symbol. Assume now that condition holds, then the definition of the first tangency implies the following inequalities between the first components of the points T_1 and Z:

$$a > |b| + 1 \tag{3.226}$$

$$\begin{cases} 1 < Z_x < T_{1x} \\ \mathcal{L}^{-2} \ (T_1)_x = \dfrac{1}{b}\mathcal{L}^{-1} \ (T_1)_y = \dfrac{x-1}{b} > 0 \end{cases} \tag{3.227}$$

The second part of (3.227) implies that the corresponding symbol for $\mathcal{L}^{-2}$ (T_1) cannot be −1, as shown in [Ishii, (1997(a))], but +1, i.e.,

$$\pi \ (T_1) = \ldots + 1 + 1 + 1 \pm 1 * +1 - 1 - 1 - 1 \ldots \tag{3.228}$$

Thus, the algebraic curve corresponding to the boundary $\partial H'$ of the parameter region H', for which the dynamic of the Lozi map (3.1) on $\Lambda_{a,b}$ is equivalent to the full shift, is given by:

$$\frac{1}{1+\dfrac{a-\sqrt{a^2+4b}}{2}} - \frac{1}{\left(a+\dfrac{a-\sqrt{a^2+4b}}{2}\right)\left(1+\dfrac{a-\sqrt{a^2+4b}}{2}\right)} = 0 \tag{3.229}$$

For the points of $\partial H'$ satisfying the conditions (3.110), only the straight line is shown. On the other hand, the first tangency point T_1 can be computed from X_1 by considering the segment $X_1\mathcal{L}^{-2} \ (X_1)$ and concluding that the first tangency of the Lozi map (3.1) occurs at the point given by the intersection of $\overline{X_1 \mathcal{L}^{-2}(X_1)}$ with the x-axis. From (3.224), we have:

$$b = 4 - 2a \tag{3.230}$$

$$\begin{cases} T_1 = \left(\dfrac{b\left(2 - a + \sqrt{a^2 + 4b}\right)\left(a(b+1) + (1-b)\sqrt{a^2 + 4b}\right)}{\chi_1 \cdot \chi_2}, 0 \right) \\[4mm] \chi_1 = (1 - a - b)\left(a - \sqrt{a^2 + 4b}\right) \\[3mm] \chi_2 = \left(2b(b-1) + a^2(1 + 2b) + a(1 - 2b)\sqrt{a^2 + 4b}\right) \end{cases} \tag{3.231}$$

For parameter values (3.230) it is easy to see that both points Z and T_1 coincide. See Fig. 6.1 in [Ishii, (1997(a))] for more details.

3.10.2 Construction of the basin of attraction of the Lozi map

In this section, we present the method of constructing the basin of attraction of the strange attractor $\Lambda_{a,b}$ of the Lozi map (3.1) given in [Baptista, et al. (2009)]. Indeed, let B be the region bounded by a polygonal line S entirely characterized by the points T_1, X_1 defined above and their successive preimages. The polygonal line S can be described in two steps: For the first step, let S_1 be the polygonal line joining the four points $\mathcal{L}^{-1}(T_1)$, $\mathcal{L}^{-1}(X_1)$, X_1 and T_1. For the second step, let us consider two polygonal lines, one defined from the point $\mathcal{L}^{-1}(T_1)$ and the second one defined from the point T_1. Let us consider the set S_2 as the polygonal line joining the points: $\mathcal{L}^{-1}(T_1) \rightarrow \mathcal{L}^{-3}(X_1) \rightarrow \mathcal{L}^{-3}(T_1) \rightarrow \mathcal{L}^{-5}(X_1) \rightarrow \mathcal{L}^{-5}(T_1) \rightarrow \dots \rightarrow \mathcal{L}^{-(2n+1)}(X_1) \rightarrow \mathcal{L}^{-(2n+1)}(T_1)$ and let S_3 be the polygonal line joining the points: $T_1 \rightarrow \mathcal{L}^{-2}(X_1) \rightarrow \mathcal{L}^{-2}(T_1) \rightarrow \mathcal{L}^{-4}X_1 \rightarrow \mathcal{L}^{-4}(T_1) \rightarrow \dots \rightarrow \mathcal{L}^{-2n}(X_1) \rightarrow \mathcal{L}^{-2n}(T_1)$. We have that S_2 and S_3 do not intersect because they are on opposite sides of the stable manifold $W^s(X)$. Also, for any positive integer n, all the points $\mathcal{L}^{-(n+1)}(X_1)$ and $L^{-n}(T_1)$, belong to the first quadrant, because T_1 and $\mathcal{L}^{-2}(X_1)$ are points of the first quadrant above or belonging to the unstable manifold $W^u(X)$. Thus, the points corresponding to odd iterates of the inverse map given by (3.3) are on the left half-plane defined by $W^s(X)$ and the points corresponding to even iterates are on the right half-plane. The reason is that the larger eigenvalues in absolute value $\lambda_{X,2} = \dfrac{-a - \sqrt{a^2 + 4b}}{2}$ are negative. Therefore, the polygonal lines S_2 and S_3 do not intersect. This result follows from the fact that the first one being totally on the left half-plane and the second totally on the right half-plane since the distance from these points to $W^s(X)$ goes to zero, as $n \to \infty$. In this case, the polygonal line S can be seen as the *concatenation* of S_1, S_2 and S_3. These sets define a region of the plane denoted B. In fact, the region B is the set of points of the plane whose successive iterations by the Lozi map (3.1) tend to $\Lambda_{a,b}$, i.e., B is the basin of attraction of $\Lambda_{a,b}$. To prove this result, we must look to the successive forward iterations of its points, i.e., to show that some iterations of these points inevitably ends on the polyhedra $\mathcal{P}$, with vertices at the points $\mathcal{L}^{-1}(T_1)$, $\mathcal{L}^{-1}(X_1)$, X_1 and T_1 and prove that $\mathcal{L}(\mathcal{P}) \subset \mathcal{P}sl$. Let us consider the linear maps R and L defined in the second part of (3.192) as $R = \mathcal{L}|_{x \geq 0}$ and $L = \mathcal{L}|_{x \leq 0}$. Let R^{-1} be the inverse linear map of R, hence, it is easy to show that for points of the first quadrant, we have $R^{-1} = \mathcal{L}^{-1}$. Let Δ_n be the sequence of triangles defined by:

$$\Delta_n = \mathcal{L}^{-(n+2)}(T_1)\, \mathcal{L}^{-(n+2)}(X_1)\, \mathcal{L}^{-n}(T_1),\ n = 0, 1, 2, \dots \qquad (3.232)$$

We have that $R^{n+1}(n) = R(\Delta_0)$ is the triangle with vertices at the points $\mathcal{L}{-1}(T_1)\mathcal{L}^{-1}(X_1)\mathcal{L}(T_1)$ because $R(\Delta_n) = \Delta_{n-1}$. This result implies that for any non-negative integer n, the inclusion $\mathcal{L}^{n+1}(\Delta_n) \subset \mathcal{P}$ holds. Note that for $(a, b) \in \partial H'$, the point T_1 and its successive backward iterations belong to the unstable manifold $W^u(X)$. This means that the only points of B outside the polyhedral $\mathcal{P}$, are those of Δ_n. On the other hand, let us consider the triangles defined by:

$$\nabla_n = \mathcal{L}^{-n}(T_1)\, \mathcal{L}^{-(n+2)}(T_1)\, X,\ n = 0, 1, 2, \dots \qquad (3.233)$$

Triangles ∇_n are dynamically related, i.e.,

$$R(\nabla_n) = \nabla_{n-1} \qquad (3.234)$$

and they have a common vertex, the fixed point X. These triangles are used to examine the rest of the points of the region B that are not in the polyhedra $\mathcal{P}\,sl$. Relation (3.234) implies that for all positive integer n, the set $R^{n+1}(\nabla_n) = R(\nabla_0)$ is the triangle with vertices at $\mathcal{L}^{-1}(T_1)$, $\mathcal{L}(T_1)$ and X. Since T_1 is a point on the line segment $X_1\, \mathcal{L}^{-2}(X_1)$, we conclude that $\mathcal{L}(T_1) \in \mathcal{L}^{-1}(X_1)\, L(X_1)$. This means that $\mathcal{L}^n(\nabla_n) \subset \mathcal{P}sl$. Finally, we split $\mathcal{P}$ into two triangles, the first one is the left half-plane triangle $\mathcal{L}^{-1}(T_1)\mathcal{L}^{-1}(X_1)X_1$ and the second one is the right half-plane triangle $\mathcal{L}^{-1}(T_1)\, X_1 T_1$. The image of the first one by the linear map L is the triangle with vertices at T_1, X_1, and $\mathcal{L}(X_1)$. The image of the second one by the linear map R is the triangle defined by the points T_1, $\mathcal{L}(X_1)$, and $\mathcal{L}(T_1)$, where the point $\mathcal{L}(X_1)$ is the intersection of the line segment $\mathcal{L}^{-1}(X_1)\, \mathcal{L}(X_1)$ with the x-axis. To this end, it suffices to prove that $\mathcal{L}(B) \subset B$ to conclude that $\mathcal{L}(\mathcal{P}) \subset \mathcal{P}$. Indeed, let $P_1 \notin B$, then its iterates $\mathcal{L}^n(P_1)$ diverge as

$n \to \infty$. To see this property, let us consider different regions, corresponding to different behavior of their points by the Lozi map (3.1).

(1) If P_1 is on the left half-plane below $W^s (Y)$, then $\mathcal{L}^n (P_1)$ diverges, as $n \to \infty$. (2) If P_1 is on any right half-plane point below the line defined by X_1 and $\mathcal{L}^{-2} (X_1)$, then $\mathcal{L}^n (P_1)$ diverges, as $n \to \infty$. The reason is that the image of this point by the linear map R is on the second quadrant triangular sector below $W^s (Y)$. (3) If P_1 is on a second quadrant point above the line defined by $\mathcal{L}^{-1} (X_1)$ and $\mathcal{L}^{-3} (X_1)$, then $\mathcal{L}^n (P_1)$ diverges, as $n \to \infty$. The reason is that the image of this point by the linear map L is a point on either the third or fourth quadrant below the line defined by X_1 and T_1, i.e., a point on the cases (1) and (2). (4) If P_1 is on the first quadrant point above the line defined by $\mathcal{L}^{-1} (T_1)$ and $\mathcal{L}^{-3} (X_1)$, then $\mathcal{L}^n (P_1)$ diverges, as $n \to \infty$. The reason is that the image of this point R is on the first quadrant triangular sector below the line defined by T_1 and $\mathcal{L}^{-2} (X_1)$, whose forward iteration was already studied in case (2). (5) If P_1 is on the first quadrant triangular sectors $\angle_n$ with vertex at the point $\mathcal{L}^{-n} (T_1)$ and defined by the line passing by $\mathcal{L}^{-n} (X_1)$ and $\mathcal{L}^{-(n+2)} (X_1)$, for any $n > 1$, then $\mathcal{L}^n (P_1)$ diverges, as $n \to \infty$. The reason is that $R(\angle_n) = \angle_{n-1}$ and $R^{n-1} (\angle_n) = R(\angle_2)$, i.e., for any point, $P_1 \in \angle_n$, the point $\mathcal{L}_n (P_1)$ belongs to the triangular sector with vertex $\mathcal{L}^{-1} (T_1)$ and defined by the line passing by $\mathcal{L}^{-1} (X_1)$ and $\mathcal{L}^{-3} (X_1)$, whose successive iteration were already studied in cases (3) and (4). Finally, based on the previous construction, the following theorem was proved in [Baptista, et al. (2009)]:

Theorem 45 *Given a Lozi map (3.1) for parameter values satisfying the condition (3.110), the region B is the basin of attraction of $\Lambda_{a,b}$.*

3.11 Relations between the forwards limit sets and the strange attractor

In this section, we will discuss the possible relations between the *forwards limit sets* of the Lozi map (3.1) with its strange attractor proved to be exist in Section 3.5. Indeed, it was proved in [Kiriki (2004)] that the Lozi family (3.1) admits an open set $\mathcal{O}$ in the parameter space such that, for almost every (a, b) in $\mathcal{O}$ the forwards limit set of a point in the y-axis coincides with the strange attractor $\Lambda_{a,b}$ defined by (3.127). In other words, let $\mathcal{Y}_{a,b}$ be the segment of the y-axis in the set F defined in Section 3.5 and consider the forward orbits of the singularities in $\mathcal{Y}_{a,b}$. Their ω-limit sets coincide with $\Lambda_{a,b}$ for almost every (a, b) in some parameter region $\mathcal{O}$, i.e., the following result:

Theorem 46 *There exists an open set $\mathcal{O} \subset \mathcal{M}$ given by (3.110) whose closure contain $(2, 0)$ such that, for Lebesgue almost every*

$$((a, b), z) \in \{ ((a, b), z) : (a, b) \in \mathcal{O}, z \in \mathcal{Y}_{a,b} \} \tag{3.235}$$

the ω-limit set $\omega(z, \mathcal{L})$ coincides with the Lozi attractor $\Lambda_{a,b}$.

Theorem 46 is an extension to the 2-dimensional context of the results (See Theorem 47(c) below) about turning orbits in the dynamical core of tent maps (3.6) on $\mathbb{R}$ proved in [Brucks &Misiurewicz (1996)]. The proof of Theorem 42 is based on two notions: the first one is the *estimations of parameter dependence* and the second one is *usefulness* and *maturity* for parameter intervals. This last concept was defined for the tent maps (3.6) in [Brucks & Misiurewicz (1996)] and it was extended to the Lozi maps in [Kiriki (2004)]. As we note, the Lozi map (3.1) is equivalent to the tent maps (3.6) when $b = 0$. The following results were proved for the tent maps (3.6):

Theorem 47 *(a) The tent maps (3.6) have a turning point $x = 0$ whose for-ward orbit can not escape from it's dynamical core defined by:*

$$\Lambda(T) = [T^2 (0), T (0)] \text{ for } 1 < a < 2. \tag{3.236}$$

(b) *There exists a G_δ-dense subset of $a \in \left[\sqrt{2}, 2 \right]$ such that the forward orbit of $x = 0$ is dense in $\Lambda(T)$. [Brucks, et al. (1991)].*

(c) *For almost every $a \in \left[\sqrt{2}, 2 \right]$ we have $\omega (0, T) = \Lambda (T)$. [Brucks & Misiurewicz (1996)].*

(d) *The complement of the set $a \in \left[\sqrt{2}, 2 \right]$ is σ-porous. [Brucks & Buc-zolich (2000)] .*

(e) *For almost every parameter value, the turning orbit is typical for an absolutely continuous invariant probability measure. [Bruin (1998), Brucks & Bruin (2004)].*

The corresponding results of Theorem 47 for the Lozi map (3.1) are not known.

3.12 Topological entropy of the Lozi maps

In this section, we will discuss the topological entropy of the Lozi maps. The most important result in this topic is the monotonicity of this entropy in the vertical direction around $a = 2$ and in some other directions for $1 < a \leq 2$ as shown in [Yildiz, (2011)]. See Fig. 3.23.

The connection of topological entropy to the pruning idea can be seen from the fact that the pruned region in symbolic space decreases entropy increases. Also, we note that Lap number entropy formula of Ishii and Sands given in [Ishii and Duncan (2007)] can be applied to Lozi maps (3.1) or (3.2) as well as any piecewise-affine homeomorphism of the plane. In this case, the topological entropy of a piecewise projective homeomorphism of the 2-sphere is given by the growth rate of the number of projective pieces under iteration. For higher dimensions and for non-invertible maps only an inequality was obtained. See also the valuable work of Buzzi in [Buzzi, (2009)]. Buzzi's results can not be directly applied to Lozi maps (3.1) or (3.2) and therefore lower semi continuity of topological entropy is still open but upper semi continuity does not hold by the results of Yildiz [Yildiz, (2012)]. It was shown in this work that the topological entropy for Lozi maps (3.1) can jump from zero to a value above 0.1203 as one crosses a particular parameter and this property still holds in a small neighborhood of this parameter along a line segment. Thus, it is not upper semi-continuous in general.

Also, in [Yildiz, (2011)], some rigorous and numerical results for the parameters at which the Lozi family (3.1) has zero entropy, i.e., for $a = 1$ and $b = 0.5$. See Figs. 3.24 and 3.25. We note that when the values of a and

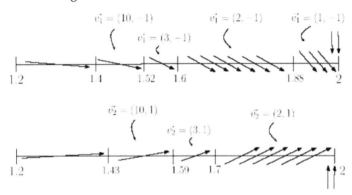

Figure 3.23: The topological entropy is non-decreasing in the direction of arrows. Reused with permission from Yildiz, Nonlinearity. Copyright 2011, IOP Publishing.

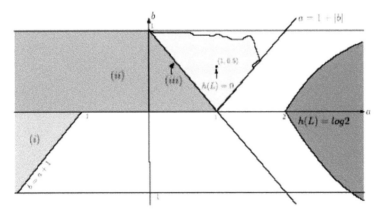

Figure 3.24: The sets (i), (ii) and (iii) where the entropy is zero. Also, the light grey region with complicated boundary has zero entropy. Reused with permission from Yildiz, Nonlinearity. Copyright 2011, IOP Publishing.

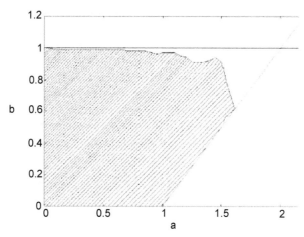

Figure 3.25: The shaded region gives the numerically observed parameters for which the entropy of the Lozi mapping (3.1) is zero. Reused with permission from Yildiz, Nonlinearity. Copyright 2011, IOP Publishing.

b move away from a neighborhood of $(a, b) = (1, 0.5)$ a homoclinic point is created and the entropy become positive. This is a result of the fact that the unstable manifold of the right fixed point intersects with the stable manifold of the same fixed point. In this case the boundary of the zero entropy locus is expected to be piecewise algebraic.

4

Dynamical Properties of Modified and Generalized Lozi Mappings

In this chapter, we discuss the most interesting results concerned with maps obtained from the idea of generalizing the Lozi maps (3.1) or (3.2) to different forms with different purposes. Some simple piecewise linear models for the zones of instability are presented in Section 4.1. In Section 4.2, we present a special case of the Lozi maps where it is possible to calculate rigorously their fractal dimensions using the usual covering. The 3-D piecewise linear noodle map is presented in Section 4.3. This map can display additional phenomena expected from four-dimensional flows and not possible in two-dimensional maps, i.e., hyperchaos phenomenon[1]. A class of systems called *generalized hyperbolic attractors* is discussed in Section 4.4. This class includes the 2-D hyperbolic attractors of Belykh and Lozi mappings (3.1) or (3.2) and the Lorenz attractor given by (1.49) and have further ergodic properties such as Smale spectral decomposition and some important statistical properties. In Section 4.5, the global periodicity property of some generalized Lozi mappings is discussed. The *generalized discrete Halanay inequality* and the global stability are presented in Section 4.6. Some global behaviors of some (max) difference equations are discussed in Section 4.7. We note that in Sections 4.5, 4.6 and 4.7, each difference equation can be reduced to a special case of the Lozi mapping. In Section 4.8, a class of piecewise-linear area-preserving plane maps is presented and discussed. The main issue here is to look to the properties of *rotation number* and the

---

[1] If more than two Lyapunov exponents are positive, then the resulting attractor is called *hyperchaotic*. The importance of these attractors is that are more non-regular, and the iteration points are seemingly *almost* full of the considered space, which explains one of applications of chaos in fluid mixing.

associated circle map. Several analytical results are obtained and discussed. In Section 4.9, a smooth version of the Lozi mappings (3.1) is presented in order to make a comparison with piecewise dynamics. In Section 4.10, a map with *border-collision period doubling scenario* is introduced. This route to chaos is relatively new in the literature and usually not taken for the Lozi map (3.1). In Section 4.11, a rigorous proof of chaos in a 2-D piecewise linear map is based on the equivalence of the matrices defining its formula. Since, only one route to chaos was observed for the Lozi map (3.1), we present in Section 4.12 a 2-D map with two different routes to chaos, namely, period doubling and border collision bifurcations. Section 4.13 is devoted to the generation of multifold chaotic attractors, this including C^1-multifold chaotic attractors in Section 4.13.1 and C^∞-multifold chaotic attractors in Section 4.13.2. In Section 4.14, a simple 2-D piecewise linear map is presented. This map differs from the Lozi map in that it has a much wider variety of attractors. Another 2-D piecewise linear map called the *discrete hyperchaotic double scroll* is presented in Section 4.15. This map is obtained from the Lozi map (3.2) using the characteristic function of the Chua circuit (1.50)-(1.51). Since all the generalized Lozi mappings are piecewise linear, we present in Section 4.16, a short description of the theory of 2-D piecewise smooth maps along with the non-invertible piecewise smooth case. Some relevant results about the normal form and the occurrence of *robust chaos* are also discussed in some detail. Finally, we note that notations in this chapter are not unified in the whole text because the subject of each section is different from the others. So the unification of notations is done only for each section of this chapter.

4.1 Simple piecewise linear models for the zones of instability

In this section, we present two examples of piecewise linear maps that serve as models for the zones of instability. The idea for these maps comes from similar results in the smooth case studied in [Moser (1973)]. Indeed, in [Devaney (1984(a))] the global behavior of the piecewise linear area-preserving transformation of the plane given by:

$$G(x, y) = \begin{pmatrix} 1 - y + |x| \\ x \end{pmatrix} \tag{4.1}$$

was studied. For this map, there are infinitely many invariant polygons surrounding an elliptic fixed point of map (3.1). The regions between these polygons are called *annular zones*, and they serve as models for the zones of instability in the corresponding smooth case [Moser (1973)]. The annular zones contain only finitely many elliptic islands and the map (3.1) exhibits stochastic behavior on the complement of these islands because it is hyperbolic (almost uniformly hyperbolic) with the property that unstable periodic points are dense in this region.

Some of the properties of map (4.1) are listed here:

(1) G is piecewise linear and continuous, and each of its two pieces is area preserving.

(2) Its elliptic fixed point $(1, 1)$ satisfies the hypotheses of the *Moser Twist Theorem* [Moser (1973)] and it is surrounded by infinitely many invariant polygons of arbitrarily large radius.

(3) G is ergodic (See Section 1.1.6) on the remaining region of positive measure.

(4) G is conjugate to its linear part given by the matrix $A = \begin{pmatrix} 1 & -1 \\ 1 & 0 \end{pmatrix}$ on a neighborhood of $(1, 1)$ and we have $A^6 = I_2$, it follows that $G^6 = Id$.

Now, let H be the maximal region about $(1, 1)$ on which $G^6 = Id$, i.e., the hexagon with vertices at $(0, 0)$, $(1, 0)$, $(2, 1)$, $(2, 2)$, $(1, 2)$ and $(0, 1)$ shown in Fig. 4.1. Thus, all points in H except the fixed point $(1, 1)$ have period 6.

The main property of map (4.1) is as follow: *Under iteration of G, the orbits of points in H lie in the right halfplane and all other orbits visit both half planes.* This means that these points are acted upon by both linear maps which comprise G. Thus, chaotic behaviors are possible for map (4.1). Indeed, Fig. 4.1 shows the orbits of several points in the plane. Each orbit is constrained in an annular region between two polygons and avoid hexagonal regions within this annulus. Also, these orbits are dense in the annulus minus the hexagonal islands as shown in Fig. 4.2.

The following result was proved in [Devaney (1984(a))] about some of the above properties of map (4.1). Namely, the existence of infinitely many invariant annuli surrounding H and the boundedness of periodic points of G in the plane.

Proposition 4.1 *(a) For each integer $k \geq 1$, there is an G-invariant polygon α_k of period $27k - 3$ passing through $(4k, 4k)$.*

(b) For each integer $l \geq 0$, there is an G-invariant polygon β_l of period $27l + 6$ passing $(4l + 2, 4l + 2)$.

(c) Periodic points of G are dense in the plane.

In what follows, we discuss the structure of ergodic regions inside the zones of instability of the map (4.1). For this purpose, we need to define some regions as follow: Let A_i and B_i be the annular regions that served as models for the zones of instability of map (4.1). Hence, there are a finite number of elliptic islands in each together with an *ergodic sea*. The dynamics of map (4.1) in these regions for $i \geq 1$ is essentially the same. This property was proved in [Devaney (1984(a))] using the notion of a *first return map* defined by:

Definition 52 *Let B_i be a fixed annular zone with $i \geq 1$. Let Γ_i be the region in B_i between $y = \frac{1}{2}x$ and $y = \frac{1}{2}x + 2$. Thus, Γ_i consists of two strips, h_i and e_i, separated by the line $y = \frac{1}{2}x + 1$. Let $\Phi_i: \Gamma_i \to \Gamma_i$ be an application defined by $\Phi_i(p) = G^k(p)$ where k is the least positive integer for which $G^k(p)$ belongs to Γ_i. Hence, Φ_i is the first return map on Γ_i.*

It was shown in [Devaney (1984(a))] that all Φ_i are conjugate. Thus, let $\Phi_i = \Phi^2$ and $\Gamma_i = \Gamma$. Let e_0 be the triangle in e bounded above by $y = 6$. Let h_0 be the triangle in h bounded below by $y = 2x + 14$. Let $e_1 = e - e_0$ and $h_1 = h - h_0$. Thus, the following result was proved in [Devaney (1984(a))]:

Proposition 4.2 *Let $(x, y) \in \Gamma$.*
  (1) *If $(x, y) \in e_0$, then*
  $\Phi(x, y) = G^5(x, y) = (4 + 2x - 3y, 1 + x - y)$,
  (2) *if $(x, y) \in h_0$, then*
  $\Phi(x, y) = G^{23}(x, y) = (y + 8, -3 - x + 3y)$,
  (3) *if $(x, y) \in e_1$, then*
  $\Phi(x, y) = G^{14}(x, y) = (8 + 2x - 3y, 1 + x - y)$,
  (4) *if $(x, y) \in h_1$, then*
  $\Phi(x, y) = G^{14}(x, y) = (y + 4, -3 - x + 3y)$.

The proof of Proposition. 4.2 is based on straightforward computations. The elliptic point of Φ has a G-period 14 since it lies in e_1. Let U be the hexagonal island with vertices at $(11, 6)$, $(12, 6)$, $(14, 7)$, $(15, 8)$, $(14, 8)$ and $(12, 7)$, then the elliptic point of Φ is surrounded by U called *hexagonal island of stability*.

The following properties of map (4.1) can be seen from the above analysis:

1. The stable and unstable manifolds of Φ have points of transverse intersection.
2. G is ergodic in the annular zones B_i minus the G-images of the hexagonal island U. This done by showing that Φ is hyperbolic on $\Gamma-U$.
3. The restriction of map (4.1) to the invariant polygons is periodic.

---

[2] Note that Φ is a discontinuous piecewise linear map.

4. There are only finitely many elliptic islands in the zones B_{μ} and the map has a simple linear structure nearby.
5. It is possible to compute rigorously the stable and unstable manifolds of the map (4.1) as in [Devaney (1984(b))].

Now, in [Aharonov, et al. (1997)] a second example of simple piecewise linear models for the zones of instability was studied. In particular, the *island structure* and the chaotic behaviors. This model is given by:

$$F(x, y) = \begin{pmatrix} 1 - y - |x| \\ x \end{pmatrix} \tag{4.2}$$

Also, the portion of the dynamics that persists when the map (4.2) is approximated by the real analytic map of the form:

$$F_{\varepsilon}(x, y) = \begin{pmatrix} 1 - y - f_{\varepsilon}(x) \\ x \end{pmatrix} \tag{4.3}$$

was studied, where $f_{\varepsilon}(x)$ is a real analytic function close to $|x|$ for small values of f_{ε}. For the case of map (4.2), the evolution of every orbit was described explicitly and completely. The analysis of the map (4.3) is simpler than map (4.2), in which calculations are considerably shorter to obtain a more refined and complete analysis. An orbit of a single point under the iteration of F is shown in Fig. 4.3. An orbit of the same point for map (4.3) with $f_{\varepsilon}(x) = \sqrt{x^2 + \varepsilon}$, $\varepsilon = 0.01$, along the orbits of two points in the hexagonal islands are shown in Fig. 4.4.

The main remark here is that, if ε is sufficiently small, then the islands of F seem to persist for F_{ε} and that orbits of points in these islands seem to be invariant circles. This assertion was proved in [Aharonov, et al. (1997)]:

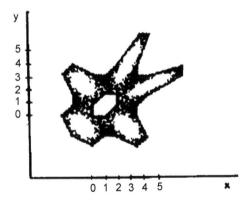

Figure 4.1: The ergodic sea is an invariant gingerbreadman: iterates of a single point under the mapping (4.1) 10, 000 iterates of a single point in the ergodic region B_0. The central hexagonal region is the region h surrounding the fixed point. Reused with permission from Devaney, Physica D, (1984). Copyright 1984, Elsevier Limited.

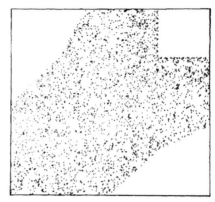

Figure 4.2: A blow-up of a portion of Fig. 4.1. Reused with permission from Devaney, Physica D, (1984). Copyright 1984, Elsevier Limited.

Theorem 48 *(a) In the complement of the hexagonal islands (these islands are periodic under F and that F^{13} is a linear rotation of period 6 on each island) in the polygonal annuli (bounded between two closed invariant polygons), F is non uniformly hyperbolic.*

(b) Consider the area preserving map (4.3) with $f_\varepsilon(x) = g^{-1}(g(x) + \varepsilon)$, $\varepsilon > 0$, and $g : (-\infty, \infty) \to [0, \infty)$ is a smooth even function, $g(0) = 0$ and g^{-1} is the branch which maps $[0,\infty)$ to $[0,\infty)$. (1) If $g', g'' > 0$ on $(0,\infty)$, where g', g'' are the derivatives of g, then for sufficiently small values of ε, F_ε has infinitely many closed invariant curves inside the triangle elliptic island $\Delta = \{x \geq 0, y \geq 0, 1 - x - y \geq 0\}$.

(2) If $\frac{1}{g'} > 0$, $\left(\frac{1}{g'}\right)' < 0$, $\left(\frac{1}{g'}\right)'' > 0$, then the same holds for all 13 hexagonal elliptic island in the annulus F_1 (when $\varepsilon = 0$).

The chaotic regions described in Theorem 44(a) are called *Birkhoff instability rings* and they are shown in Figs. 4.3, 4.5, 4.6.

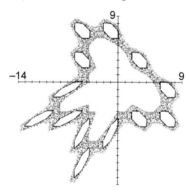

Figure 4.3: Chaotic region in annulus F_1 of the map (4.2).

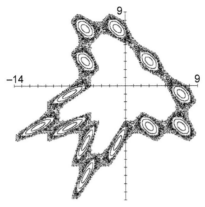

Figure 4.4: Invariant curves in the islands of F_1 of the map (3.3) with $f_\varepsilon(x) = \sqrt{x^2 + 0.01}$.

4.2 Rigorous calculation of fractal dimension of a strange attractor

In [Tél (1983(b))] the fractal dimension of the strange attractor of the following map:

$$x' = ax - sgn\,(x) + bz, z' = x. \qquad (4.4)$$

was calculated rigorously by using the construction of the unstable manifolds of period-two points. The symbol $sgn(x)$ denotes the sign of x. In this case, map (4.4) displays a strange attractor consists of parallel straight

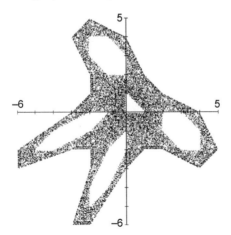

Figure 4.5: Iterates of $(1.01, 0)$ of the map (4.2).

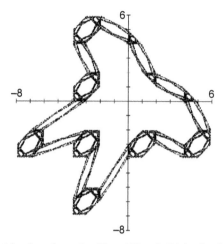

Figure 4.6: Big elliptic islands and two families of 38 and 47 islands of the map (4.2).

line segments only as shown in Fig. 4.7(a). This property is a result of the equations describing the invariant curves of periodic points of map (4.4) given in [Tél (1983(a))] as follows:

$$f^*_{i+1}(z) = f(z) + b \, (f^*_i)^{-1} \, (z), \, i = 1, \, ..., \, n \tag{4.5}$$

Eq. (4.5) indicates that the invariant curve $x = f^*_i(z)$ around the period n point G_i is mapped onto $x = f^*_{i+1}(z)$, the invariant curve around G_{i+1}, the next element of the n-cycle. For the map (4.4), the fixed points are $H_{1(2)}$: $\left(x^* = \pm \, (a+b-1)^{-1}, x^* \right)$. These points are located outside the strange attractor as shown in Fig. 4.7(b). Also, the period-two points for the map (4.4) are $F_{1(2)}$: $\left(x_\pm = \pm \, (a-b+1)^{-1}, - \, x_\pm \right)$ and they are located inside the strange attractor. In this case, the branch of the invariant manifold going through F1(2) has the form:

$$f^*_{1(2)} \, (z) : x = x_\pm + \lambda \, (z + x_\pm) \tag{4.6}$$

where the upper index belongs to f^*_1, from (4.5) the values of $\lambda^u(s)$ are:

$$\lambda^u(s) = \frac{\left[a \pm (a^2 + 4b)^{\frac{1}{2}} \right]}{2} \tag{4.7}$$

From (4.7), if $a > 1 - b$, then $\lambda^u > 1$ and $|\lambda^s| < 1$. These results imply that the branch with λ^u is located in the unstable manifold of $F_1(2)$ and the one with λ^s belongs to the stable one. The inverse of the first branch has a slope $\dfrac{1}{\lambda}$, and Eq. (4.5) generates new straight line segments with a slope $a + \dfrac{b}{\lambda}$, and the fact that $\lambda = a + \dfrac{b}{\lambda}$ implies that the new branches run parallel to the first one.

This result implies that the unstable (stable) manifolds of the fixed points and other periodic points consist of straight line segments with slope λ^u (λ^s) only. Note that the strange attractor of map (4.4) is related to the unstable manifolds of the period-two points $F_{1(2)}$ as shown in Fig. 4.7(b). It is clear to see a few branches of $W^u (F_1)$. Now, the following result was proved in [Tél (1983(b))] for the expression of the fractal dimension of map (4.4):

Theorem 49 *The fractal dimension of the map (4.4) is given by:*

$$d = \lim_{n \to \infty} \frac{\ln N_n}{\ln \dfrac{1}{e_n}} = \frac{2 \ln\left(\dfrac{a}{2} + \sqrt{\dfrac{a^2}{4} + b}\right) - \ln b}{\ln\left(\dfrac{a}{2} + \sqrt{\dfrac{a^2}{4} + b}\right) - \ln b} \tag{4.8}$$

The proof is based on the behaviors of the stable and unstable manifolds of map (4.4). Indeed, the first branch going through the period-two point F_1 (resp. F_2) intersects the x-axis at $P_{1,0}$: $(c^*, 0)$ (resp. $P_{1,0}$: $(-c^*, 0)$) where:

$$c^* = (1 + \lambda^u) \, x_+ = 1 + \frac{c * b}{\lambda^u} \tag{4.9}$$

   Let $\overline{P}$ denote the inverse of the point P and $P_{1,n}$ denote the nth image of $P_{1,0}$. Since the map (4.4) is linear for $z \leq 0$, then (4.6) is restricted to $z \leq 0$ in the case of f^*_1 and to $z \geq 0$ in the case of f^*_2 and the inverse functions are thus defined for $z \leq c^*$ and $z \geq -c^*$. Hence, Eq. (4.5) implies that the right endpoint of the branch going through F_2 is $P_{1,1}$: $(ac^* - 1, c^*)$, and the left end-point of the other one is $\overline{P}_{1,1}$ and it is remarkable that further steps do not create segments outside these branches. Thus, these branches are the outermost lines of the strange attractor of the map (4.4). It is easy to show that Eq. (4.5) generates two new branches $\overline{P}_{1,2} \, P_{2,0}$ and $\overline{P}_{2,0} \, P_{1,2}{}^3$ where the x-coordinate of $P_{2,0}$ is given by $1 - \dfrac{c * b}{\lambda^u}$. These branches

intersect the x axis at $\pm \left(1 \pm \dfrac{1 - \left(\dfrac{c * b}{\lambda^u}\right) b}{\lambda^u}\right)$. Hence, it is also easy to show

that the region outside the band $P_{1,0} \overline{P}_{1,1} P_{1,2} P_{2,0}$ and its inverted image cannot belong to the strange attractor of map (4.4). Now, Eq. (4.9) implies that the width of the bands is $\dfrac{2c * b}{\lambda^u \sqrt{1 + (\lambda^u)^2}}$, $q = \dfrac{b}{\lambda^u}$ times the width of the parallelogram $P_{1,0} \overline{P}_{1,1} P_{1,2}$

---

[3] They are the innermost lines of the strange attractor of map (4.4).

$P_{2,0}$ and their area is $2c^{*2}b$, b times the area of the band $P_{1,0}\overline{P_{1,1}P_{1,2}}P_{2,0}$. The fact that the image of the forbidden region is also forbidden, implies that the bands covering the strange attractor after this step have a width $\dfrac{2c^*}{\sqrt{1+(\lambda^u)^2}}q^2$ and an area $2c^{*2}b^2$.

Finally, after n steps, one has bands of width $\dfrac{2c^*}{\sqrt{1+(\lambda^u)^2}}q^n$ and of a total area $2c^{*2}b^n$. Thus, $N_n \sim \left(\dfrac{b}{q^2}\right)^n$ small squares of side e_n, $e_n \sim q^n$ can be taken to cover the strange attractor of map (4.4).

From Eq. (4.8) it is easy to see that for small values of b, the dimension d behaves as $1-\dfrac{\ln a}{\ln b}$. Result similar to Theorem 49 can be obtained for the more general case of the form:

$$x' = \frac{f(-x)}{b} + \frac{z}{b}, z' = x \qquad (4.10)$$

Figure 4.8 shows $d_c = d(a = a_c = 2\sqrt{1-b})$ as a function of b. We note that above a_c transverse heteroclinic points appear, and the strange attractor ceases to exist. When $a = a_c$, the strange attractor possesses its greatest fractal dimension at ac and its behaviors corresponds to the 2-D extension of the *diadic map* (or Bernoulli shift) $x' = 2x \,(mod\,1)$. Finally, the Lyapunov numbers of map (4.4) can be calculated easily since the strange attractor consists of parallel straight line segments only. In this case, we have $\Lambda_1 = \ln \lambda^u$ and $\Lambda_2 = \ln |\lambda^s|$, and the fractal dimension is $d = 1 - \dfrac{\Lambda_1}{\Lambda_2}$ in agreement with the well know results in [Kaplan & Yorke (1979)].

4.3 The piecewise linear noodle map

In this section, we present a three-dimensional piecewise linear invertible map. This map is called the *smooth noodle map* studied in [Rôssler, et al. (1983)] in the same relationship as the piecewise-linear two-dimensional Lozi map (4.1) does to the Hénon diffeomorphism (2.1). This map cannot be reduced to two-dimensional submaps and it can display additional phenomena expected from four-dimensional flows and not possible in two-dimensional maps. The smooth noodle map is given by:

$$\begin{cases} x_{n+1} = ax_n(x_n^2 - 1) - y_n - z_n \\ y_{n+1} = \beta x_n \\ z_{n+1} = \gamma(x_n^2 - 0.33) + \delta z_n \end{cases} \qquad (4.11)$$

It is well known that there is a qualitative difference between the solutions of high- and low-dimensional dynamical systems. Especially when studying

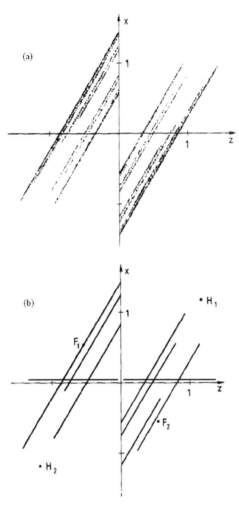

Figure 4.7: (a) The strange attractor of map (4.4) obtained in a numerical simulation after 3000 points, $a = 1.35$, $b = 0.5$. (b) The first branches of the unstable manifolds of F_1 as obtained after 3 steps of construction. Reused with permission from Tél, Physics Letters A, (1983). Copyright 1983, Elsevier Limited.

the topology, the algebraic geometry, or measures of a such a solution. Many of these differences are the occurrence of hyperchaos and persistent chaos. On one hand, if more than two Lyapunov exponents are positive, then the resulting attractor is called *hyperchaotic*. The importance of these attractors is that they are more non-regular, and the iteration points are seemingly *almost* full of the considered space, which explains one of applications of chaos in fluid mixing. On the other hand, *persistent chaos* offer a conjecture about the geometric mechanism characterizing this persistence in higher

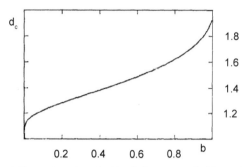

Figure 4.8: The fractal dimension given by Eq. (4.8) at $a_c = 2\sqrt{1-b}$ as a function of b. Reused with permission from Tél, Physics Letters A, (1983). Copyright 1983, Elsevier Limited.

dimensions. In other words, dynamical persistence means that a behavior type, i.e., equilibrium, oscillation, chaos or hyperchaos does not change with functional perturbations or parameter variations. See [Albers, et al. (2006)] for more details. Now, the map (4.11) is a global diffeomorphism just like the two-dimensional Hénon (2.1) because it is locally invertible since its Jacobian determinant is a constant$= \beta\delta \neq 0$ and its right-hand side is analytic and there exists an explicit inverse. The map (4.11) is almost chaotic since its Lyapunov exponents are positive, negative, and negative respectively for almost all initial conditions [Rosen (1970)]. In this case, the well known formula conjectured by Kaplan and Yorke in [Kaplan & Yorke (1979)] implies that the fractal dimension of the topologically 1-D attractor becomes larger then 2. Generally, the analytical study of nonlinear maps is not an easy task and the piecewise linear maps are more amenable to this kind of analysis. For this reason, a piecewise linear analogue of the map (4.11) was studied in [Hudson & Rôssler (1984)]. This map is given by:

$$
\begin{cases}
x_{n+1} = \begin{cases} 3.3\,(x_n - 1) - 2, \text{if } x_n \geq 1 \\ -2x_n, \text{if } -1 \leq x_n \leq 1 \\ 3.3\,(x_n + 1) + 2, \text{if } x_n < -1 \end{cases} - y_n - z_n \\
y_{n+1} = \beta x_n \\
z_{n+1} = \gamma(|\,x_n\,| - 1) + \delta z_n
\end{cases}
\tag{4.12}
$$

The map (4.12) is invertible just like map (4.11) and it consist of several linear diffeomorphisms glued together continuously in a C^0 manner. Noodle maps of the form (4.12) occur in realistic systems, for example, see [Shaw (1984)] for more information.

In Figs. 4.9 and 4.10 two numerical calculations performed on map (4.12) were shown. In Fig. 4.9(a) the first 1600 iterates of the origin and a box around it that contains all these iterates are shown stereoscopically. The edges (the first iterate of this very box) of the box are shown in Fig. 4.9(b) with 200 points per edge and corresponding sides are marked by arrows. The same two calculations in a different projection are shown in Fig. 4.10. In this case, the first iterate stays always within the original box as shown in Fig. 4.9(b). For the initial conditions closer to the final attractor shown in Figs. 4.9(a) and .4.10(a), the first iterate of this complicated region does not protrude outside its original domain nor, outside the box of Figs. 4.9 and .4.10. In this case, the whole shape of the attractor shown in Figs. 4.9(a) and 4.10(a) and the first iterate of the box shown in Figs. 4.9(b) and .4.10(b) is like two letter V's glued together. Because in the first case, the shape is like a letter S made up from straight line segments, in xy-plane, and in the second case, the shape is like the small Greek letter α made up from straight line segments in xz-plane.

Finally, the full shape of the attractor in three-dimensional view resembles a letter-Z shaped noodle whose two half loops belong to two different planes. In fact, the map (4.12) contains a rotational component and there exists a Cantor-set like structure not just in two directions across the attractor. At this end, we note that the map (4.12) covers a rather wide range of qualitative behaviors, i.e., it can generate hyperchaos. The inverse of map (4.12) is a *folded-towel type map* studied in [Rôssler (1980)] when the Jacobian determinant is taken to be larger than unity. In this case, no fully invertible piecewise linear map of this class can exist. The case where this map is almost invertible with one discontinuity on the right-hand side was studied in [Lozi (1983)]. Finally, no attractor remains when the Jacobian determinant of map (4.12) is equal to 1, i.e., the volume-preserving case. If there are no regions (in the bifurcation parameters space) of bounded recurrent behavior persist as appears in [Devaney (1984(a))] for the Lozi map (4.1) (See also, Section 4.1), then the resulting attractors should share some properties with the two arbitrarily close-by existing attractors [Rôssler (1980)]. Also, the map (4.12) and some chaotic attractors generated from the special case of *many-particle Hamiltonian systems* possess a non-uniform trajectorial density in the sense of classical) entropy [Farmer (1982)] in recurrent domains of volume-conserving systems.

4.4 Generalized hyperbolic attractors

In this section, we present piecewise hyperbolic maps or generalized Lozi maps, followed by a class of generalized hyperbolic attractors and then a definition of two-dimensional (2-D) generalized hyperbolic attractors is given along with some ergodic properties.

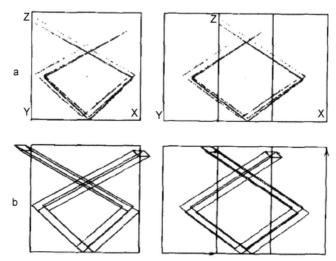

Figure 4.9: A numerical calculation of map (4.12). (a) The chaotic attractor.1600 iterates of the initial point $(0, 0, 0)$ near the middle of the picture) are shown. The box is the same as in (b). (b) The shape of the map. An initial box is shown along with its first iterate (200 points per edge). Corresponding sides are marked by arrows. Note that at the left-hand (diamondshaped) end, the first iterate (inset) corresponds to the left side of the box, and the right-hand diamond to the right side. The same parallel projections are shown in (a) and (b). Calculations performed on an HP9845B desk-top computer with peripherals. Parameters $\beta = \gamma = 0.1$, $\delta = 0.05$. Axes -2.475 to 2.343 for x, -0.249 to 0.235 for y, -0.1052 to 0.1543 for z. Reused with permission from Hudson and Rôssler, Physica D, (1984). Copyright 1984, Elsevier Limited.

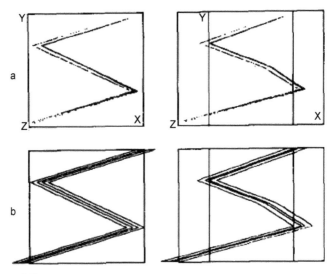

Figure 4.10: A different view of the attractor. (a) and the first iterate, (b) that were shown in Fig. 4.9. Reused with permission from Hudson and Rôssler, Physica D, (1984). Copyright 1984, Elsevier Limited.

4.4.1 Piecewise hyperbolic maps or generalized Lozi maps

In this section, we present some ergodic properties of a class of piecewise hyperbolic maps that makes a natural generalization of the Lozi maps (3.1). Indeed, in [Young (1985)] the existence of invariant Bowen-Ruelle measures with absolutely continuous conditional measures on unstable manifolds of a class of piecewise C^2 Lozi-like maps was proved. The main purpose of this study is a partial answer to the following question: *Suppose a compact neigh-borhood is mapped into itself and the map displays some chaotic behavior. Is there a strange attractor, or more specifically, is there a Bowen-Ruelle mea-sure?*. One can remark that Axiom A systems introduced in Section 1.2.1 cannot gives answers to these questions. A positive answer to this problem is given for the class of maps called *generalized Lozi maps* defined by the following manner: Let $f : R \to R$ be a continuous injective map where $R = [0, 1] \times [0, 1]$. Assume that f or some iterate of f takes R into its interior. Let $0 < a_1 < ... < a_q < 1$ and let $S = \{a_1, ..., a_q\} \times [0, 1]$. Assume that $f|_{(R-S)}$ is a C^2 diffeomorphism onto its image with $|Jac\,(f)| < 1$ and that both $f|_{(R-S)}$ and $f^{-1}|_{(R-S)}$ have bounded second derivative. Assume the following conditions on $f|_{(R-S)}$ with their geometric interpretations given in Fig. 4.11:

(H1) (Df preserves cones making $< 45^0$ with the x-axis), i.e.,

$$\inf \left\{ \left(\left| \frac{\partial f_1}{\partial x} \right| - \left| \frac{\partial f_1}{\partial y} \right| \right) - \left(\left| \frac{\partial f_2}{\partial x} \right| + \left| \frac{\partial f_2}{\partial y} \right| \right) \right\} \geq 0 \tag{4.13}$$

(H2) (restricted to these cones the action of Df when projected onto the x-axis is uniformly expanding), i.e.,

$$\inf \left\{ \left(\left| \frac{\partial f_1}{\partial x} \right| - \left| \frac{\partial f_1}{\partial y} \right| \right) \right\} = u > 1 \tag{4.14}$$

(H3) (horizontal expansion dominates the action of Df on vertical vectors), i.e.,

$$\sup \left\{ \left(\frac{\left| \frac{\partial f_1}{\partial y} \right| + \left| \frac{\partial f_2}{\partial y} \right|}{\left| \frac{\partial f_1}{\partial x} \right| - \left| \frac{\partial f_1}{\partial y} \right|} \right) \right\} < 1 \tag{4.15}$$

(H4) f expands horizontally more than it folds, i.e.,

$$\exists N \in \mathbb{Z}^+ \text{such that } u^N > 2 \text{ and } f^k\,(S) \cap S = \varnothing \text{ for } 1 \leq k \leq N \tag{4.16}$$

It is possible to verify that these hypotheses are indeed satisfied by many Lozi maps (3.1) which can be rewritten in the form (3.2) by using a change of coordinates. In this case, for open intervals of a and b, the map f takes some square $[c, c'] \times [c, c']$ into itself and satisfies hypotheses (**H1**)-(**H4**). This result was confirmed by the one given in [Collet & Levy (1984)]. See Section 2.2.

In order to state the main result of [Young (1985)], we need the following definitions about conditional measures:

Definition 53 *A Borel probability measure μ on R is said to have absolutely continuous conditional measures on unstable manifolds if there exist measurable partitions $P_1 \subset P_2 \subset \dots$ of R and measurable sets $V_1 \subset V_2 \subset \dots$such that:*

  (a) *$\mu V_n \uparrow 1$ as $n \to \infty$.*
  (b) *each element of $P_n \mid V_n$, is an open subset of some unstable manifold.*
  (c) *if $\{\mu_c, c \in P_n \mid_{V_n}\}$, denotes the system of conditional measures on elements of $P_n \mid_{V_n}$ and mc denotes Riemannian measure on c, then for almost every $c \in P_n \mid_{V_n}$, we have $\mu_c \ll m_c$.*
  (d) *A Borel probability measure μ on R is called a Bowen-Ruelle measure [Bowen (1975)] if there is a set $U \subset R$ of positive Lebesgue measure such that for every continuous function:*

$$\varphi: R \to R, \frac{1}{n}\sum_{i=0}^{n-1} \varphi(f^i(x)) \to \varphi d\mu \qquad (4.17)$$

for Lebesgue-a.e. $x \in U$.

The main result about the existence of invariant Bowen-Ruelle measures for the above family of maps was proved in [Young (1985)]:

Theorem 50 *If $f: R \to R$ is a generalized Lozi map, then f has an invariant Borel probability measure μ such that: (a) Local unstable manifolds exist at μ-a.e. point. (b) μ has absolutely continuous conditional measures on unstable manifolds.*

The proof of Theorem 50 is purely constructive and based on the following Lemma proved also in [Young (1985)]:

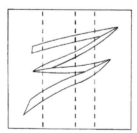

Figure 4.11: Example of a generalized Lozi map. Reused (Fig. 1) with permission from Young, Trans. Amer. Math. Soc (1985). Copyright 1985, American Mathematical Society.

Lemme 4.1 *Under the hypotheses of the Theorem 50, there exists an invariant Borel probability measure μ and a function $g : [0, 1] \rightarrow [0,\infty)$ of bounded variation such that:*

$$d(p^*\mu) = gdm \qquad (4.18)$$

and the following ideas:

1. The unstable manifolds (when they exist) are piecewise smooth curves *zigzagging* across the set R.
2. The construction of an invariant measure μ that behaves nicely on neighborhoods of the singularity set S is done using a combination of the methods used in [Sinai (1968), Lasota & Yorke (1973)].
3. The fact that all turns are created by passing through S, allows the construction of a non invariant measure $\hat{\mu}$ equivalent to μ on an arbitrarily large set, in which its conditional measures on unstable manifolds are absolutely continuous.

If f is a generalized Lozi map, then the following results were also proved in [Young (1985)]:

(a) Let μ be constructed as in Lemma. 4.1. Then there are measurable sets E_1, E_2, ... $\subset R$ such that $f^{-1}(E_i) = E_i$, $\mu(E_i) > 0 \,\forall i$, $\mu(E_i) = 1$ and for each i, $f \mid_{E_i}$: $(E_i, \mu \mid_{E_i}) \rightarrow (E_i, \mu \mid_{E_i})$ is ergodic. (b). Let μ be constructed as in Lemma. 4.1. Then corresponding to each E_i in Corollary 7(a), $\mu \mid E_i$ normalized is a Bowen-Ruelle measure. (c). Let $f : R \rightarrow R$ be a generalized Lozi map. Then f has an invariant Borel probability measure μ such that:

$$h_\mu(f) = \lambda_1(x)\, d\mu(x) \qquad (4.19)$$

where $\lambda_1(x)$ is the positive μ-exponent of f at x.

The idea of the proof of Corollary 7(a) follows the one given in [Pesin (1977)] with the fact that on most $c \in P_n \mid_{Vn}$, μ_c is equivalent to m_c with the use of the notion of absolute continuity of stable manifolds given in [Katok, et al. (1986)]. This last reference can be considered also as a guide to the proof of Corollary 7(b). The proof of Corollary 7(c) is based essentially on the work given in [Ledrappier & Strelcyn (1982)]. At the end of this section, we note that the statistical properties of dynamical systems with some hyperbolicity were studied in [Young (1998(b))].

4.4.2 Class of generalized hyperbolic attractors

In this section, we present some statistical properties of a class of hyperbolic attractors introduced in [Pesin (1992)] called *generalized hyperbolic attractors*. This class includes the 2-D hyperbolic attractors of Belykh, the Lozi mappings (3.1) and the Lorenz attractor given by (1.49). The Smale spectral decomposition with countably many components was established for this type of systems. Each component is ergodic with respect to any Gibbs *u*-measure called SRB-measures where every component is decomposed into finitely many subsets which are cyclically permuted with the property that the corresponding iteration of the map is of *mixing* and Bernoulli type. The number of ergodic components is finite as shown [Sataev (1992)] and the generalized Lozi attractors satisfy the assumption required in [Pesin (1992)]. Thus, the generalized Lozi attractors are ergodic-even-mixing- for some value of their parameters as shown in [Misiurewicz (1983)] for the Lozi map (3.1). See Section 3.5.2.

4.4.3 Definition of 2-D generalized hyperbolic attractors

This class can be defined as follows: Let M be a smooth two-dimensional (2-D) manifold equipped by a Riemannian metric ρ and Riemannian volume $vol(.)$. Let $U \subset M$, be an open connected subset with compact closure, and $\Gamma \subset U$, a closed subset. Assume that $S^+ = \Gamma \subset \partial U$ consists of a finite number of compact smooth curves and the set $U \setminus \Gamma$ consists of a finite number of open connected components. Let $f : U \setminus \Gamma \rightarrow f(U \setminus \Gamma) \subset U$ be a C^2-diffeomorphism that is twice differentiable up to the boundary $\partial (U \setminus \Gamma) = S^+$ which is the singularity set of it. This means that the boundary $\partial (f(U \setminus \Gamma)) = S^{-4}$ is a finite union of compact smooth curves. Hence, the inverse map f^{-1} is twice differentiable up to S^- and the first and the second partial derivatives of both f and f^{-1} are uniformly bounded because $\overline{U}$ is a compact set in M. Let S^+_m be the union of the curves on which the map f^m is singular. Let Γ_ε be the ε-neighborhood of the set Γ and v the Lebesgue measure on M. Hence, we have the following definition:

Definition 54 *A smooth curve γ in U is unstable (resp. stable) if its tangent line belongs in the cone C^u (z) (resp., C^s (z)) defined by (4.23) below at any $z \in \gamma$.* Now, let us define the two sets U^+ and D as follow:

$$\left\{ \begin{array}{c} U^+ = \{x \in U : f^n (x) \notin S^+, n = 0, 1, 2, ...\} \\ D = \cap_{n \geq 0} f^n (U^+) \end{array} \right. \tag{4.20}$$

The set D is invariant under both f and f^{-1} and its closure $\Lambda = \overline{D}$ is called the *attractor for f*, for systems with:

$$Vol (\Lambda) = 0 \tag{4.21}$$

and this case includes also piecewise linear toral automorphisms attractors like the Lozi map (3.1), i.e., when $Vol (\Lambda) > 0$. Property (4.21) can be assured if $f (U \setminus \Gamma) \subset U$ as shown in [Sataev (1992)].

---

[4] which is the set of singularities for f^{-1}.

In order to define a hyperbolic structure for the map f, let $z \in U$ be any point and P any line lying in the tangent plane $T_z M$ and any real number $\alpha > 0$. Let $C(z, \alpha, P)$ the cone set defined by:

$$C(z, \alpha, P) = \{v \in T_z M : \angle(v, P) \le \alpha\} \quad (4.22)$$

Assume that for each point $z \in U \backslash S^+$ there are two cones:

$$\begin{cases} C^u(z) = C(z, \alpha^u(z), P^u(z)) \\ C^s(z) = C(z, \alpha^s(z), P^s(z)) \end{cases} \quad (4.23)$$

with the following three properties:

1. The angle between $C^u(z)$ and $C^s(z)$ is uniformly bounded away from zero.
2. The following inclusions hold:

$$\begin{cases} df(C^u(z)) \subset C^u(f(z)) \text{ for any } z \in (U \backslash S^+) \\ df^{-1}(C^s(z)) \subset C^s(f^{-1}(z)) \text{ for any } z \in f(U \backslash S^+) \end{cases} \quad (4.24)$$

3. There exist constants $C > 0$ and $\lambda \in (0, 1)$ such that for any integer $n > 0$, we have:

(a) if $z \in U^+$ and if $v \in C^u(z)$, then

$$\| df^n v \| \ge C \lambda^{-n} \| v \| \quad (4.25)$$

(b) if $z \in f^n(U^+)$ and if $v \in C^s(z)$, then

$$\| df^{-n} v \| \ge C \lambda^{-n} \| v \| \quad (4.26)$$

4. The two cones $C^u(z)$ and $C^s(z)$ depend continuously on $z \in U^+$.

Note that the Lozi map (3.1) belongs to this class of mappings satisfying the above properties (1)-(3) and it is possible to choose $C = 1$, because it was shown in [Sataev (1992)] that there exist an integer $m \ge 1$ so that f^m enjoys the properties (1)-(3). Thus, the following definition is given in [Pesin (1992)]:

Definition 55 *An attractor Λ is called a generalized hyperbolic attractor if the two families of cones given by (4.23) exist.*

For $z \in D$, then properties (1)-(3) yield families of invariant subspaces E^u_z and E^s_z in the tangent space $T_z M$ with the following two properties:

$$\begin{cases} E^u_z \subset C^u(z) \text{ and } E^s_z \subset C^s(z) \\ df(E_z^{u,s}) = E_{f(z)}^{u,s}. \end{cases} \quad (4.27)$$

To guarantee the uniform hyperbolic structure for the map f, it suffices to assume the following property given in [Pesin (1992)]: From property 4, it follows that for any $z \in \Gamma$, the two limit cones $C^{u,s}(z) = \lim_{z' \to z} C^{u,s}(z')$ exist on both sides of Γ, and the angle between the tangent line Γ at z and the unstable limit cone $C^u(z)$ is uniformly bounded away from zero. We note also that the singularities of f and f^{-1} are *mild* and they are concentrated on a finite union of smooth compact curves, and the first and second derivatives of f and f^{-1} have one-sided limits. To guarantee that expansion and contraction prevail over discontinuities, it suffice to assume the following conditions given in [Afraimovich, et al. 1995]:

Condition A1: There exists an integer $\tau \geq 1$ such that $f^{-k}(\Gamma) \cap \Gamma = \varnothing$ for $k = 1, 2, ..., \tau$ and $\lambda^{-\tau} > 2$, where $\lambda^{-1} > 1$ is, as before, the minimal factor of expansion of vectors in the unstable cones $C^u(z)$ at all points $z \in U \backslash S^+$. Moreover, there is a neighborhood of the attractor in which the smooth components of Λ do not intersect one another.

Condition A2: There exist constants $C_0 > 0$ and $K_0 < \lambda^{-1}$ such that for any integer $m \geq 1$ no more than $C_0 K_0^n$ smooth components of the union $\cup_{l=0}^m = S^+_l$ can meet at any point $z \in U$.

Condition A3: There exist constants $B > 0$, $\beta > 0$, and $\varepsilon_0 > 0$ such that for any integer $n \geq 1$ and any $\varepsilon \in (0, \varepsilon_0)$ one has $v(f^{-n}\Gamma_\varepsilon) < B\varepsilon^\beta$.

Condition A4: There is a constant $\varepsilon_0 > 0$ such that for any unstable curve W^u there exist an integer $n_0 = n_0(W^u)$ and a constant $B_0 = B_0(W^u)$ such that for any $\varepsilon \in (0, \varepsilon_0)$ one has: (a) $v^u(W^u \cap f^{-n}\Gamma_\varepsilon) < \varepsilon^\beta v^u(W^u)$ for all integers $n > n_0$, (b) $v^u(W^u \cap f^{-n}\Gamma_\varepsilon) < B_0\varepsilon^\beta v^u(W^u)$ for all integers $n \geq 1$. It is important to set some remarks about these conditions: First, we note that some (but not all) Lorenz, Lozi and Belykh attractors satisfy **Condition A1**, but in some cases weaker assumption than **Condition A1** is sufficient [Pesin (1992), Afraimovich, et al. (1995)]. Second, **Condition A1** implies **Condition A2** with $K_0 = 2^{\frac{1}{\tau}}$. When $K_0 = 1$, **Condition A2**, in a more stringent form was used in [Bunimovich, et al. (1990), Chernov, et al. (1993)]. Hence, **Condition A2** holds for a single, sufficiently large value of $m \geq 1$. Two more conditions were adduced in [Sataev (1992)]. Third, **Condition A3** and a weaker version of **Condition A4** were also assumed in [Pesin (1992)]. Also, it is very hard to check **Conditions A3** and **A4** for particular examples. In fact, it was shown in [Afraimovich, et al. (1995)] that **Condition A2** implies **Conditions A3** and **A4**. See Proposition. 4.3 below:

Proposition 4.3 *If a generalized hyperbolic attractor* Λ *satisfies* **Condition A2**, *then it satisfies* **Condition A3** *and* **A4**.

In [Pesin (1992)], let $\varepsilon > 0$ and $l = 1, 2, ...$, thus, it is possible to define the following subsets:

$$
\begin{cases}
\widehat{D}^{+}_{\varepsilon,l} = \{z \in U^{+} : \rho(f^{n}(z), S^{+}) \geq l^{-1}e^{-\varepsilon n}, n=0,1,...\}, \\
D^{-}_{\varepsilon,l} = \{z \in D : \rho(f^{-n}(z), S^{-}) \geq l^{-1}e^{-\varepsilon n}, n=0,1,...\}, \\
\qquad\qquad D^{+}_{\varepsilon,l} = \widehat{D}^{+}_{\varepsilon,l} \cap \Lambda, \\
\qquad\qquad D^{0}_{\varepsilon,l} = D^{+}_{\varepsilon,l} \cap D^{-}_{\varepsilon,l}, \\
\qquad\qquad D^{\pm}_{\varepsilon,l} = \cup_{l \geq 1} D^{\pm}_{\varepsilon,l} \\
\qquad\qquad D^{0}_{\varepsilon} = \cup_{l \geq 1} D^{0}_{\varepsilon,l}.
\end{cases}
\qquad (4.28)
$$

The subset $D^{+}_{\varepsilon,l}$ $(D^{-}_{\varepsilon,l})$ consists of points that do not approach the singularity set too rapidly in the future (resp., in the past). The sets $\widehat{D}^{+}_{\varepsilon,l'}$ $D^{\pm}_{\varepsilon,l}$ and D^{0}_{ε} are closed, $D^{0}_{\varepsilon} = D^{+}_{\varepsilon} \cap D^{-}_{\varepsilon}$, the set D^{+}_{ε} is f-invariant, D^{-}_{ε} is f^{-1}-invariant and D^{0}_{ε} is both f and f^{-1} invariant, $D^{0}_{\varepsilon} \subset D$ for any $\varepsilon > 0$.

Definition 56 *(Regular attractor) [Pesin (1992)] The attractor Λ is said to be regular if $D^{0}_{\varepsilon} \neq \emptyset$ for all sufficiently small $\varepsilon > 0$.*

In fact, it was proved in [Pesin (1992)] that the attractor Λ is regular under weaker assumptions than **Condition A3** and **Condition A4**, which means that **Condition A2** implies that the attractor is regular.

Proposition 4.4 *There exists an $\varepsilon > 0$ such that for any point $z \in D^{+}_{\varepsilon,l}$ $(z \in D^{-}_{\varepsilon,l})$ there is a local stable fiber, LSF, denoted by $V^{s}(z)$ (resp., a local unstable fiber, LUF, denoted by $V^{u}(z)$). An LSF (LUF) (In some cases, such as the linear toral automorphism the LUF's and LSF's have infinite length and hence, LUF's and LSF's were redefined for a large $L > 0$ and denote by $V^{s,u}(x)$ a segment of the LUF (LSF) at the point x that has length $L > 0$ and is centered at x) is a C^{1}-curve in M. It is tangent to the line E^{s}_{z} (resp., to E^{u}_{z}) at z. The ρ-distance of the point z from the endpoints of that fiber is at least $\delta_{l} = \dfrac{1}{l}$, a quantity determined by l and independent of z.*

Let $V^{s,u}(z)$ be the maximal smooth local stable and unstable fiber passing through z. We have $V^{u}(z) \subset D^{-}_{\varepsilon}$ for any $z \in D^{-}_{\varepsilon}$ and the Gibbs u-measures on Λ are defined as follow:

Definition 57 *Let $J^{u}(z)$ be a one-step expansion factor in E^{u}_{z}, i.e., the Jacobian of the map $df|_{E^{u}_{z}}$ at z. Then, for any $z \in D^{+}_{s,l}$ such that $V^{u}(z)$ exists, define for all $y \in V^{u}(z)$:*

$$
k(z, y) = \lim_{\pi \to \infty} \Pi^{n}_{j=1} [J^{u}(f^{-j}(z))] \cdot [J^{u}(f^{-j}(y))]^{-1} \qquad (4.29)
$$

Limit (4.29) exists, positive, continuous on $D^{-}_{s,l}$ and uniformly bounded away from zero and infinity on D^{-}_{ε} because f is smooth up to S^{+}.

A measure μ on Λ is called a Gibbs u-measure, or a Bowen-Ruelle-Sinai measure (BRS measure) if: (a) it is f-invariant; (b) $\mu\,(D^0_\varepsilon) = \mu\,(\Lambda) = 1$ for some $\varepsilon > 0$; (c) the conditional measure on LUF's $V^u\,(z)$ induced by μ has a density with respect to the Lebesgue measure on $V^u\,(z)$ proportional to $k\,(z, y)$.

Hence, the Gibbs u-measures are constructed as follows:

(a) Consider a point $z \in D^-_\varepsilon$ and take the normalized Lebesgue measure v^u on $V^u\,(z)$.
(b) Pull the normalized Lebesgue measure forward under $f: v_k = f^k_* v^u$, i.e., for any Borel set $A \subset U$ one takes $v_k\,(A) = v^u\,(f^{-k}A \cap V^u\,(z))$.
(c) The sequence of measures:

$$\mu'_n = \frac{1}{n}\sum_{k=0}^{n-1} v_k \qquad (4.30)$$

has a limit point (a measure) in the weak topology and defines a Gibbs u-measure[5] for any $z \in D^-_\varepsilon$.

We have the following result:

Proposition 4.5 (a) *If Vol(Λ) = 0, then any Gibbs u-measure is singular with respect to the Lebesgue measure on U and with respect to the Lebesgue measure on any LSF.*
  (b) *A Gibbs measure μ has no atoms, and any particular LUF or LSF has μ measure zero.*

On the other hand, it was shown in [Sataev (1992)] that there is always one measure satisfying three additional properties:

(e) For every $l > 0$ the sets $\Lambda^j_i \cap D^-_{\varepsilon,l}$ are closed.
(f) For every $l > 0$, every $i = 1, ..., r$ and every open subset $Q \subset U$ such that $Q\,\Lambda^j_i \cap D^-_{\varepsilon,l} \neq \varnothing$, we have $\mu_i\,(Q \cap \Lambda^j_i \cap D^-_{\varepsilon,l}) > 0$.
(g) If $z \in \Lambda^j_i$, then $V^u(z) \subset \Lambda^j_i$.

4.4.4 Smale spectral decomposition for generalized hyperbolic attractors

The following proposition called *Smale spectral decomposition* was proved in [Pesin (1992)] for generalized hyperbolic attractors defined in Section 1.2.1:

Proposition 4.6 *There are subsets Λ_i, i = 0, 1, .. and Gibbs u-measures μ_i, i $\geq$ 1 such that:*

---

[5] One can take any measure equivalent to the Lebesgue measure on LUF. Thus, a Gibbs u-measure is not unique.

(a) $\Lambda = \bigcup_{i \geq 0} \Lambda_i$ and $\Lambda_i \cap \Lambda_j = \varnothing$ for $i \neq j$;

(b) for $i \geq 1 : \Lambda_i \subset D, f(\Lambda_i) = \Lambda_i, \mu(\Lambda_i) = 1$ and $f|_{\Lambda i}$ is ergodic with respect to μ_i;

(c) for $i \geq 1$: there exists a finite decomposition $\Lambda_i = \cup^{ri}_{j=1} \Lambda^j_i$, where $\Lambda^{j'}_i \cap \Lambda^{j'}_i = \varnothing$ for $j \neq j', f(\Lambda^j_i) = \Lambda^{j+1}_i$ and $f(\Lambda^{ri}_i) = \Lambda^1_i$, and $f^{ri}|\Lambda^1_i$ is a Bernoulli automorphism;

(d) any Gibbs u-measure μ is a weighted sum $\mu = \sum_{i \geq 1} \alpha_i \mu_i$ with some $\alpha_i \geq 0$ and $\sum \alpha_i = 1$. In particular, $\mu(\Lambda_0) = 0$.

More detailed study of this class is done in [Afraimovich, et al. (1995)] where some statistical properties of 2-D generalized hyperbolic attractors were studied such as *stretched* exponential bound on the decay of correlations and the central limit Theorem. The method of analysis is based on the Markov approximation to hyperbolic dynamical systems developed for hyperbolic billiards and similar models in [Bunimovich, et al. (1990), Chernov (1994)].

4.4.5 Statistical properties of generalized hyperbolic attractors

Let H denote the class of Hôlder continuous (HC) functions on the attractor (Recall Definition 11). To estimate the decay of correlations discussed in Section 1.1.9 for generalized hyperbolic attractors, it suffices to take an arbitrary subcomponent $\Lambda_* = \Lambda^j_i$ of any ergodic component Λ_i of the attractor, and define $f_* = f^{ri}|_{\Lambda_*}$, μ_* is the normalized measure $\mu_i|_{\Lambda_*}$, $r_* = r_i$ and denote by $\langle . \rangle$ the expectation with respect to μ_*. Thus, Proposition. 4.6, implies that the triple (Λ_*, f_*, μ_*) is a Bernoulli dynamical system, that is mixing. Thus, the following results about decay of correlations, central limit theorem, and relaxation to equilibrium distribution for generalized hyperbolic attractors, were proved in [Afraimovich, et al. (1995)]. The method of analysis is based on the use of Markov approximation[6] to (Λ_*, f_*, μ_*) for Theorem 51(a) and Theorem 51(b) below:

Theorem 51 (a) (Decay of correlations) Let $F(x)$ and $G(x)$ be two HC or PHC functions on M. Then, for any integer N, we have:

$$\left| \left\langle \left(F \circ f^N_* \right) . G \right\rangle - \langle F \rangle \langle G \rangle \right| \leq c(F, G) \, \alpha \sqrt{|N|} \tag{4.31}$$

where $c(F, G) > 0$ depends on F and G and $\alpha < 1$ is determined by the subcomponent $\Lambda_* = \Lambda^j_i$ and the class of HC or PHC functions under consideration.

---

[6] The so called Markov sieves developed in [Bunimovich, et al. (1990-1991), Chernov (1992), Chernov, et al. (1993), Chernov (1994)].

(b) *(Central limit theorem) Again, let F(x) be an HC or a PHC function. Assume that $\langle F \rangle = 0$. Then, the quantity:*

$$\sigma_F^2 = \sum_{n=-\infty}^{\infty} \left\langle \left(F \circ f_*^N\right).F\right\rangle \tag{4.32}$$

is finite and nonnegative. If $\sigma_F \neq 0$, then the sequence:

$$\frac{F(x) + F(f_*x) + \dots + F(f_*^{N-1}x)}{\sqrt{\sigma_F^2 N}} \tag{4.33}$$

converges in distribution to the standard normal law as $N \to \infty$.

(c) *(Relaxation to equilibrium distribution) For any integers $k \geq l > 1$ and $1 \leq i_1 < i_2 < \dots < i_k \leq N$ there is a subset $R_* = R_*(i_1, \dots, i_k) = \mathcal{L}^{k-l+1}$ of $(k-l+1)$-tuples of indices such that:*

(i) *if $(j_1, \dots, j_k) \in R_*$, then*

$$\left\{ \begin{array}{c} \sum\limits_{j_1,\dots,j_{l-1}=0}^{l} |\mu_*(A_1) - \mu_*(A_2)| \leq \Delta \\ A_1 = f_*^i A_{j_1} \cap \dots \cap f_*^{i_{l-1}} A_{j_{l-1}} / f_*^{i_l} A_{j_l} \cap \dots \cap f_*^{ik} A_{j_k} \\ A_2 = f_*^{i_1} A_{j_1} \cap \dots \cap f_*^{i_{l-1}} A_{j_{l-1}} \end{array} \right. \tag{4.34}$$

(ii) *one has*

$$\sum_{(j_1,\dots,j_k)\in R_*}^{l} \left| \mu_* \left(f_*^{i_l} A_{j_l} \cap \dots \cap f_*^{i_k} A_{j_k} \right) \right| \geq 1 - \Delta \tag{4.35}$$

where

$$\Delta = \max\left\{ c_4 \alpha_4^n, \left(1 - \frac{g_1}{2}\right)^{\left[\frac{L}{2}\right]} \right\} \text{ with } L = \left[\frac{i_l - i_{l-1}}{g_0 n}\right]. \tag{4.36}$$

We note that Theorem 51(c) is still true for the *reverse time*, i.e., if $1 \geq i_1 > i_2 > \dots > i_k \geq N$ and it means that the conditional distributions relax to equilibrium exponentially fast in the parameter $|i_l - i_{l-1}|$, which represents the *interval* between the *future* and the *past*, at least as long as that interval is less than *const.n²*.

4.5 Global periodicity property of the generalization Lozi mappings

In this section, we will discuss the *global periodicity property* of the generalization Lozi mappings given by:

$$y_{n+1} = a \, | \, y_n \, | + by_{n-1} + \gamma, \, n = 0, 1, 2, \ldots, y_{-1}, y_0 \in \mathbb{R} \qquad (4.37)$$

For this purpose, we need to define the global periodicity property with period p of the general difference equation given by:

$$y_{n+1} = f \, (y_n, \ldots, y_{n-k+1}), \, n = 0, 1, 2, \ldots, k, \, k \geq 2 \qquad (4.38)$$

where $y_{n-k+1}, \ldots, y_0 \in D$, a non-degenerate interval of real numbers, and f: $D^k \to D$ is a single-valued function.

Definition 58 *Eq. (4.38) is said to possess the global periodicity property if all its solutions are periodic of the same period p. When p is the smallest period that works for all solutions, we call it the prime period[7].*

Several necessary and/or sufficient conditions were established for the global periodicity property [Abu-Saris (1999-2000(a-b)), Feuer & Janowski (2000), Golumb (1992)]. In [Janowski, et al. (1995)] necessary and/ or sufficient conditions on the parameters of certain classes of functions were developed. In [Kurshan & Gobinath (1974), Mestel (2003)] necessary conditions on the characteristic roots were developed. In [Sivak (1997), Abu-Saris (2000(a-b))] necessary conditions on the asymptotic behavior were developed. In [Mestel (2003), Abu-Saris & Al-Jubouri (2004)] functional equations criterion were used to obtain necessary and sufficient conditions for second-order difference equations. This approach was generalized to any order in [Abu-Saris & Al-Hassan (2003(a-b))]. This approach requires only the function being single-valued and it can be generalized to any topological space. Due to some difficulties[8] arising from this method, practical conditions at the minimum possible level were investigated in [Abu-Saris (2006)] by adding the continuity of f. Thus, the global periodicity of one-dimensional discrete dynamical systems governed by a continuous function $f : D \to D$ was investigated. The results obtained can be applied to some generalizations of the Lozi mappings (3.2). Indeed, the following results were proved in [Abu-Saris (2006)]:

Theorem 52 *(a) Every solution of the first-order difference equation:*

$$z_{n+1} = g(z_n), \, n = 0, 1, 2, \ldots, z_0 \in D \qquad (4.39)$$

---

[7] In this case, p need not be the prime (minimal) period for an individual solution.
[8] More computations when the period p is large.

is periodic of prime period p if and only if $g^{-1}(x) = g(x)$ for all $x \in D$. In other words, either $p = 1$ or 2.

(b) *Suppose that all solutions of higher order difference equations (4.38) are periodic of prime period p. Then for $x = (x, ..., x) \in D^{k-1}$, either $f(x, x_k) = x_k$ or $f(x, x_k)$ is self-inverse in x_k, i.e., $f((x, f(x, x_k))) = x_k$ for all $x, x_k \in D$.*

The proof of Theorem 52 is based on several results about some inequalities verified by Eq. (4.39). Theorem 52(a) can be applied to the generalization of the Lozi map (4.37). In [Kulenovic & Ladas (2002)] the authors raised an open problem about the global periodicity of (4.37) according to the values of the parameter γ. The case where $\gamma = 0$ was solved in [Abu-Saris (1999)] but the case $\gamma \neq 0$ is still an open problem. Finally, Theorem 52(b), implies that a necessary condition for global periodicity of Eq. (4.37) is $b = -1$.

4.6 Generalized discrete Halanay inequality and the global stability of Lozi mapping

In this section, we discuss the global stability of Lozi mapping $\mathcal{L}_1$ given by (3.2) using the so called *generalized discrete Halanay inequality* introduced in [Liz, et al. (2003)] where a discrete Halanay-type inequalities was derived with some applications to the study of the dynamics of nonlinear difference equations involving the *maximum functional*. In particular, the asymptotic stability of max functional differential equations was the subject of many works as for example in [Halanay (1966), Ivanov, et al. (2002), Liz & TroBmchuk (2000), Liz & Ferreiro (2002)]. In fact, it was shown in [Liz & Ferreiro (2002)] that the asymptotic stability of generalized difference equations given in [Mohamad & Gopalsamy (2000)] can be studied in an accurate manner by using some discrete versions of these max inequalities. Indeed, the first result was proved in [Liz & Ferreiro (2002), Theorem 1] and it is expressed as:

Theorem 53 *(The discrete Halanay Lemma) Let $r > 0$ be a natural number, and let $\{x_n\}_{n \geq -r}$ be a sequence of real numbers satisfying the inequality:*

$$\Delta x_n = x_{n+1} - x_n \leq -ax_n + b \max\{x_n, x_{n-1}, x_{n-r}\}, n \geq 0 \qquad (4.40)$$

If $0 < b < a \leq 1$, then there exists a constant $\lambda_0 \in (0, 1)$ such that:

$$x_n \leq \max\{0, x_0, x_{-1}, ..., x_{-r}\} \lambda_0^n, n \geq 0. \qquad (4.41)$$

Moreover, λ_0 can be chosen as the root in the interval $(0, 1)$ of the equation:

$$\lambda^{r+1} + (a - 1)\lambda^r - b = 0 \qquad (4.42)$$

By a simple use of Theorem 53, the following statement was proved in [Liz & Ferreiro (2002), Theorem 2]:

Theorem 54 *Assume that $0 < a \leq 1$ and that there exists a positive constant $b < a$ such that*

$$|f(n, x_n, ..., x_{n-r})| \leq b \, \|(x_n, ..., x_{n-r})\|_\infty, \, \forall \, (n, x_n, ..., x_{n-r}) \in \mathbb{R}^{r+1} \qquad (4.43)$$

Then there exists $\lambda_0 \in (0, 1)$ such that:

$$|x_n| \leq \left(\max_{-r \leq i \leq 0} \{|x_i|\} \right) \lambda_0^n, n \geq 0 \qquad (4.44)$$

for every solution $\{x_n\}$ of

$$\Delta x_n = -ax_n + f(n, x_n, ..., x_{n-r}), a > 0 \qquad (4.45)$$

where λ_0 can be calculated in the form established in Theorem 53, i.e., Eq. (4.42).

In [Liz, et al. (2003)] a proof was given for a generalization of Theorem 54 (a generalization of the discrete Halanay lemma), which gives some conditions for the asymptotic stability of a family of difference equations (including the Lozi mapping $\mathcal{L}_1$ given by (3.2)) by introducing a discrete analog of the Yorke condition [Kuang (1993), Section 4.5]. Let us consider the following inequalities:

$$\Delta u_n \leq -Au_n + B\tilde{u}_n + Cv_n + D\hat{v}_n, n \geq 0 \qquad (4.46)$$

$$u_n \leq (1-A)^n u_0 + \sum_{i=0}^{n-1} (1-A)^{n-i-1} [B\tilde{u}_i + Cv_i + D\hat{v}_i], n \geq 0 \qquad (4.47)$$

$$v_n \leq Eu_n + F\tilde{u}_n, n \geq 0 \qquad (4.48)$$

where

$$\begin{cases} \Delta u_n = u_{n+1} - u_n, \\ \tilde{u}_n = \max\{u_n, ..., u_{n-r}\}, \\ \hat{v}_n = \max\{v_{n-1}, ..., v_{n-r}\}, r \geq 1, \end{cases} \qquad (4.49)$$

and A, B, C, D, E, F are real constants. Denote $u = \{u_n\}_{n \geq -r}, v = \{v_n\}_{n \geq -r}$. We remark that if $A \leq 1$, then by induction, we have that if the pair (u, v) satisfies inequality (4.46), then it also satisfies (4.47). The following result was proved in [Liz, et al. (2003)]:

Theorem 55 *Assume that (u, v) satisfies the system of inequalities (4.44)–(4.48). If*

$$\begin{cases} B, C, D, E, F \geq 0 \\ FD + B > 0, E + F > 0 \\ B + (E + F)(C + D) < A \leq 1 \end{cases} \qquad (4.50)$$

then there exist constants $K_1 \geq 0, K_2 \geq 0$, and $\lambda_0 \in (0, 1)$ such that:

$$u_n \leq K_1 \lambda^n_0 , v_n \leq K_2 \lambda^n_0, n \geq 0. \qquad (4.51)$$

Moreover, λ_0 can be chosen as the smallest root in the interval $(0, 1)$ of the equation $h(\lambda) = 0$, where

$$h(\lambda) = \lambda^{2r+1} - (1 - A + CE)\lambda^{2r} - (B + FC + ED)\lambda^r - FD \qquad (4.52)$$

To show the global asymptotic stability of some forms of difference equations, Theorem 55 was applied to the following generalized difference equation:

$$\Delta x_n = -ax_n - bf (n, x_n, x_{n-1}, ..., x_{n-r}), b > 0 \qquad (4.53)$$

Obviously, given $r + 1$ points $\{x_{-r}, x_{-r+1}, ..., x_0\}$ there is a unique solution[9] $\{x_n\}$ of Eq. (4.53). To apply Theorem 55, we need to assume that f satisfies the following hypotheses:

$$\begin{cases} \textbf{(H1)}: |f (n, x_n, ..., x_{n-r})| \leq \|(x_n, ..., x_{n-r})\|_\infty \\ \textbf{(H2)}: |f (n, x_n, ..., x_{n-r}) - x_n| \leq r \|(\Delta x_{n-1}, ..., \Delta x_{n-r})\|_\infty \end{cases} \qquad (4.54)$$

Then, it is easy to prove the following result:

Lemme 4.2 *(a) Hypotheses (H1) and (H2) are satisfied for some important linear and nonlinear difference equations because:*

$$\begin{cases} |\max\{x_n, ..., x_{n-r}\}| \leq \max \{|x_n|, ..., |x_{n-r}|\} \\ x_n - x_i = \sum_{j=i}^{n-1} \Delta x_j \leq r \max\{\Delta x_{n-1}, ..., \Delta x_{n-r}\}, \forall i = n - r, ..., n. \end{cases} \qquad (4.55)$$

(b) Hypotheses (H1) and (H2) hold if the following condition is satisfied:

$$\min\{x_n, ..., x_{n-r}\} \leq f(n, x_n, ..., x_{n-r}) \leq \max\{x_n, ..., x_{n-r}\} \qquad (4.56)$$

Condition (4.56) is an analog of the so-called *Yorke condition for functional differential equations* [Ivanov, et al. (2002), Kuang (1993)]. Thus, some conditions for the asymptotic stability of Eq. (4.53) using Theorem 55 were obtained in [Liz, et al. (2003)] as follow:

Theorem 56 *Assume that f satisfies hypotheses (H1) and (H2). If either*

(a)

$$\begin{cases} 0 \leq a \leq -b \\ 0 < br < 1 \end{cases} \qquad (4.57)$$

---

[9] Can be explicitly calculated by iterations.

or

(b)

$$\begin{cases} a < 0 \\ 0 < br < (a + b)(-a + b)^{-1} \end{cases} \qquad (4.58)$$

holds, then there exist $K > 0$ and $\lambda_0 \in (0, 1)$ such that for every solution $\{x_n\}$ of (4.53), we have:

$$|x_n| \leq K \lambda_0^n, n \geq 0 \qquad (4.59)$$

where λ_0 can be calculated in the form established in Theorem 55. As a consequence, the trivial solution of Eq. (4.53) is globally asymptotically stable.

We remark that the discrete Halanay lemma can only applies when $|b| < a$, which imply that Theorem 56 show the global asymptotic stability in (4.53) for some values of $b > a > 0$ and also for $a < 0$.

To study the global asymptotic stability of the Lozi map (4.1), Theorem 50 was used to obtain some results: Indeed, define $L_1 (x, y) = 1 - a |x| + by$, then the iterations of the Lozi mapping $\mathcal{L}_1$ given by (3.2) can be written as:

$$\begin{cases} x_{n+1} = L_1(x_n, y_n) \\ y_{n+1} = x_n, n \geq 0. \end{cases} \qquad (4.60)$$

where $(x_0, y_0) \in \mathbb{R}^2$ is a given initial condition. The map (4.60) is equivalent to the second-order difference equation:

$$x_{n+1} = L_1(x_n, x_{n-1}) = 1 - a |x_n| + bx_{n-1} \qquad (4.61)$$

Obviously, the positive equilibrium $x_* = (1 + a - b)^{-1}$ exists if and only if a + 1 > b. The change of variable $z_n = x_n - x_*$ transforms (the convergence of x_n to x_* is equivalent to the convergence of z_n to 0) (4.61) into:

$$z_{n+1} = -a |z_n + x_*| + ax_* + bz_{n-1}. \qquad (4.62)$$

that is,

$$\begin{cases} \Delta z_n = -z_n + f(z_n, z_{n-1}) \\ f(z_n, z_{n-1}) = -a |z_n + x_*| + ax_* + bz_{n-1} \end{cases} \qquad (4.63)$$

and it is straightforward to check that:

$$|f(z_n, z_{n-1})| \leq (|a| + |b|)\max\{|z_n|, |z_{n-1}|\} \qquad (4.64)$$

Thus, an application of Theorem 50 provides the following corollary: If

$$|a| + |b| < 1 \qquad (4.65)$$

then

$$|x_n - x_*| \le \max\{|x_0 - x_*|, |x_1 - x_*|\}(|a| + |b|)^{\frac{n}{2}}, n \ge 0 \qquad (4.66)$$

for all solutions $\{x_n\}$ to Lozi mapping (4.60).

In particular, if (4.65) holds, then the point (x_*, x_*) is the global attractor of the Lozi mappings (4.60). This result is strong when $a > 0, b > 0$, because the equilibrium (x_*, x_*) is a saddle point for $a > 1 - b$.

4.7 Global behaviors of some max difference equations

In [Papaschinopoulos & Hatzifilippidis (2001)] the existence of generalized invariants and the periodicity of the positive solutions of the following max equations:

$$x_{n+1} = \frac{\max\left\{a_n\left(\prod_{i=n-k+1}^{n} x_i\right), b_n\right\}}{\prod_{i=n-k}^{n} x_i}, n = 0, 1, \dots \qquad (4.67)$$

was studied, where a_n, b_n are sequences of positive numbers, $x_{-k}, x_{-k+1}, x_0 \in (0, \infty)$ and $k \in \{2, 3, \dots\}$. From (4.67) it follows that for $a_n = 1, b_n = A_n$, for $n = 0, 1, \dots$ and $k = 1$, Eq. (4.67) reduces to Eq. (4.69). In [Feuer, et al. (2000)] the asymptotic behavior, oscillatory character, and periodic nature of the solutions of the max equation:

$$x_{n+1} = \frac{\max\{x_n, A\}}{x_n x_{n-1}}, n = 0, 1, \dots \qquad (4.68)$$

was investigated. Here A is a real constant and x_{-1}, x_0 are nonzero constants. Note that (4.68) can be reduced to an equation that is a special case of the Lozi map (3.1).

In [Briden, et al. (2000)] the periodic nature, the boundedness and persistence of the solutions of the non autonomous max equation:

$$x_{n+1} = \frac{\max\{x, A_n\}}{x_n x_{n-1}}, n = 0, 1, \dots \qquad (4.69)$$

were studied, where A_n is a sequence of positive numbers and x_{-1}, x_0 are positive numbers.

Definition 59 *According to [Grove, et al. (1998)], we say that a non autonomous difference equation:*

$$x_{n+1} = F(n, x_{n-k}, x_{n-k+1}, x_n) \qquad (4.70)$$

$n = 0, 1, ...,$ *where* $F : \mathbb{N} \times (0,\infty) \times ... \times (0,\infty) \rightarrow (0,\infty)$ *is a continuous function on* $(0,\infty) \times ... \times (0,\infty)$, *has a generalized invariant if there exists a function* $I : \{v_0, v_0 + 1, ...\} \times (0,\infty) \times ... \times (0,\infty) \rightarrow (0,\infty)$ *(where* $v_0 \geq 0$*) continuous on* $(0,\infty) \times ... \times (0,\infty)$ *such that for every positive solution of Eq. (4.70), we have:*

$$I_n = I(n, x_{n-k}, x_{n-k+1}, x_n) \leq M, n \geq v_0 \tag{4.71}$$

where M is a positive number.

Definition 60 *A solution* $(x_n)_n$ *of (4.70) is bounded and persists if there exist positive constants C and D such that for all* $n \in \mathbb{N}$, *we have:*

$$C \leq x_n \leq D \tag{4.72}$$

To study the existence of generalized invariants for (4.67) we need the following definition:

Definition 61 *According to [Grove, et al. (1998)] and [Papaschinopoulos & Schinas (1999)], we say that* a_n *and* b_n *in (4.67) satisfy hypothesis* **H1** *if there exists a sequence* $(\Delta_n)^{\infty}_{n=-1}$ *of positive numbers such that the following hold:*

(i) $(\Delta_n)^{\infty}_{n=-1}$ *is non increasing function.*

(ii) *There exists a number* $\Delta \in (0,\infty)$ *such that* $\lim_{n\to\infty}\Delta_n = \Delta$.

(iii) $\Delta_{n-1}a_n \geq \Delta_n a_{n+1}, \Delta_{n-1}b_n \geq \Delta_n b_{n+1}$ *for all* $n \geq 0$.

We say that a_n *and* b_n *satisfy hypothesis* **H2** *if there exists a sequence* $(\Delta_n)^{\infty}_{n=1}$ *of positive numbers satisfying conditions (i) and (ii) of Definition 61 and*

$$\begin{cases} \Delta_{n-1}a_n \geq \Delta_n a_{n+k+1} \\ \qquad\qquad\qquad\qquad , n \geq 0 \\ \Delta_{n-1}b_n \geq \Delta_n b_{n+1} \end{cases} \tag{4.73}$$

Thus, the following result was proved in [Papaschinopoulos & Hatzifilippidis (2001)]:

Proposition 4.7 *(a) Suppose that* a_n *and* b_n *satisfy hypothesis* **H1**. *Then the following statements are true:*

$$\begin{cases} (i)\ \Delta^2_{n-(k+2)}a_{n-(k+1)} \geq \Delta^2_{n-1}a_n. \\ (ii)\ \Delta^2_{n-(k+2)}a^2_{n-(k+1)} \geq \Delta^2_{n-1}a^2_n. \\ (iii)\ \Delta^2_{n-2}a_{n-1}b_{n-1} \geq \Delta^2_{n-1}a_n b_n. \end{cases} \tag{4.74}$$

(b) *Suppose that* a_n *and* b_n *satisfy hypothesis* **H1**. *Then Eq. (4.67) has a generalized invariant* $I \{k + 2, k + 3, ...\} \times (0,\infty) \times...\times (0,\infty) \rightarrow (0,\infty)$ *given by:*

$$I(n, y_1, y_2, ..., y_{k+1}) = \max \{\eta_1, \eta_2\} \tag{4.75}$$

where

$$\begin{cases} \eta_1 = \left\{ \Delta^2_{n-(k+2)} a_{n-(k+1)}\, y_1,\; \Delta^2_{n-(k+1)} a_{n-k} y_2, ..., \Delta^2_{n-2} a_{n-1}\, y_{k+1} \right\} \\[2em] \eta_2 = \left\{ \dfrac{\Delta^2_{n-(k+2)} a_{n-(k+1)}}{y_1},\; \dfrac{\Delta^2_{n-(k+1)} a_{n-k}}{y_2}, ...,\; \dfrac{\Delta^2_{n-2} a_{n-1}}{y_{k+1}},\; \dfrac{\Delta^2_{n-2} a_{n-1} b_{n-1}}{\prod\limits_{i=1}^{k+1} y_i} \right\} \end{cases} \tag{4.76}$$

(c) *Suppose that a_n and b_n satisfy hypothesis **H2**. Then Eq. (4.67) has a generalized invariant I $\{k + 2, k + 3, ...\} \times (0,\infty) \times ... \times (0,\infty) \to (0,\infty)$ given by:*

$$I\,(n,\, y_1,\, y_2,\, ...,\, y_{k+1}) = \max\,\{\eta_3,\, \eta_4\} \tag{4.77}$$

where

$$\begin{cases} \eta_3 = \left\{ \Delta^2_{n-(k+2)} y_1,\; \Delta^2_{n-(k+1)} y_2, ..., \Delta^2_{n-2}\, y_{k+1} \right\} \\[2em] \eta_4 = \left\{ \dfrac{\Delta^2_{n-(k+2)} a_{n-(k+1)}}{y_1},\; \dfrac{\Delta^2_{n-(k+1)} a_{n-k}}{y_2}, ...,\; \dfrac{\Delta^2_{n-2} a_{n-1}}{y_{k+1}},\; \dfrac{\Delta^2_{n-2} b_{n-1}}{\prod\limits_{i=1}^{k+1} y_i} \right\} \end{cases} \tag{4.78}$$

Necessary and sufficient conditions for which a positive solution of Eq. (4.67) is bounded and persists were given in the following result proved in [Papaschinopoulos & Hatzifilippidis (2001)]:

Proposition 4.8 *Consider Eq. (4.67). Then the following statements are true:*

I. *Suppose that Eq. (4.67) has a positive bounded solution. Then the sequences a_n and b_n are also bounded.*

II. *Suppose that a_n and b_n satisfy hypothesis **H1** and a_n persists. Then every positive solution of Eq. (4.67) is bounded and persists.*

III. *Suppose that a_n and b_n satisfy hypothesis **H2**. Then the following conditions hold: (a) Every positive solution of Eq. (4.67) is bounded. (b) If a_n persists, then every positive solution of Eq. (4.67) is bounded and persists.*

The periodicity of the positive solutions of Eq. (4.67) was given in the following result proved in [Papaschinopoulos & Hatzifilippidis (2001)]:

Proposition 4.9 *Consider Eq. (4.67), where*

$$a_n = 1 \text{ and } b_n = A, n = 0, 1, ... \tag{4.79}$$

where A is a positive constant. Let x_n be a positive solution of Eq. (4.67) with initial conditions satisfying:

$$\begin{cases} (i) \ \prod_{i=-k+1}^{0} x_i \leq A \\[2ex] (ii) \ \prod_{i=-k}^{-1} x_i \leq A \\[2ex] (iii) \ x_i x_{i-1} \geq 1, \ for \ i = -k+1, ..., -1, 0, \end{cases} \qquad (4.80)$$

where $k \geq 2$ and $x_{-k}, ..., x_0$ are real positive numbers. Then x_n is periodic with prime period equal to $k + 2$.

Consider Eq. (4.67) such that:

$$a_n = 1, 0 < b_n \leq A, n = 0, 1, ... \qquad (4.81)$$

where A is a constant satisfying $0 < A < 1$. Define the set:

$$R_2^{k+1} = \left\{ (y_1, y_2, ..., y_{k+1}) \in \mathbb{R}_+^{k+1} : y_1, y_2, ..., y_{k+1} \in \left[\sqrt[k]{A}, \frac{1}{\sqrt[k]{A}} \right] \right\}, k > 2. (4.82)$$

Then R_2^{k+1} is an invariant set. Furthermore, if x_n is a nontrivial solution of Eq. (4.67) with $x_{-k}, x_{-k+1}, ..., x_0 \in R_2^{k+1}$, then x_n is periodic with prime period $2k + 2$. Consider Eq. (4.67), where

$$a_n = 1, n = 0, 1, ... \qquad (4.83)$$

Suppose that b_n is periodic with prime period $k+1$ and the following conditions are true:

$$\begin{cases} (i) \ \prod_{i=-k}^{-1} x_i \leq \min\left\{ \dfrac{b_0^2}{b_k}, \dfrac{b_1^2}{b_k}, ..., \dfrac{b_{k-1}^2}{b_k}, b_k \right\} \\[2ex] (ii) \ \prod_{i=-k+1}^{0} x_i \leq \min\left\{ \dfrac{b_1^2}{b_0}, \dfrac{b_2^2}{b_0}, ..., \dfrac{b_k^2}{b_0}, b_0 \right\} \\[2ex] (iii) \ x_i x_{i+1} \geq \max\left\{ \dfrac{b_{k+i}}{b_{k+i+1}}, \dfrac{b_{k+i+1}}{b_{k+i}}, ..., \dfrac{b_{k+i} b_{k+i+1}}{b_{k+i+2}^2}, ..., \dfrac{b_{k+i} b_{k+i+1}}{b_{2k+i}^2} \right\} \end{cases} \qquad (4.84)$$

where $k \geq 2$ and $i \in \{-k, -k + 1, ..., -1\}$. Then x_n is periodic with prime period $(k + 2)(k + 1)$.

In particular, the above results hold for some Lozi mappings.

4.8 Generalized piecewise-linear area-preserving plane maps

In this section, we discuss the dynamics of some special cases of the following generalized piecewise-linear area-preserving plane maps:

$$f(x, y) = (\alpha_0 + \alpha_1 F(x) + \alpha_2 y, \alpha_3 x) \tag{4.85}$$

First, we remark that the Lozi mappings (3.1) and (3.2) are special cases of system (4.85). Second, we give in this section, some results concerning the dynamics of the maps of the plane $\mathbb{R}^2$ given by:

$$T_{ab}(x, y) = (F_{ab}(x) - y, x) \tag{4.86}$$

studied deeply in the three papers of Lagarias & Rains [Lagarias & Rains (2005(a-b-c))]. The function $F_{ab}(x)$ is given by:

$$F_{ab}(x) = \begin{cases} ax \text{ if } x \geq 0 \\ bx \text{ if } x < 0 \end{cases} \tag{4.87}$$

The map (4.86) with the function (4.85) is equivalent to the following nonlinear difference equation:

$$\begin{cases} x_{n+2} = \mu \, | \, x_{n+1} \, | + v x_{n+1} - x_n \\ \mu = \dfrac{1}{2}(a - b), v = \dfrac{1}{2}(a + b) \end{cases} \tag{4.88}$$

The maps (4.86) have the parameter space $(a, b) \in \mathbb{R}^2$ and they are areapreserving homeomorphisms of $\mathbb{R}^2$ that map rays from the origin into rays from the origin. This action on rays define an auxiliary map $S_{ab} : S^1 \to S^1$ of the circle, which has a well-defined *rotation number*. In [Beardon, et al. (1995)] the periodic orbits of the maps (4.88) are studied along the set of parameter values Ω_B for which all orbits are bounded and it was shown that $\Omega_R = \Omega_{B'}$ where Ω_R is the set of maps topologically conjugate to a rotation of the plane. This result follows from results of Herman in [Herman (1986), VIII.2.4].

Let Ω_{SB} be the set of all parameter values for which the map T_{ab} has at least one nonzero bounded orbit. Let Ω_Q be the set of parameter values (a, b) for which $T_{a,b}$ has a piecewise conic invariant circle of the type given in Theorem 63(d), i.e., parameter values such that the S_{ab}-orbit of $(0, 1)$ contains $(0, -1)$, and S_{ab} has irrational rotation number. Thus, the following result was proved in [Lagarias & Rains (2005(b))]:

Theorem 57 *The set* $\Omega_P \cup \Omega_Q$ *has Hausdorff dimension 1.*

and we have the following inclusions:

$$\Omega_P \subset \Omega_P \cup \Omega_Q \subseteq \Omega_R = \Omega_B \subseteq \Omega_{SB} \tag{4.89}$$

with the following conjecture.

Conjecture 58 *The set* $\Omega_P \cup \Omega_Q$ *is a dense subset of* Ω_{SB}.

This conjecture was discussed with some examples in [Lagarias & Rains (2005(b))]. More results on the set Ω_{SB} can be found in Section 4.8.4 below.

The map (4.86) can be viewed as a solution to the one-dimensional nonlinear difference equation of Schrôdinger type:

$$x_{n+2} + 2x_{n+1} - x_n + V_\mu(x_{n+1})x_{n+1} = Ex_{n+1} \tag{4.90}$$

where the *potential* $V_\mu(x)$ and the energy value E are given by:

$$\begin{cases} V_\mu(x) = \begin{cases} \mu \text{ if } x \geq 0, \\ -\mu \text{ if } x < 0. \end{cases} \\ E = 2 - v \end{cases} \tag{4.91}$$

In [Lagarias & Rains (2005(a))] the possible dynamics (where the map T_{ab} is a periodic map or the set Ω_p of purely periodic maps) under iteration of T_{ab} when the auxiliary map S_{ab} has rational rotation number was characterized, i.e., the induced dynamics on rays, which is described by a circle map S_{ab} defined by:

$$S_{ab}(e^{i\theta}) = \frac{T_{ab}(e^{i\theta})}{\|T_{ab}(e^{i\theta})\|}, 0 \leq \theta \leq 2 \tag{4.92}$$

where $\| T_{ab}(e^{i\theta}) \|$ is the Euclidean norm on $\mathbb{R}^2$, and $e^{i\theta} = x + iy \in \mathbb{C}$ with $(x, y) \in \mathbb{R}^2$. Formula (4.92) is well defined since the map T_{ab} is homogeneous, so sends rays $[v] = \{\lambda v : v \geq 0\}$ to rays $T_{ab}([v])$. In particular, the dynamics of the map T_{ab} can be understood by two things: the first one is the study of the dynamics of the circle map S_{ab}, and the second one is the study of the motion of points inside the individual rays. The circle map S_{ab} has a well-defined rotation number $r(S_{ab})$ as shown in [Sutherland & Kohmoto (1987)][10] given by:

$$r(S_{ab}) = \lim_{n \to \infty} \frac{\widetilde{S}_{ab}^{(n)}(x) - x}{n} \tag{4.93}$$

where $\tilde{S}_{ab}$ is any lift of S_{ab} to $\mathbb{R}$ (2π lifts to1). Here $r(S_{ab})$ is independent of the initial value x (*mod*1) and of the choice of lift. Generally, it was assumed that positive rotation number corresponds to counter-clockwise rotation. If S_{ab} has a periodic point, then $r(S_{ab})$ is rational and conversely, as well. More properties of the map (4.86) can be found in Section 4.8.2. For one-parameter families maps and cases where the map T_{ab} is purely periodic, the rotation number can be determined exactly. See [Lagarias & Rains (2005(a)), Beardon, et al. (1995)].

---

[10] In fact, every orientation-preserving homeomorphism $S : S^1 \to S^1$ has a well-defined rotation number.

The following result proved in [Lagarias & Rains (2005(a))] determine the allowed range of the rotation number as follows:

Theorem 59 *For fixed real a, and* $-\infty < b < \infty$, *the rotation number* $r\ (S_{ab})$ *is continuous and non increasing in b, and completely fills out the following intervals:*
(a) For a < 0, we have:

$$r\ (S_{ab}) \in \left[0, \frac{1}{2}\right] \tag{4.94}$$

(b) Let $0 \le a < 2$. *Then for each integer* $n \ge 2$, *on the interval*

$$2\cos\frac{\pi}{n} \le a < 2\cos\frac{\pi}{n+1} \tag{4.95}$$

$$r\ (S_{ab}) \in \left[0, \frac{1}{n+1}\right] \tag{4.96}$$

(c) For a ≥ 2, *we have:*

$$r\ (S_{ab}) = 0 \tag{4.97}$$

The range of the rotation number in (μ, v)-space, for fixed μ can be given by the following result proved also in [Lagarias & Rains (2005(a))]:

Theorem 60 *For fixed real* μ *and* $-\infty \le v < \infty$ *the values of* $r\ (S_{\mu v})$ *are continuous and non increasing in* μ *and completely fill out the closed interval:*

$$r\ (S_{\mu v}) \in \left[0, \frac{1}{2}\right] \tag{4.98}$$

4.8.1 Rational rotation number

A characterization of the values of (a, b) where T_{ab} is periodic is given in the next result proved in [Beardon, et al. (1995)]. In fact, The maps S_{ab} exhibit a *modelocking* behavior which means that there are open sets in the parameter space for which $r\ (S_{ab})$ takes a fixed rational value (for certain rationales) as in Theorem 55(c) above.

Theorem 61 T_{ab} *is of finite order if and only if the orbit of* $(0, 1)$ *is periodic.*

Theorem 61 implies that the parameter values at which T_{ab} is a periodic map fall in a countable number of one-parameter families, plus a countable number of isolated values, described by the period and possible symbolic dynamics of the orbits.

The characterization of the dynamics in all cases of rational rotation number was proved in [Lagarias & Rains (2005(a))]:

Theorem 62 *If the rotation number $r(S_{ab})$ is rational, then S_{ab} has a periodic orbit, and one of the following three possibilities occurs: (a) S_{ab} has exactly one periodic orbit. Then T_{ab} has exactly one periodic orbit (up to scaling) and all other orbits diverge in modulus to $+\infty$ as $n \to \pm\infty$. (b) S_{ab} has exactly two periodic orbits. Then T_{ab} has no periodic orbits. All orbits of T_{ab} diverge in modulus to $+\infty$ as $n \to \pm\infty$, with the exception of orbits lying over the two periodic orbits of S_{ab}. These exceptional orbits have modulus diverging to $+\infty$ in one direction and to 0 in the other direction, with forward divergence for one, and backward divergence for the other. (c) S_{ab} has at least three periodic orbits. Then Tab is of finite order, i.e., $T_{ab}^{(k)} = I$ for some $k \geq 1$, and all its orbits are periodic.*

Theorem 62 answers a question raised in [Beardon, et al. (1995), p. 671]: Can there be a nonperiodic T_{ab} having two disjoint orbits of rays on each of which $T_{ab}^q = I$?. This corresponds to case Theorem 62(b), which means that the answer is: no. The proof of Theorem 62 uses the following auxiliary lemma that confirms essentially the uniqueness assertion. This lemma was proved in [Lagarias & Rains (2005(a))]:

Lemme 4.3 *For every rational $r = \dfrac{p}{q}$ with $0 < r < \dfrac{1}{2}$ and each fixed $\mu \in \mathbb{R}$, the following holds: For each $v = e^{i\theta} \in S^1$ there is a unique parameter value $v \in \mathbb{R}$ such that v is a periodic point of $S_{\mu v}$ of rotation number r.*

4.8.2 Properties of the associated circle map

In this section, we present some properties of the associated circle map (4.92). Indeed, it was shown in [Lagarias & Rains (2005(a))] that the circle map $S_{\mu v}(\theta)$ in the (μ, v)-coordinates has the following properties:

Theorem 63 *(a) Each $S_{\mu v}(\theta)$ is an orientation-preserving homeomorphism from S^1 to itself. The derivative $\dfrac{d}{d\theta}(S_{\mu v}(\theta))$ is continuous and of bounded variation. (b) (i) For fixed b, the rotation number $r(S_{ab})$ is nonincreasing in a, and for fixed a it is nonincreasing in b. (ii) (c) For fixed μ, the rotation number $r(S_{\mu v})$ is nonincreasing in v. (d) For each fixed a and $-\infty \leq b \leq \infty$ the values of $r(S_{ab})$ completely fill out the following intervals, including their endpoints: (i) For a < 0, the rotation number satisfies:*

$$r(S_{ab}) \in \left[0, \frac{1}{2}\right] \tag{4.99}$$

(ii) Let $0 \leq a \leq 2$. For each $n \geq 2$ on the interval (4.95) the rotation number satisfies:

$$r(S_{ab}) \in \left[0, \frac{1}{n+1}\right] \tag{4.100}$$

(iii) For a ≥ 2, the rotation number satisfies:

$$r\,(S_{ab}) = 0 \tag{4.101}$$

(e) Let n ≥ 1 and suppose that:

$$S^{(n)}_{ab}\,(0,\,1) = (0,\pm 1)\,,\,or\,S^{(n)}_{ab}\,(0,-1) = (0,\pm 1) \tag{4.102}$$

Then one of the following two relations holds:

$$T^{(n)}_{ab}\,(0,\,1) = (0,\,\lambda) \tag{4.103}$$

$$T^{(n)}_{ab}\,(0,-1) = (0,-\lambda^{-1}) \tag{4.104}$$

where λ is a nonzero real number. (i) If λ > 0, then both relations above hold. In addition, we have:

$$T^{(n)}_{ab}\,(-1,\,0) = (-\lambda,\,0)\,and\,T^{(n)}_{ab}\,(1,\,0) = (\lambda^{-1},\,0) \tag{4.105}$$

The rotation number r (S_{ab}) is rational. (ii) If λ < 0, then necessarily λ = −1. In the first case, we have:

$$T^{(n)}_{ab}\,(0,\,1) = (0,-1)\,and\,T^{(n)}_{ab}\,(-1,\,0) = (1,\,0) \tag{4.106}$$

while in the second case, we have:

$$T^{(n)}_{ab}\,(0,-1) = (0,\,1)\,and\,T^{(n)}_{ab}\,(1,\,0) = (-1,\,0) \tag{4.107}$$

The rotation number r (S_{ab}) can be irrational or rational.

4.8.3 Irrational rotation number

In this section, we discuss different results obtained about the dynamics of map (4.86) when the rotation number $r\,(S_{ab})$ is irrational, in which the dynamics becomes more complicated. In this case, a result of Herman given in [Kotani (1990), VIII.2.4] implies a dichotomy: *either the map T_{ab} is topologically conjugate to a rotation of the plane, or else it has a dense orbit.* In fact, it was proved in [Lagarias & Rains (2005(c))] that if the map (4.86) has a dense orbit then it also possesses a bounded orbit. On the other hand, smoothness properties of the circle maps S_{ab}, leads to the following result proved in [Lagarias & Rains (2005(a))]:

Theorem 64 *(a) If the rotation number r (Sab) = r is irrational, then there exists a homeomorphism h : $S^1 \to S^1$ such that:*

$$h\,o\,S_{\mu\nu}\,o\,h{-}1 = \Theta_r \tag{4.108}$$

where $\Theta_r : S^1 \to S^1$ is rotation by r (θ → θ + 2πr).

(b) Let T : $\mathbb{R}^2 \to \mathbb{R}^2$ be any homeomorphism which is scale-invariant; that is T (λv) = λT (v) for all λ ≥ 0, all v ∈ $\mathbb{R}^2$. Let

$$S(v) = \frac{T(v)}{\|T(v)\|} \qquad (4.109)$$

be its associated circle map, and suppose that S is topologically conjugate to an irrational rotation. If T has a forward orbit $O^+(v_0) = \{v_n : n \geq 0\}$, such that:

$$\frac{1}{C} \leq |v_n| \leq C, n \geq 0 \qquad (4.110)$$

for some positive constant $C > 1$.
   Then T has an invariant circle, and all orbits are bounded.

We note that Theorem 61 means that the existence of an orbit with elements of modulus bounded away from 0 and ∞ implies that there exists an invariant circle. Theorem 61 means that an invariant circle is necessarily preserved under the reflection symmetry $R(x, y) = (y, x)$. The application of Theorem 60(b) to T_{ab}, gives a criterion when it has invariant circles, i.e., the symmetry is manifest in plots of invariant circles [Lagarias & Rains (2005(a))]:

Theorem 65 *If the map T_{ab} has associated irrational rotation number r (S_{ab}), then any invariant circle of T_{ab} is symmetric under the involution $R(x, y) = (y, x)$.*

More results about irrational rotation number can be found in [Lagarias & Rains (2005(b))]. Indeed, the existence of special parameter values was obtained for which the map T_{ab} has every nonzero orbit contained in an invariant circle with an irrational rotation number, with invariant circles that are piecewise unions of arcs of conic sections, i.e., there exist parameter values (a, b) with irrational rotation number for which T_{ab} has invariant circles with a *striking* structure as shown by the following result proved and tested numerically in [Lagarias & Rains (2005(b))]. The method of analysis is based on the construction of the *first-return maps* to suitable sectors of the plane.

Theorem 66 *Suppose that the rotation number $r(S_{ab})$ is irrational and that the S_{ab} orbit of (0, 1) contains (0,−1). Then the following hold:*
   (a) The T_{ab} orbit of (0, 1) contains (0,−1).
   (b) The closure of every (nonzero) orbit of T_{ab} is an invariant circle, which is a piecewise union of arcs of conic sections. The conic sections occurring in such an invariant circle are all of the same type, either ellipses, hyperbolas or straight lines.

By considering those first-return maps of maps in the irrational rotation case to suitable sectors of the plane, the following results were proved in [Lagarias & Rains (2005(b))]:

Theorem 67 *Suppose that the rotation number $r(S_{ab})$ is irrational. Then we have:*

(a) *For any half-open sector $J = R+ [v, v')$ the first return map $T^{(1)}J : J \to J$ of T_{ab} to J is piecewise linear with at most five pieces.*

(b) *Let m_+, m_-, n_+, n_- be nonnegative integers, and set*

$$\begin{cases} O = O_1 \cup O_2 \cup O_3 \cup O_4 \\ O_1 = \{T_{ab}^{(1+i)} (0, 1) : 0 \le i < m_+\} \\ O_2 = \{T_{ab}^{(-i)} (0, 1) : 0 \le i < m_-\} \\ O_3 = \{T_{ab}^{(1+i)} (0,-1) : 0 \le i < n_+\} \\ O_4 = \{T_{ab}^{(-i)} (0,-1) : 0 \le i < n_-\} \end{cases} \qquad (4.111)$$

Let $J = R^+ [v, v')$ be any half-open sector determined by two elements $v, v' \in O$ such that J contains no points of O in its interior. Then the first return map $T_J : J \to J$ is piecewise linear with at most three pieces.

(c) *Suppose that S_{ab} has irrational rotation number, and that $(0, 1)$ and $(0,-1)$ are in the same T_{ab}-orbit. Let*

$$O' = \{T_{ab}^{(j)} (v_0) : 1 \le j \le m\} \qquad (4.112)$$

be any finite segment of that orbit which includes all of the points strictly between $(0, 1)$ and $(0,-1)$, as well as whichever of $(0, 1)$ and $(0,-1)$ comes last in the orbit. If $J = R+ [v, v')$ is a half-open sector determined by two members of O' containing no elements of O' in its interior, then the first return map T_J on J is piecewise linear with exactly two pieces, and the linear transformations corresponding to the two pieces commute.

It was shown in [Lagarias & Rains (2005(b))] that if the *Tab*-orbit of $(0, 1)$ contains $(0,-1)$, then sectors on which the first return map is piecewise linear with at most two pieces can be obtained:

Theorem 68 *Suppose that S_{ab} has irrational rotation number and that the S_{ab} orbit of $(0, 1)$ contains $(0,-1)$. Then the T_{ab}-orbit of $(0, 1)$ contains $(0,-1)$, and T_{ab} has a piecewise conic invariant circle. If $S_{ab}^{(n)} (0, 1) = (0,-1)$ for $n \in \mathbb{Z}$, then the number of conic pieces of the invariant circle is at most $|n|$, and all conic pieces are of the same type, either arcs of ellipses, arcs of hyperbolas or line segments, respectively.*

The proof of Theorem 68 is based essentially on the following well-known lemma proved in Arnold and Avez [Arnold & Avez (1968), Appendix 27]:

Lemme 4.4 *Let $M = \begin{pmatrix} a & b \\ c & d \end{pmatrix} \in SL(2,\mathbb{R})$, and suppose that $M \ne \pm I$. Then the map $(x, y) \to M (x, y)$ leaves invariant the quadratic form $Q(x, y) = cx^2 + (d - a)xy - by^2$, and the only quadratic forms it leaves invariant are scalar multiples of $Q(x, y)$. The level sets $cx^2 + (d-a)xy - by^2 = \lambda$ for real λ are invariant sets, and fill the* *plane. Furthermore: (1) If | Tr(M) | < 2, then the nonempty level sets are ellipses, or else a single point.*

(2) *If | T r(M) | > 2, then the level sets are hyperbolas, except for λ = 0, where they are their asymptotes, consisting of two straight lines through the origin.*

(3) *If | T r(M) | = 2, then the nonempty level sets are either two parallel lines whose vector sum is a fixed parallel line through the origin; or, for λ = 0, this fixed line. Any matrix M' commuting with M also preserves the quadratic form Q.*

As a first example of one-parameter families where (0, 1) and (0,−1) are in the same orbit, we consider the case where $1 < a < \sqrt{2}$, and $b = \dfrac{2(a^2 - 1)}{a(a^2 - 2)}$, in which the map T_{ab} is either periodic or has a piecewise elliptical invariant circle[11]. In this case, the explicit formula for the rotation number $r(S_{ab})$ is given by:

$$
\begin{cases}
r(S_{ab}) = \dfrac{3\pi - 7\theta}{14\pi - 32\theta} \\[2mm]
a = 2\cos\theta, \dfrac{\pi}{4} < \theta < \dfrac{\pi}{3}
\end{cases}
\tag{4.113}
$$

Formula (4.113) implies that $r(S_{ab})$ is rational if and only if θ is a rational multiple of π. In the second example, we consider the case where $0 < a < 1$, and $b = \dfrac{1}{2}\dfrac{3a^2 - 3 + \sqrt{a4 - 2a2 + 9}}{a(a^2 - 2)}$, in which the map T_{ab} is either periodic or has

a piecewise conic invariant circle. In this case, all three cases of piecewise ellipses, straight lines or hyperbolas occur[12].

4.8.4 The set of nonzero bounded orbit

In [Lagarias & Rains (2005(c))] a study of the set Ω_{SB} of nonzero bounded orbit was given. This study corresponds to l_∞-eigenfunctions of the difference operator (4.90). In fact, it was shown that for transcendental μ given by (4.88), the set of energy values E having a bounded solution is a Cantor set that has positive (one-dimensional) measure for all real values of μ. Recall that a Cantor set in $\mathbb{R}$ is a perfect totally disconnected set. The first result was proved in [Lagarias & Rains (2005(c))]:

$$
Spec_\infty[\mu] = \{E = 2 - v : (\mu, v) \in \Omega_{SB}\}
\tag{4.114}
$$

---

[11] These parameter values were obtained from the assumption that $T^{(8)}_{ab}(0,-1) = (0, 1)$, with appropriate sign conditions.

[12] The choice of parameter values is based on the relation $T^{(10)}_{ab}(0, 1) = (0,-1)$ with appropriate sign conditions.

Theorem 69 *The set Ω_{SB} is a closed set. It consists of all parameter values (μ, v) for which the associated values (a, b) satisfy one of the conditions below:*
(a) $r(S_{ab})$ is rational, and T_{ab} has a periodic orbit.
(b) $r(S_{ab})$ is irrational.

To prove Theorem 69(b), the (μ, v)-parametrization for constant μ must be studied to show the following facts. (Theorem 70 below):

(1) The set of values v with $r(S_\mu) = r$ with $0 < r = \dfrac{p}{q} < \dfrac{1}{2}$ is either a single

point v (then $T_{\mu v}$ is periodic) or a closed interval $[v_1, v_2] = [v^-(r), v^+(r)]$ (then no $T_{\mu v}$ is periodic and the only $T_{\mu v}$ with a periodic orbit are the endpoints $v = v^-(r)$ and $v^+(r)$).

(2) The set of values v with $r(S_\mu) = r$ irrational with $0 < r < \dfrac{1}{2}$ is a single point v.

The proof of these items can be done by obtaining a bounded orbit in case (2) above by a limiting procedure using bounded orbits in case (1) with suitable rational rotations $\dfrac{p_n}{q_n}$ approaching r. We note that the endpoints of rotation intervals of positive length in Theorem 69(a) give points in the set Ω_{SB} that are not in the set Ω_B. The second result was also proved in [Lagarias & Rains (2005(c))] and gives informations on the location of the set Ω_{SB} viewed in the (μ, a) parameter space:

Theorem 70 *The set of values for which T_μ has a bounded orbit, i.e., $(\mu, v) \in$ SB, in the range $\mu \geq 0$ lies inside the cylinder $-2 \leq a \leq 2$, where $a = v + \mu$. Furthermore:*

(a) For $-2 \leq a < 0$, one has:

$$0 \leq \mu \leq \frac{2}{|a|} - \frac{|a|}{2} \tag{4.115}$$

(b) For $0 \leq a \leq 2$, set $a = 2 \cos \theta$. Then for each $n \geq 2$ and $\dfrac{\pi}{n+1} \leq \theta < \dfrac{\pi}{n}$, we have:

$$0 \leq \mu \leq \cos \theta - \frac{\sin n\theta + \sin \theta}{\sin(n+1)\theta} \tag{4.116}$$

We note that a weaker bound for the set Ω_B was obtained in [Beardon, et al. (1995), Theorem 1.1(iii)]. In some cases, namely for $0 \leq \mu \leq 6$, the region $\mathcal{R}$ of Theorem 70 becomes unbounded. Thus, the following result was proved also in [Lagarias & Rains (2005(c))]:

Theorem 71 *The set Ω_B consists of those parameter values (μ, v) for whose associated (a, b) the map T_{ab} is conjugate to a rotation of the plane. This occurs if and only if one of the following occur.*

(a) $r(S_{ab})$ is rational and T_{ab} is of finite order, i.e., $T^{(k)}_{ab} = I$ for some finite k.
(b) $r(S_{ab})$ is irrational, and T_{ab} contains an invariant circle.

The following theorem is the main result of [Lagarias & Rains (2005(c))]:

Theorem 72 *For $\mu_0 \in \mathbb{R}$ the set*

$$Spec_\infty [\mu] = \{E = 2 - v : (\mu, v) \in \Omega_s\} \tag{4.117}$$

is a Cantor set if μ is not in a countable exceptional set $\mathcal{E}$ that consists entirely of algebraic numbers.

We note that the set $\mathcal{E}$ contains $\mu = 0$, and it is the only point in $\mathcal{E}$. If so, then the set Ω_{SB} would have the structure: (Cantor set) × (half-line) in the (μ, a)-parameter space, in the region $-2 \leq a \leq 2$ and $\mu > 0$. But, numerical evidence supports the assertion that $Spec_\infty[\mu]$ has positive one-dimensional Lebesgue measure for all real values of μ.

In what follows we discuss the existence of bounded orbits for the map (4.86). Indeed, if one consider the set Ω_{SB} of parameter values having at least one bounded orbit, then the following result was proved in [Lagarias & Rains (2005(c))]:

Theorem 73 *The set Ω_{SB} is a closed set. Each $(\mu, v) \in \Omega_{SB}$ the associated T_{ab} possesses a bounded orbit $O(v_0) = \{v_n : k \in Z\}$ such that:*

$$\|v_0\| = \sup_{n \in Z} \left\| T^{(n)}_{ab}(v_0) \right\| \tag{4.118}$$

The proof of Theorem 73 is done by shifting the orbit appropriately and by using the compactness property of the unit ball in $\mathbb{R}^2$. Also, the proof of Theorem 69 needs the following preliminary result proved in [Lagarias & Rains (2005(c))] by using the (μ, v)-parameters.

Theorem 74 *Let $\mu \in \mathbb{R}$ be fixed and let v vary over $-\infty < v < \infty$.*

(a) *Let $0 < r < \dfrac{1}{2}$ be rational. Then the set $I_\mu(r)$ of values v such that $r(S_\mu) = r$ is either a point $v^\pm(r)$ or an interval $[v^-(r), v^+(r)]$. In the point case, $T_{\mu\pm}$ is of finite order. In the interval case, T_μ is never of finite order, and T_μ for $v \in [v^-(r), v^+(r)]$ contains a periodic orbit if and only if v is one of the endpoints of the interval $v = v^-(r)$ or $v^+(r)$.*

(b) *Let $0 < r < \dfrac{1}{2}$ be irrational. Then the set $I_\mu(r)$ of values v such that $r(S_\mu) = r$ consists of a point $v^\pm(r)$.*

(c) *The set $I_\mu(0)$ of values v with $r(S_\mu) = 0$ is a half-infinite interval $[v^-(0), +\infty)$. In this interval, T_μ is never periodic, and T_μ has a periodic orbit only for $v = v^-(0)$.*

(d) *The set* $I_\mu\left(\dfrac{1}{2}\right)$ *of values* v *with* $r\,(S_\mu) = \dfrac{1}{2}$ *is a half-infinite interval* $\left(-\infty, v^+\left(\dfrac{1}{2}\right)\right]$. *In this interval,* T_μ *is never periodic, and* $T_{\mu h}$ *as a periodic orbit only for* $v = v^+\left(\dfrac{1}{2}\right)$.

We note that the continuity and nonincreasing properties of $S_{\mu v}$ are used to prove Theorem 74. The following result was proved in [Lagarias & Rains (2005(c))] and it makes a relation between the fact that $r(S_{ab})$ is irrational and the existence of bounded orbit for the map T_{ab}.

Theorem 75 *If the rotation number* $r(S_{ab})$ *is irrational, then* T_{ab} *has a bounded orbit.*

Theorem 75 above implies that if the map T_μ had a localized orbit, then S_μ has irrational rotation number, and T_μ is not conjugate to a rotation. In fact, it was observed in [Herman (1986)] that any such map necessarily has a dense orbit and it was conjectured in [Herman (1986), VIII.2.4] that there are some elements of a family of maps having this property, i.e., the case of the *Froeschlé group of homeomorphisms of the plane* generated by $SL(2, \mathbb{R})$ and given by:

$$G_{ab} = \begin{pmatrix} 1 & 0 \\ F_{ab}(x) & 1 \end{pmatrix} \qquad (4.119)$$

for real a, b, viewed as acting on column vectors. In [Beardon, et al. (1995)] the possibility that this occurs for some maps T_{ab} was proved for the parameter line $v = \dfrac{1}{2}(a+b) = 0$. Numerical evidence (for $0 \le \mu \le 1$ and $-2 \le a \le 2$) show that the width of rational rotation numbers $r = \dfrac{1}{n}$ appears to be nondecreasing in μ. This means that in the limit as the denominator cutoff $N \to \infty$, for $\mu = 1$ there is an uncovered set of positive Lebesgue measure (approximately= 2). Thus, the monotonicity hypothesis imply that the two-dimensional Lebesgue measure of Ω_{SB} in the region $0 \le \mu \le 1$ in the (μ, a)-plane is between 2 and 4.

To this end, we discuss the infinitely many of the rational rotation number intervals that have positive length for the family $T_{\mu v}$ for fixed μ and variable v. For $\mu = 0$, the maps $T_{\mu v}$ are linear and the following results were proved in [Lagarias & Rains (2005(c))]:

Theorem 76 (a) *For* $\mu = 0$, $T_{\mu v}$ *has a nontrivial bounded orbit if and only if* $v \in [-2, 2]$, *all orbits are bounded if and only if* $v \in (-2, 2)$.

(b) *Let $q \geq 3$ be prime. For each rational $r = \dfrac{j}{q}$ with $1 \leq j \leq \dfrac{q-1}{2}$, the set $E\left(\dfrac{j}{q}\right)$ of all $\mu \geq 0$ for which the rotation interval of rotation number $\dfrac{j}{q}$ is degenerate is a finite set of algebraic numbers.*

Thus, the set $Spec_\infty\,[\mu]$ defined by (4.117) is a totally disconnected set, because Theorem 76(b) guarantees that the rotation interval $I_\mu\left(\dfrac{j}{q}\right)$ is nontrivial, and by Theorem 55 its interior is not in Ω_{SB}. Also, the set $Spec_\infty\,[\mu]$ is perfect for all $\mu \in \mathbb{R}$ because for a given $v \in Spec_\infty\,[\mu]$, the rotation number $r\,(S_{\mu v})$ can be approximated both from above and below by sequences $(\mu, v^+_{\ n}), (\mu, v^-_{\ n}) \in \Omega_{OB}$. These sequences have irrational rotation numbers converging to $r\,(S_{\mu v})$. Thus, we conclude that at least one of $\lim_{n \to \infty} v^-_{\ n} = v$ or $\lim_{n \to \infty} v^+_{\ n} = v$ holds, and both hold if $r\,(S_{\mu v})$ is irrational. Two different numerical experiments suggest that the measure is positive for at least some non-zero μ. First, the calculation at $\mu = 1$ of the uncovered area (as a function of denominator) suggested that positive measure will persists. Second, numerical random selection of parameter values in the allowed interval seemed to produce invariant circles with positive probability. Finally, the following conjecture was formulated in [Lagarias & Rains (2005(c))]:

Conjecture 77 *The set Ω_{SB} has positive two-dimensional Lebesgue measure. Furthermore, for each real μ the l_∞-spectrum (3.117) $Spec_\infty\,[\mu]$ has positive one-dimensional Lebesgue measure.*

This conjecture was formulated via some numerical evidences on the dynamics of the map (4.86), i.e., the set Ω_{SB} has positive Lebesgue measure, in the region $0 \leq \mu \leq 1$ and $-2 \leq a \leq 2$.

4.9 Smooth versions of the Lozi mappings

In [Aziz-Alaoui, et al. (2001)] a smooth version $L_{a,b,\varepsilon}$ of the piecewise linear Lozi map (3.1) was studied analytically and tested numerically. This version is given by:

$$L_{a,b,\varepsilon}\,(x, y) = (1 - aS_\varepsilon\,(x) + y,\ bx) \qquad (4.120)$$

where $a, b \in \mathbb{R}, b \geq 0$ and $S_\varepsilon\,(x)$ is defined by:

$$S_\varepsilon : \mathbb{R} \to \mathbb{R},\ x \to S_\varepsilon\,(x) = \begin{cases} |x|, \text{if} \, |x| \geq \varepsilon \\ g_\varepsilon(x) = \dfrac{x^2}{2\varepsilon} + \dfrac{\varepsilon}{2}, \text{if} \, |x| \leq \varepsilon \end{cases} \qquad (4.121)$$

where ε satisfies the conditions:

$$0 < \varepsilon < 1 \text{ and } \lim_{\varepsilon \to 0} S_\varepsilon(x) = |x| \qquad (4.122)$$

It was found that map (4.120)-(4.121) still possesses some characteristics of Hénon-type dynamics. In particular, if ε measures the degree of smoothness, then it was proved that if $\varepsilon \to 0$, then the stability and the existence of the fixed points are the same for both maps. General properties of $L_{a,b,\varepsilon}$ can be seen from the following result proved in [Aziz-Alaoui, et al. (2001)]:

Proposition 4.10 *As $\varepsilon \to 0$, the number, existence and stability of the fixed points of both Lozi map (3.1) and $L_{a,b,\varepsilon}$ mappings (4.120)-(4.121) are the same (in almost all the (a, b)-plane).*

Hence, the importance of the results obtained for the map (4.120)-(4.121) lies in the similarity of those obtained for the Lozi map (3.1). Indeed, maps (4.120)-(4.121) and (3.1) have identical behaviors for small ε and almost all the (a, b)-plane. In particular, the chaotic dynamics for the two maps is the same both in the form (of the attractor) and of a chaotic saddle as shown in Fig. 4.12 and 4.13.

Other examples about generating Hénon-Lozi mappings dynamics can be found in [Dobrynskiy (2008)]. In this case, for a Hénon–Lozi mapping F, sufficient conditions were derived such that there exists a domain U on the plane such that its closure is mapped by F strictly inside U. This result confirms the existence of a compact invariant set V in U. Also, it was proved that there exists an open set of parameter values for which V contains a zero-dimensional locally maximal topologically transitive hyperbolic Markov set V' such that the $F\,|_{V'}$ is topologically conjugate to the shift automorphism in the space of sequences of two symbols. Thus, $V = \overline{W^s}(P)$, where P a fixed point of F lying in V which is a topologically indecomposable one-dimensional continuum. The above results were found to be true for the Hénon mapping (3.1) for some parameter values, i.e., the existence of a parameter range in which the invariant set of the Hénon mapping (2.1) is a one-dimensional topologically indecomposable *Brauer–Janiszewski continuum* was proved. This set contains a zero-dimensional locally maximal set and lies in the attraction domain of itself. Other smooth versions of the Lozi map (3.1) can be found in Section 4.13.

4.10 Maps with border-collision period doubling scenario

Power electronics is an area with wide practical application. In particular, power converters exhibit several nonlinear phenomena such as border-collision bifurcations, coexisting attractors and chaos. All these phenomena are created by switching elements. Recently, several researchers have

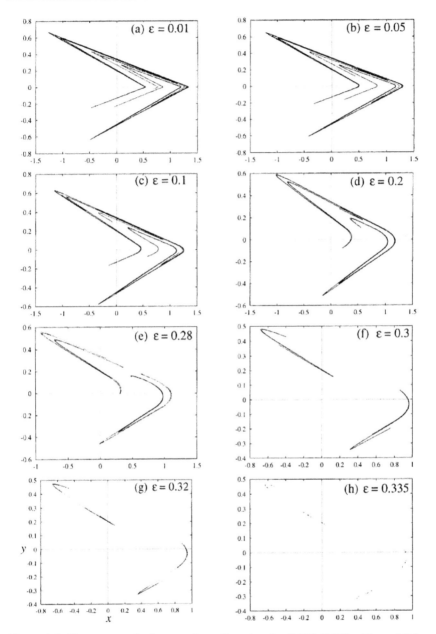

Figure 4.12: The attractors for $L_{a,b,\varepsilon}$ transformation given by (4.120)-(4.121) with $a = 1.7$, b = 0.5. The parameter ε is given in each figure. Reused with permission from M.A. Aziz-Alaoui, Carl Robert and Celso Grebogi, Chaos, Solitons and Fractals, (2001). Copyright 2001, Elsevier Limited.

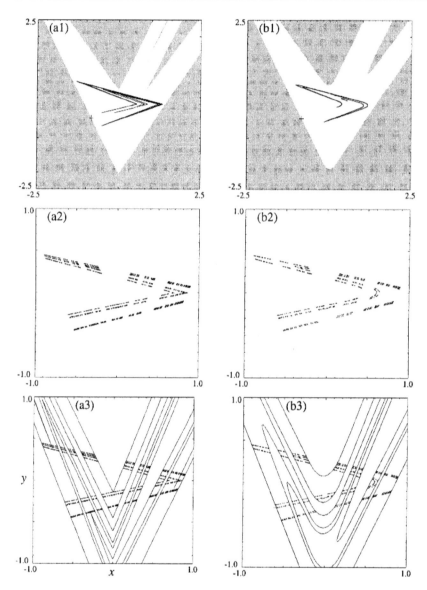

Figure 4.13: Comparison of (a) $L_{a,b,0}$ and (b) $L_{a,b,\varepsilon}$ given by (4.120)-(4.121) before a = 1.7, b = 0.5 and after a = 2.2, b = 0.5 the boundary crisis. Before crisis: chaotic attractor with its basin of attraction, see (a1) and (b1). After crisis: only a chaotic saddle remains, see (a2) and (b2). Chaotic saddle with part of the stable manifold of the fixed point to show how the gaps are formed in the saddle, see (a3) and (b3). Note: only part of the stable manifold is being computed in order to help visializing the mechanism. Reused with permission from M.A.Aziz-Alaoui, Carl Robert and Celso Grebogi, Chaos, Solitons and Fractals, (2001). Copyright 2001, Elsevier Limited.

studied border-collision bifurcations in piecewise-smooth systems. These systems can exhibit classical smooth bifurcations, but if the bifurcation occurs when the fixed point is on the border, there is a discontinuous change in the elements of the Jacobian matrix as the bifurcation parameter is varied. A variety of such border-collision bifurcations have been reported in this situation.

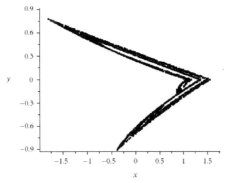

Figure 4.14: A Lozi-type chaotic attractor obtained from system (4.123) for $a = -1.8$, $b = 0.5$, Zeraoulia (2005).

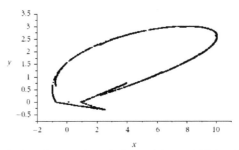

Figure 4.15: A typical orbit of system (4.123) obtained for $a = 1.35$, $b = 0.3$, Zeraoulia (2005).

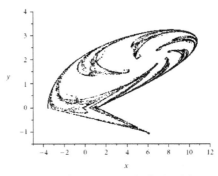

Figure 4.16: The chaotic attractor of system (4.123) obtained for $a = 1.4$, $b = 0.3$, Zeraoulia (2005).

It was proved numerically in [Zeraoulia (2005)] that the following map:

$$f(x, y) \rightarrow \begin{pmatrix} 1 + a\left(|x| - y^2\right) + y \\ bx \end{pmatrix} \tag{4.123}$$

can display two different chaotic attractors, where one of them shown in Figs. 4.16 was obtained via a *border-collision period-doubling scenario*[13] shown in Fig. 4.17. This route to chaos is different from the classical *period-doubling bifurcation* as shown in [Avrutin & Schanz (2004)]. The system (4.123) has the same complexity as the Lozi system (3.1) and the two models are topologically not equivalent because the Lozi system (3.1) is a piecewise-linear map, but model (4.123) is a piecewise nonlinear system. Also, a straightforward proof can show that a non-singular homeomorphism that transforms each system to other does not exist. It is well known that the Hénon attractor (3.1) is obtained via a period-doubling bifurcation route to chaos as a typical future, unless for Lozi map (3.1), no period doubling route to chaos is allowed, and the attractor goes directly from a border collision bifurcation developed from a stable periodic orbit. For system (4.123) the chaotic attractor is obtained from a border-collision period doubling bifurcation scenario. This scenario shown in Fig. 4.17 is formed by a sequence of pairs of bifurcations, whereby each pair consists of a border-collision bifurcation and a *pitchfork bifurcation*.

4.11 Rigorous proof of chaos in a 2-D piecewise linear map

In [Zeraoulia (2009)] a rigorous proof of chaos in a discontinuous piecewise linear planar map was given by using the standard Lyapunov exponents method (Recall Definition. 1.8). An analytic formula tested numerically for the dynamics of this map is also presented in terms of a single bifurcation parameter. The 2-D piecewise linear planar map is defined by:

$$\begin{cases} x_{n+1} = -ah\left(y_n\right) + x_n \\ y_{n+1} = x_n \end{cases} \tag{4.124}$$

where the function h is given by:

$$h\left(y\right) = \begin{cases} by - c, \text{ if } y \geq 0 \\ by + c, \text{ if } y < 0 \end{cases} \tag{4.125}$$

where $a, b, c \neq 0$, are positive bifurcation parameters.

---

[13]The dynamics of piecewise smooth maps is the first reason for some investigations.

For $b = 0.5$, $c = 0.3$, and $a \geq 0$ vary and the initial condition $x = y = 0.01$, the map (4.124)-(4.125) exhibits the following dynamical behaviors: (i) For $0 \leq a < 2$, the map (4.124)-(4.125) converges to a stable fixed point. (ii) For $a = 2$, the map (4.124)-(4.125) converges to a 5-periodic orbit (depends on the choice of the initial condition $x_0 = y_0 = 0.01$). (iii) For $a > 2$, the map (4.124)-(4.125) converges to a chaotic attractor as shown in Fig. 4.18. This chaotic attractor is obtained via border-collision bifurcation from a stable 5-periodic orbit to a fully developed chaotic regime as shown in Fig. 4.19.

The map (4.124)-(4.125) can be rewritten as follow:

$$X_{n+1} = f(X_n) = \begin{cases} AX_n + B, & \text{if } y_n \geq 0 \\ AX_n - B, & \text{if } y_n < 0 \end{cases} \qquad (4.126)$$

where $A = \begin{pmatrix} 1 & -ab \\ 1 & 0 \end{pmatrix}$, $B = \begin{pmatrix} ac \\ 0 \end{pmatrix}$, $X_n = \begin{pmatrix} x_n \\ y_n \end{pmatrix}$. The plane can be divided into two linear regions for the map (4.124)-(4.125) as follow: $D_1 = \{(x, y) \in \mathbb{R}^2 / y \geq 0\}$ and $D_2 = \{(x, y) \in \mathbb{R}^2 / y < 0\}$. For all values of the parameter $a \geq 0$, $c > 0$, and $b > 0$, the map (4.124)-(4.125) has two fixed points given by: $P_1 = \left(\frac{c}{b}, \frac{c}{b}\right)$ and $P_2 = \left(-\frac{c}{b}, -\frac{c}{b}\right)$. The Jacobian matrix of the map (4.124)-(4.125) evaluated at the fixed points P_1 and P_2 is the same and given by: $J_{1,2} = \begin{pmatrix} 1 & -ab \\ 1 & -0 \end{pmatrix}$. Some calculations lead to the result that the two equilibrium points P_1 and P_2 have the same stability type, namely, if $0 \leq a \leq \frac{1}{b}$, then both fixed points are asymptotically stable. If $a > \frac{1}{b}$, then both fixed points are unstable. If $0 \leq a \leq \frac{1}{4b}$, then the eigenvalues λ_1 and λ_2, are given by:

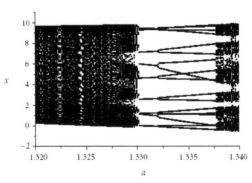

Figure 4.17: Border-collision period-doubling scenario route to chaos observed for system (4.123), Zeraoulia (2005).

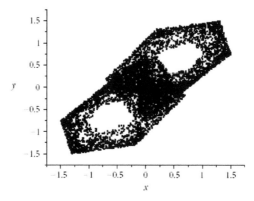

Figure 4.18: The piecewise linear chaotic attractor obtained from map (4.124)-(4.125) for: $a = 2.5$, $b = 0.5$, $c = 0.3$, and the initial condition $x = y = 0.01$. Zeraoulia (2009).

$$\lambda_1 = \frac{1+\sqrt{1-4ab}}{2}, \lambda_2 = \frac{1-\sqrt{1-4ab}}{2} \qquad (4.127)$$

and if $a > \dfrac{1}{4b}$, then the eigenvalues are complex and they are given by:

$$\lambda_1 = \frac{1+\sqrt{-(1-4ab)i}}{2}, \lambda_2 = \frac{1-\sqrt{-(1-4ab)i}}{2} \qquad (4.128)$$

Before we state the rigorous proof of chaos in map (4.124)-(4.125) using Lyapunov exponents method, we note that for a continuous map, positive Lyapunov exponents indicate chaos, negative exponents indicate fixed points, and if the Lyapunov exponent is equal to 0, then the dynamics is periodic. Generally, in the context of continuous systems it is true that Lyapunov exponent equals zero implies periodic dynamics. However, examples of works focused on systems with zero Lyapunov exponents, presenting weaker mechanisms of mixing and sensitivity to initial conditions than dynamical chaos can be found in [Dettmann, et al. (1999), Dettmann & Cohen (2000), Grassberger, et al. (2002)]. These systems sustain transport processes such as diffusion as well as some stochasticity. Thus, it is possible to have zero Lyapunov exponents without periodic dynamics. Examples are: The *fixed squares model* studied in [Dettmann & Cohen (2000)] and the limiting attractor of the logistic map following the period doubling cascade, that is the case at a bifurcation point. For the case of a discontinuous map, a zero Lyapunov exponent does not always indicate periodic behavior. This means that there are some cases (depend mainly on position of the initial conditions) where the behavior of map (4.124)-(4.125) is not periodic in spite its Lyapunov exponent is zero when $\alpha = \dfrac{1}{b}$. The following result was proved in [Zeraoulia (2009)]:

Theorem 78 *(a) The map (4.124)-(4.125) converges to a stable fixed point when* $0 \leq \alpha < \dfrac{1}{b}$.

(b)   The behavior of map (4.124)-(4.125) is not clear[14] when $\alpha = \dfrac{1}{b}$.

(c)   The map (4.124)-(4.125) converges to a chaotic attractor when $\alpha > \dfrac{1}{b}$.

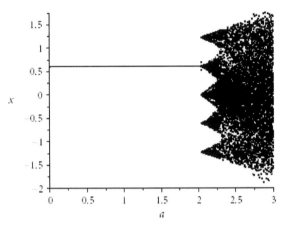

Figure 4.19: The bifurcation diagram for the map (4.124)-(4.125) for: $0 \leq a \leq 3.0$, with $b = 0.5$, $c = 0.3$, and the initial condition $x = y = 0.01$., Zeraoulia (2009).

Proof 28 *The Jacobian matrix of map (4.124)-(4.125) is* $Df(X_n) = A = \begin{pmatrix} 1 & -ab \\ 1 & 0 \end{pmatrix}$, *then the matrix* $T_n(X_0)$ *is given by* $T_n(X_0) = A^n$.

(I) If $0 \leq a \leq \dfrac{1}{4b}$, *then A has two real and distinct eigenvalues* λ_1 *and* λ_2 *given by*

(4.127). Let P be the matrix whose columns consist of the two eigen-vectors of the matrix A, hence, the Jordan normal form for the matrix A is $A = PJP^{-1}$, *where J is given by* $J = \begin{pmatrix} \lambda_1 & 0 \\ 0 & \lambda_2 \end{pmatrix}$. *We have* $A^n = PJ^nP^{-1}$, *for all* $n \in \mathbb{N}$, *then*

the eigenvalues of A^n *are the same as the eigenvalues of the matrix* J^n, *thus the*

---

[14] That is, the map generates a symbolic sequence $\hat{s} = \{\hat{s}_0; \hat{s}_1; ...; \hat{s}_j ; ...\}$ composed of symbols $\hat{s}_j$ $= i$ if $X_j = f_j(X_0) \in D_i$, $i = 1, ..., m$ (where the map can be expressed as $f(X) = A_i X + Bs_i$, if $X \in D_i$, and $s_i \in \{-1, 1\}$). Each of those symbolic sequences is called *"admissible"* and its symbols describe the order in which trajectories, starting from any initial condition X_0, visit the various sub regions D_i, $i = 1, 2$. In [Maggio, et al. (2000)] a general approach for finding periodic trajectories in piecewise-linear maps is introduced. This procedure is based on the decomposition of the initial state via the eigenvectors of their jacobian and it is applied to digital filters with two's complement overflow and modulators [Feely & Chua (1992), Sharkovsky & Chua (1993), Rajaraman, et al. (1996), Banerjee, et al. (2004)].

eigenvalues of $T^n(X_0)$ *are* $\delta_1 = \left(\dfrac{1+\sqrt{1-4ab}}{2}\right)^n$, $\delta_2 = \left(\dfrac{1-\sqrt{1-4ab}}{2}\right)^n$, *and therefore one*

has $J_1(X_0, n) = |\delta_1|^{\frac{1}{n}} = \left|\dfrac{1+\sqrt{1-4ab}}{2}\right| < 1$, *and* $J_2(X_0, n) = |\delta_1|^{\frac{1}{n}} = \left|\dfrac{1-\sqrt{1-4ab}}{2}\right|$

< 1. *The condition* $0 \le a \le \dfrac{1}{4b}$, *implies that Lyapunov exponent satisfies the following conditions:*

$$\lambda_1(X_0) = \ln\left|\frac{1+\sqrt{1-4ab}}{2}\right| < 0, \ \lambda_2(X_0) = \ln\left|\frac{1-\sqrt{1-4ab}}{2}\right| < 0 \qquad (4.129)$$

in each linear regions D_1 *and* D_2. *Thus, the map (4.124)-(4.125) converges to a stable fixed point for all* $0 \le a \le \dfrac{1}{4b}$.

(II) *If* $a > \dfrac{1}{4b}$, *then the matrix A has two conjugate complex eigenvalues* λ_1 *and* λ_2, *given by (4.128). The Jordan normal form for the matrix A is* $A = PJP^{-1}$, *where J is given by* $J = \begin{pmatrix} \alpha & \beta \\ -\beta & \alpha \end{pmatrix}$, *where* $\alpha = \dfrac{1}{2}$ *and* $\beta = \dfrac{\sqrt{4ab-1}}{2}$.

The same method shows that the Lyapunov exponents are given by:

$$\lambda_1(X_0) = \lambda_2(X_0) = \frac{1}{2}\ln(ab). \qquad (4.130)$$

in each linear regions D_1 *and* D_2. *Thus means that map (4.124)-(4.125) converges to a stable fixed point when* $\lambda_1(X_0) = \dfrac{1}{2}\ln(ab) < 0$, *i.e.,* $\dfrac{1}{4b} < a < \dfrac{1}{b}$, *since, we suppose that* $a > \dfrac{1}{4b}$, *and the behavior is not clear if* $\lambda_1(X_0) = \dfrac{1}{2}\ln(ab) = 0$, *i.e.,* $a = \dfrac{1}{b}$. *Finally, the map (4.124)-(4.125) converges a chaotic attractor if* $\lambda_1(X_0) = \dfrac{1}{2}$ $\ln(ab) > 0$, *i.e.,* $\alpha > \dfrac{1}{b}$.

The above analysis was generalized in [Zeraoulia (2007)]. Indeed, sufficient conditions for the existence of chaotic attractors in a general n-D piecewise linear discrete maps were derived, along with the exact determination of its dynamics using the standard definition of the largest Lyapunov exponent (Recall Definition 8). For this purpose, let us consider the following n-D map of the form: $f : D \rightarrow D$, $D \subset \mathbb{R}^n$, defined by:

$$x_{k+1} = f(x_k) = A_i x_k + b_i, \text{ if } x_k \in D_i, i = 1, 2, ...,m \qquad (4.131)$$

where $A_i = (a^i_{jl})_{1\le j,l\le n}$ and $b_i = (b^i_i)_{1\le i\le n}$ are respectively $n \times n$ and $n \times 1$ real matrices, for all $i = 1, 2, ..., m$, and $x_k = (x^j_k)_{1\le j\le n} \in \mathbb{R}^n$ is the state variable, and m is the number of disjoint domains on which D is partitioned. Due to the shape of the vector field f of the map (4.131) the plane can be divided into m regions denoted by $(D_i)_{1\le i\le m}$, and in each of these regions the map (4.131) is linear. The Jacobian matrix of the map (4.131) is defined by:

$$J(x_k) = \begin{cases} A_1, \text{ if } x_k \in D_1 \\ A_2, \text{ if } x_k \in D_2 \\ ... \\ A_m, \text{ if } x_k \in D_m \end{cases} \qquad (4.132)$$

The essential idea of the proof is the assumption that the matrices $(A_i)_{1\le i\le m}$ have the same eigenvalues, i.e., they are equivalent, then if one computes analytically a Lyapunov exponent (which is an eigenvalue of a matrix A_i) of the map (4.131) in a region $D_{i'}$ then one can find that these exponents are identical in each linear region $D_{i'}$ for all $i \in \{1, 2, ..., m\}$. Thus, one can consider the Jacobian matrix $J(x_k)$ of the map (4.131) as any matrix $A_{i'}$ denoted by $A = (a_{jl})_{1\le j,l\le n}$. Assume that the eigenvalues of A are listed in order as follow:

$$|\lambda_1| \ge |\lambda_2| \ge ... \ge |\lambda_n| \qquad (4.133)$$

where the notation $\lambda_i = \lambda_i((a_{jl})_{1\le j,l\le n})$, $i = 1, ..., n$ indicate that the eigenvalue λ_i depends only the coefficients $(a_{jl})_{1\le j,l\le n}$. The matrix defined in (1.15) is $T_r(x_0) = A_r$, and its eigenvalues are $\lambda^r_1((a_{jl})_{1\le j,l\le n}), ..., \lambda^r_n((a_{jl})_{1\le j,l\le n})$. The Lyapunov exponents of the map (4.131) are:

$$\omega_i(x_0) = \ln\left(\lim_{r\to+\infty} \left(|\lambda_i((a_{jl})_{1\le j,l\le n})|^r\right)^{\frac{1}{r}} \right) = \ln\left|\lambda_i((a_{jl})_{1\le j,l\le n})\right| \qquad (4.134)$$

Hence, according to (4.133) all the Lyapunov exponents are listed as follow:

$$\omega_1((a_{jl})_{1\le j,l\le n}) \ge \omega_2((a_{jl})_{1\le j,l\le n}) \ge ... \ge \omega_n((a_{jl})_{1\le j,l\le n}) \qquad (4.135)$$

Define the following subsets in $\mathbb{R}^{n^2}$ in term of the vector $(a_{jl})_{1\le j,l\le n}$ as follows:

$$
\begin{cases}
\Omega_1 = \left\{ (a_{jl})_{1 \leq j,l \leq n} \in \mathbb{R}^{n^2}, |\lambda_n| > 1 \right\} \\
\Omega_2 = \left\{ (a_{jl})_{1 \leq j,l \leq n} \in \mathbb{R}^{n^2}, |\lambda_1| < 1 \right\} \\
\Omega_3 = \left\{ (a_{jl})_{1 \leq j,l \leq n} \subset \mathbb{R}^{n^2}, |\lambda_1| = 1 \right\} \\
\Omega_4 = \left\{ (a_{jl})_{1 \leq j,l \leq n} \in \mathbb{R}^{n^2}, |\lambda_i| < 1, \ i = 2, ..., n \right\} \\
\Omega_5 = \left\{ (a_{jl})_{1 \leq j,l \leq n} \in \mathbb{R}^{n^2}, |\lambda_2| = 1 \right\} \\
\Omega_6 = \left\{ (a_{jl})_{1 \leq j,l \leq n} \in \mathbb{R}^{n^2}, |\lambda_i| < 1, \ i = 3, ..., n \right\} \\
\Omega_7 = \left\{ (a_{jl})_{1 \leq j,l \leq n} \in \mathbb{R}^{n^2}, |\lambda_i| = 1, 1 \leq i \leq K, 1 \leq K \leq n \right\} \\
\Omega_8 = \left\{ (a_{jl})_{1 \leq j,l \leq n} \in \mathbb{R}^{n^2}, |\lambda_i| < 1, K+1 \leq i \leq n \right\} \\
\Omega_9 = \left\{ (a_{jl})_{1 \leq j,l \leq n} \in \mathbb{R}^{n^2}, |\lambda_1| > 1, \ \text{and} \ \prod_{i=2}^{i=n} |\lambda_i| < 1 \right\}
\end{cases}
\tag{4.135}
$$

The following theorem was proved in [Zeraoulia (2007)]:

Theorem 79 *Let us consider the general n-D piecewise linear map of the In form:*

$$
f(x_k) = x_{k+1} = A_i x_k + b_i, \ if \ x_k \in D_i \subset \mathbb{R}^n, \ i = 1, 2, ..., m \tag{4.137}
$$

and assume the following:

(a) *The map (4.131) is piecewise linear, i.e., the integer m verify $m \geq 2$, and there exist $i, j \in \{1, 2, ..., m\}$ such that $b_i \neq 0$ and $b_i \neq b_j$.*

(b) *The map (4.131) has a set of fixed points. i.e., There is a set of integers i in $\{1, 2, ..., m\}$ such that the equations $A_i x + b_i = x$, has at least a zero x in the subregion Di.*

(c) *All the matrices A_i and A_j are equivalent. i.e., there exist invertible matrices P_{ij} such that $A_i = P_{ij} A_j P^{-1}_{ij}$, for all $i, j \in \{1, 2, ..., m\}$. Then the dynamics of the map (4.131) is known in term of the vector $(a_{jl})_{1 \leq j,l \leq n} \in \mathbb{R}^{n^2}$ in the following cases:*

(1) *if $(a_{jl})_{1 \leq j,l \leq n} \in \Omega_1$, then the map (4.131) is hyperchaotic.*

(2) *if $(a_{jl})_{1 \leq j,l \leq n} \in \Omega_2$, then the map (4.131) converges to a stable fixed point.*

(3) *if $(a_{jl})_{1 \leq j,l \leq n} \in \Omega_3 \cap \Omega_4$, then the map (4.131) converges to a circle attractor.*

(4) *if $(a_{jl})_{1 \leq j,l \leq n} \in \Omega_3 \cap \Omega_5 \cap \Omega_6$, then the map (4.131) converges to a torus attractor.*

(5) *if $(a_{jl})_{1 \leq j,l \leq n} \in \Omega_7 \cap \Omega_8$, then the map (4.131) converges to a K-torus attractor.*

(6) *if $(a_{jl})_{1 \leq j,l \leq n} \in \Omega_9$, then the map (4.131) is chaotic.*

4.12 Occurrence of chaos via different routes

In [Zeraoulia (2008)] a 2-D piecewise smooth discrete-time chaotic map with a rarely observed phenomenon that is the occurrence of the same chaotic attractor via different and distinguishable route to chaos: period doubling and border-collision bifurcations as typical futures. This phenomenon is justified by the location of system equilibria of the proposed map and the possible bifurcations types in smooth dissipative systems. This phenomenon was observed for the following map:

$$f(x, y) = \begin{pmatrix} 1 - a|x^2 - y| + y \\ bx \end{pmatrix} \tag{4.138}$$

The corresponding chaotic attractor of map (4.138) is shown in Fig. 4.20 and Fig. 4.21 show that when $-0.1 \le a \le 1.4$ and $b = 0.2$, the chaotic attractor given by the map (4.138) is obtained via border-collision bifurcation from a stable two periodic orbit to a fully developed chaotic regime, while for the variations of the parameter b one can observe a typical period doubling bifurcation route to chaos, for $-0.2 \le b \le 0.2$ and $a = 1.4$, as shown in Fig. 4.22, and the map (4.138) gives the same chaotic attractor shown in Fig. 4.20. For the case of the map (4.138), the plane can be divided into two nonlinear regions and a smooth curve between them. These regions are denoted by:

$$\begin{cases} R_1 = \{(x, y) \in \mathbb{R}^2 / x^2 < y\} \\ R_2 = \{(x, y) \in \mathbb{R}^2 / x^2 > y\} \\ \Omega = \{(x, y) \in \mathbb{R}^2 / y = x^2\} \end{cases} \tag{4.139}$$

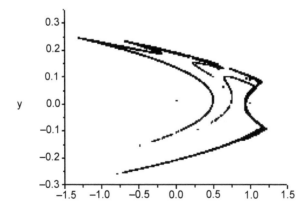

Figure 4.20: The chaotic attractor obtained from the map (4.138) for: $a = 1.4$, $b = 0.2$, and $x_0 = y_0 = 0.1$., Zeraoulia (2008).

Let us define the following components of fixed points of the map (4.138):

$$\begin{cases}
x_1 = \dfrac{0.1a - \dfrac{\sqrt{3.68a + 0.04a^2 + 0.64}}{2} - 0.4}{a}, x_2 = \dfrac{0.1a + \dfrac{\sqrt{3.68a + 0.04a^2 + 0.64}}{2} - 0.4}{a} \\[4mm]
s_1 = 0.142\,86b + 0.357\,14 + 0.357\,14\sqrt{0.8b + 0.16b^2 - 4.6} \\
s_2 = 0.142\,86b + 0.357\,14 - 0.357\,14\sqrt{0.8b + 0.16b^2 - 4.6} \\
w_1 = 0.857\,14b - 0.357\,14 + 0.357\,14\sqrt{5.76b^2 - 4.8b + 6.6} \\
w_2 = 0.857\,14b - 0.357\,14 - 0.357\,14\sqrt{5.76b^2 - 4.8b + 6.6}
\end{cases} \qquad (4.140)$$

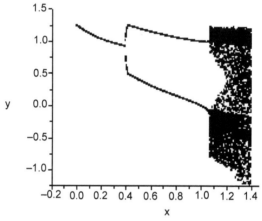

Figure 4.21: The border collision bifurcation from a stable two periodic orbit to a fully developed chaotic regime for the map (4.138) for: $-0.1 \leq a \leq 1.4$, and $b = 0.2.$, Zeraoulia (2008).

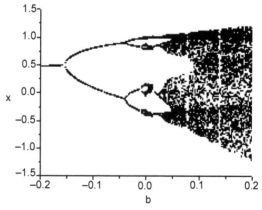

Figure 4.22: The typical period doubling bifurcation route to chaos in the map (4.138) for: $a = 1.4$ and $-0.2 \leq b \leq 0.2.$, Zeraoulia (2008).

The following result was proved in [Zeraoulia (2008)]:

Proposition 4.11 *(a) For b = 0.2, and a ∈ ℝ, there are no fixed points for the map (4.138) in the region R_1.*

(b) *For all a ∈ (−∞,−91. 826] ∪ [−0.174 24,∞) = I_1 and a ≠ 0 and b = 0.2, there exist two fixed points $(x_1, 0.2x_1)$ and $(x_2, 0.2x_2)$ for the map (4.138) in the region R_2, and they do not collide with the borderline Ω.*

(c) *For a = 0, and b = 0.2, there is one fixed point (1.25, 0.25) in the borderline Ω.*

(d) *For a = 1.4, and b ∈ (−∞,−8. 416 1] ∪ [3. 416 1,∞) = I_2, there is two fixed points (s_1, bs_1) and (s_2, bs_2) in the region R_1, and they do not collide with the borderline Ω.*

(e) *For a = 1.4, and b ∈ ℝ, there is two fixed points (w_1, bw_1) and (w_2, bw_2) in the region R_2, and they do not collide with the borderline Ω.*

(f) *If b = 0.2 and a ∈ [−0.1, 1.4], then only $(x_1, 0.2x_1)$ and $(x_2, 0.2x_2)$ exists in R_2 since [−0.1, 1.4] ⊂ I_1, and they collide with the borderline Ω at a = 0.*

(g) *If a = 1.4 and b ∈ [−0.2, 0.2], then (s_1, bs_1) and (s_2, bs_2) does not exist in R_1 since [−0.2, 0.2] ∩ I_2 =, and only (w_1, bw_1) and (w_2, bw_2) exists in R_2 which do not collide with the borderline Ω.*

Proposition 4.11 implies that if there exist a chaotic attractor of the map (4.138) then it surrounds only the two equilibria in the region R_1 or R_2 and its components are almost confined in the corresponding region where the fixed points exists, but it visits sometimes the other subregions. If a bifurcation occurs, then it is one of the generic types, namely, period doubling, saddle-node, or Hopf bifurcation [Banargee, et al. (1998)]. Thus, map (4.138) converges to the chaotic attractor shown in Fig. 4.20 via a period doubling bifurcation route to chaos as shown in Fig. 4.22.

In the case where a fixed point collides with the borderline Ω, i.e., when a = 0, there is a discontinuous *jump* in the eigenvalues of the associated Jacobian matrix of the map (4.138). This means that an eigenvalue may not cross the unit circle in a smooth way, but rather jumps over it as a parameter is varied continuously, thus, a rich variety of bifurcations have been reported in this situation, which have been called *border collision bifurcation*. See Section 4.16 for more details. In this case, the map (4.138) must converge to the same chaotic attractor shown in Fig. 4.20 via border collision bifurcation as shown in Fig. 4.21.

4.13 Generating multifold chaotic attractors

In this section, we discuss the procedures used to generate chaotic attractors with any number of *folds* from some versions of the Hénon map (2.1) or Lozi map (3.1). The first method was described in [Aziz Alaoui (2000)] using a

C^1-piecewise linear continuous *zigzagging function* and the second method was described in [Zeraoulia & Sprott (2008)] using a C^∞ mapping.

4.13.1 C^1-multifold chaotic attractors

An appropriate C^1-modification in [Aziz Alaoui (2000)] of the Lozi map (3.1) or the Hénon map (2.1) gives *multifold chaotic attractors* shown in Figs. 4.23(a1-d1). These chaotic attractors were obtained by a direct modification of the Lozi map (3.1) by replacing the term $|x|$ by a piecewise linear continuous zigzagging function $L_N(x)$ defined via the connection of N segments as shown in Figs. 4.23(a2-d2).

4.13.2 C^∞-multifold chaotic attractors

In this section, we present a C^∞-multifold chaotic attractors studied in [Zeraoulia & Sprott (2008)] where a two-dimensional (2-D) C^∞ discrete

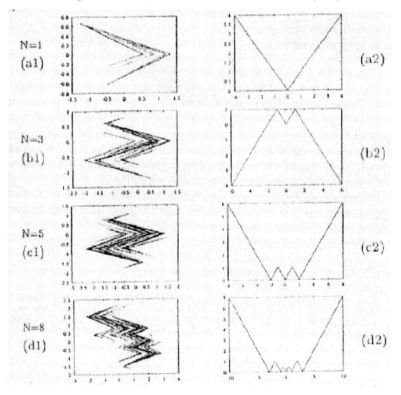

Figure 4.23: Multifold chaotic attractors obtained from map (4.138) with $L_N(x)$ for: (a) $N = 1$, (b) $N = 3$, (c) $N = 5$, (d) $N = 8$.

bounded map capable of generating multifold strange attractors via period-doubling bifurcation routes to chaos was introduced and studied for all values of its bifurcation parameters. This map is given by:

$$f(x_n, y_n) = \begin{pmatrix} x_{n+1} \\ y_{n+1} \end{pmatrix} = \begin{pmatrix} 1 - a\sin x_n + by_n \\ x_n \end{pmatrix} \qquad (4.141)$$

The smoothness of map (4.141) simplifies its study and avoids some problems related to the lack of continuity or differentiability of the map. The choice of the term $\sin x$ (or $\cos x$) in map (4.141) has an important role in that it makes the solutions bounded for values of b such that $|b| \le 1$, and all values of a, while they are unbounded for $|b| > 1$. Indeed, the following results were proved in [Zeraoulia & Sprott (2008)]:

Theorem 80 (a) The orbits of the map (4.141) are bounded for all $a \in \mathbb{R}$, and $|b| < 1$, and all initial conditions $(x_0, x_1) \in \mathbb{R}^2$, i.e., in the region of $\mathbb{R}^4$ defined by:

$$\Omega_1 = \{(a, b, x_0, x_1) \in \mathbb{R}^4 / |b| < 1, (x_0, x_1) \in \mathbb{R}^2\}. \qquad (4.142)$$

(b) The map (4.141) possesses unbounded orbits in the following subregions of $\mathbb{R}^4$:

$$\begin{cases} \Omega_2 = \left\{(a, b, x_0, x_1) \in \mathbb{R}^4 / |b| < 1, \text{ and both } |x_0|, |x_1| > \dfrac{|a|+1}{|b|-1}\right\}, \\ \Omega_3 = \{(a, b, x_0, x_1) \in \mathbb{R}^4 / |b| < 1, \text{ and } |a| < 1\}. \qquad (4.143) \end{cases}$$

Note that there is no similar proof for the following subregions of $\mathbb{R}^4$:

$$\begin{cases} \Omega_4 = \left\{(a, b, x_0, x_1) \in \mathbb{R}^4 / |b| < 1, \text{ and both } |x_0|, |x_1| \le \dfrac{|a|+1}{|b|-1}\right\} \\ \Omega_5 = \{(a, b, x_0, x_1) \in \mathbb{R}^4 / |b| < 1, \text{ and } |a| \ge 1\}. \qquad (4.144) \end{cases}$$

We note that the observed multifold chaotic attractors evolve around a large number of fixed points. It appears that the number of these points increase with increasing a when b is fixed. These multifold chaotic attractors presented in Fig. 4.24 are obtained via a period-doubling bifurcation route to chaos as shown in Fig. 4.27(a). Fig. 4.25 shows regions of unbounded, fixed point, periodic, and chaotic solutions in the ab-plane for the map (4.141), where 10^6 iterations were used for each point. For $b = 0.3$ and $-1 \le a \le 4$, the map (4.141) exhibits the dynamical behaviors as shown in Fig. 4.27. In particular, in the interval $-1 \le a \le 0.76$, the map (4.141) converges to a fixed point. For $0.76 < a \le 1.86$, there is a series of period-doubling bifurcations as shown in Fig. 4.27(a). In the interval $1.86 < a \le 2.16$, the orbit converges

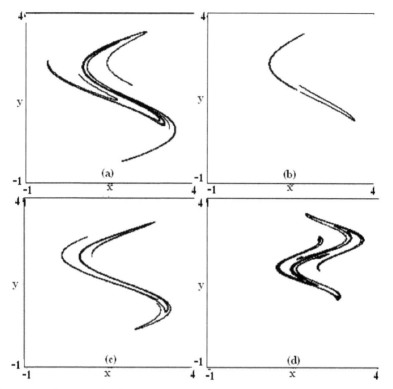

Figure 4.24: Chaotic multifold attractors of the map (4.141) obtained for (a) $a = 2.4$, $b = -0.5$. (b) $a = 2$, $b = 0.2$. (c) $a = 2.8$, $b = 0.3$. (d) $a = 2.7$, $b = 0.6$., Zeraoulia & Sprott (2008).

to a chaotic attractor. For $2.16 < a \leq 2.27$, it converges to a fixed point. For $2.27 < a \leq 2.39$, there are periodic windows. For $2.39 < a \leq 2.92$, it converges to a chaotic attractor. For $a > 2.92$, the map (4.141) is chaotic. The Lyapunov exponents for $a = 3$ and $b = 0.3$ are $\lambda_1 = 0.56186$ and $\lambda_2 = -1.76583$, giving a Kaplan-Yorke dimension of $D_{KY} = 1.31818$. There are also fixed points and periodic orbits. This map is invertible for all $b \neq 0$, especially for $|b| < 1$, and there is no hyperchaos since the sum of the Lyapunov exponents $\lambda_1 + \lambda_2 = \ln |b|$ is never positive.

Generally, for $b = 0.3$ and $-150 \leq a \leq 200$, map (4.141) is chaotic over all the range as shown in Fig. 4.28, except for the small intervals mentioned above and shown in Fig. 4.27. However, for $a = 3$ and $b \in \mathbb{R}$, the map (4.141) exhibits very complicated dynamical behaviors as shown in Fig. 4.25 where a large fraction of the region has chaotic attractors. There is a very wide variety of possible multifold chaotic attractors with different numbers of folds, only some of which are shown in Figs. 4.24, 4.26, and 4.31. For these cases, their basins of attraction include the entire xy-plane for $|b| < 1$. For $|b| > 1$, the map (4.141) does not converge as shown by Theorem 76(a).

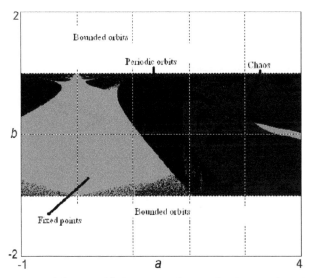

Figure 4.25: Regions of dynamical behaviors in *ab*-space for the map (4.141), Zeraoulia & Sprott (2008).

Coexisting attractors occur as shown in black in Fig. 4.30, both in the regular and chaotic regimes. Indeed, for $a = 2$ and $b = -0.6$, a two-cycle (1.314326, −0.584114) coexists with a period-3 strange attractor. Similarly, for $a = 2.2$ and $b = -0.36$, there is a strange attractor surrounded by a second period-3 strange attractor as shown in black in Fig. 4.31 with their corresponding basins of attraction.

4.14 A new simple 2-D piecewise linear map

In [Zeraoulia & Sprott (2010(a))] a simpler non dissipative 2-D map was studied. Chaotic attractors displayed by this map are obtained via border collision bifurcation with many coexisting chaotic attractors. This map is given by:

$$f(x, y) = \begin{pmatrix} 1 - a \, | \, y \, | + bx \\ x \end{pmatrix}, \qquad (4.145)$$

The map (4.145) differs from the Lozi map (3.2) in that it has a much wider variety of attractors[15]. All the attractors for the map (4.145) with regular basins of attraction, have either dimension 2 (if chaotic) or 0 (if not) for both the Kaplan-Yorke dimension and the correlation dimension except perhaps for a set of measure zero in the parameter space at the bifurcation boundaries.

---

[15] We note that the attractors of the Lozi mappings are qualitatively similar.

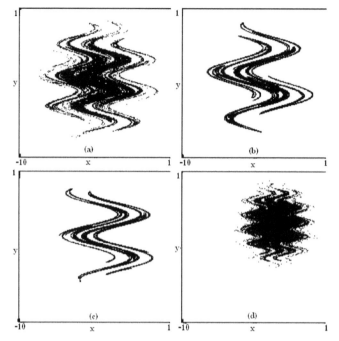

Figure 4.26: Chaotic multifold attractors of the map (4.141) obtained for (a) $a = 3.4$, $b = -0.8$. (b) $a = 3.6$, $b = -0.8$. (c) $a = 4$, b = 0.5. (d) $a = 4$, $b = 0.9$., Zeraoulia & Sprott (2008).

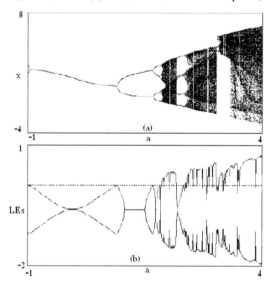

Figure 4.27: (a) Bifurcation diagram for the map (4.141) obtained for $b = 0.3$ and $-1 \leq a \leq 4$. (b) Variation of the Lyapunouv exponents of map (4.141) over the same range of a., Zeraoulia & Sprott (2008).

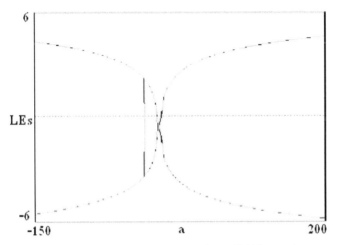

Figure 4.28: Variation of the Lyapunouv exponents of map (4.141) over the range $-150 \leq a \leq 200$ with $b = 0.3$, Zeraoulia & Sprott (2008).

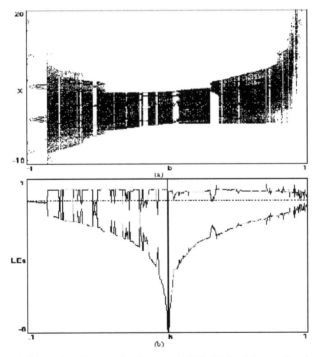

Figure 4.29: (a) Bifurcation diagram for the map (4.141) obtained for $a = 3$ and $-1 \leq b \leq 1$. (b) Variation of the Lyapunouv exponents of map (4.141) for the same range of b, Zeraoulia & Sprott (2008).

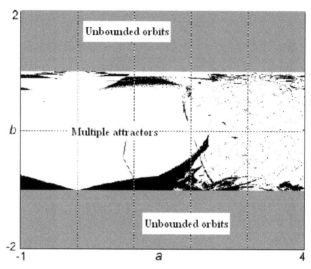

Figure 4.30: The regions of *ab*-space for the map (4.141) where multiple attractors are found (shown in black), Zeraoulia & Sprott (2008).

The map (4.145) has the same fixed points as for the Lozi map (3.1) but with different stability types, due to differences in their Jacobian matrices. The different chaotic attractors (with their basins of attraction in white) of map (4.145) are shown in Figs. 4.32 and 4.34. These attractors result from a stable period-1 orbit to a fully developed chaotic regime, i.e., a border collision bifurcation route to chaos as shown in Fig. 4.33(a), and it is the only observed scenario.

Note that the noisy region just above the $b = -1$ line for the Lozi map (3.1) in Fig. 4.34(b) is actually a region of multiple attractors. For $a = -0.9$ and $b = -0.9$, a period-5 attractor coexists with a fixed point (at $x = 1.0$); and with $a = 1$ and $b = -0.9$, a period-4 attractor coexists with a fixed point (at $x = 0.344828$). From Fig. 4.36 we have that coexisting attractors are evident for map (4.145) in the chaotic region just above the line $a = 1$. Also, map (4.145) displays a variety of hyperchaotic attractors as shown in Figs. 4.32 and 4.35 for the cases $a = 1.1$, or $b = 1.1$. Coexisting chaotic attractors for map (4.145) can be seen at $a = 1.2$ and $b = -0.6$ as shown in black in Fig. 4.37. The large chaotic attractor at the center is surrounded by a period-3 chaotic attractor at its periphery, with their fractal basins of attraction.

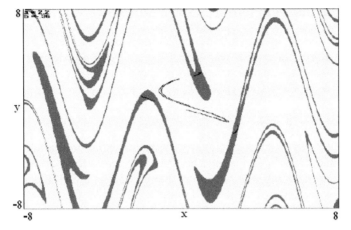

Figure 4.31: Two coexisting attractors occur for $a = 2.2$ and $b = -0.36$, for the map (4.141) where a strange attractor is surrounded by a second period-3 strange attractor with their corresponding basins of attraction, Zeraoulia & Sprott (2008).

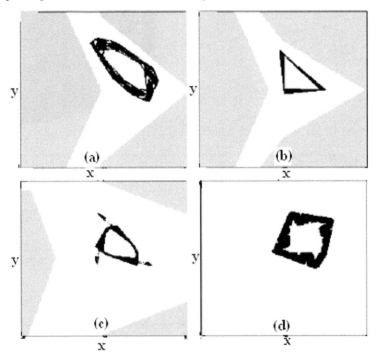

Figure 4.32: Chaotic attractors of the map (4.145) with their basins of attraction (white) for $a = 1.1$ and (a) $b = -1.4$, (b) $b = -1.1$, (c) $b = 0.8$, (d) $b = 0.2$., Zeraoulia & Sprott (2010(a)).

4.15 The discrete hyperchaotic double scroll

In [Zeraoulia & Sprott (2009)] a modified version of the Lozi map (3.1) capable of generating chaotic attractors with one and two scrolls called *discrete hyperchaotic double scroll* was proposed as follow:

$$f(x, y) = \begin{pmatrix} x - ah(y) \\ bx \end{pmatrix} \qquad (4.146)$$

where h is the characteristic function of the so-called double scroll attractor [Chua, et al. (1986)], and m_0 and m_1 are respectively the slopes of the inner and outer sets of the original Chua circuit (1.150)-(1.151). The map (4.146) has the same nonlinearity as used in the well-known Chua circuit given by (1.150)-(1.151) [Chua, et al. (1986)]. A rigorous proof of the hyperchaoticity[16]

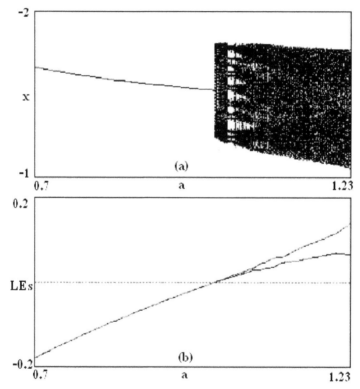

Figure 4.33: (a) The bifurcation diagram for the map (4.145) obtained for $b = 1.1$ and $0.7 \le a \le 1.23$. (b) Variation of the Lyapunouv exponents of map (4.145) versus the parameter $0.7 \le a \le 1.23$, with $b = 1.1$., Zeraoulia & Sprott (2010(a)).

---

[16]The corresponding strange attractor is characterized by two positive Lyapunov exponents.

of this attractor was given in [Zeraoulia & Sprott (2009)] based on Theorem 2. The main idea of this map is that hyperchaotic attractors make robust tools for some real-world applications. The idea comes from the fact that Chua's circuit (1.150)-(1.151) does not exhibit hyperchaos because of its limited dimensionality. Several works have focused on the hyperchaotification of Chua's circuit using several techniques such as coupling many Chua circuits as in [Kapitaniak, et al. (1994)] where a 15-D dynamical system is obtained. However, the resulting system is very complicated and difficult to construct. A simpler method introduces an additional *inductor* in the canonical Chua circuit as given in [Thamilmaran, et al. (2004)], to obtain

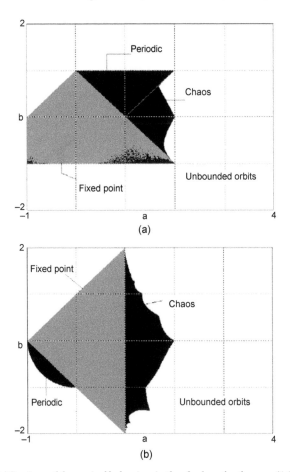

Figure 4.34: (a) Regions of dynamical behaviors in the *ab*-plane for the map (4.145). (b) Regions of dynamical behaviors in the *ab*-plane for the Lozi map (2.1), Zeraoulia & Sprott (2010(a)).

a 4-D dynamical system that converges to a hyperchaotic attractor via a border collision bifurcation [Banerjee & Grebogi (1999)]. In [Suneel (2006)] some techniques employed in the circuit realization of smooth systems were extended to other systems such as piecewise linear or piecewise smooth maps. Thus, it seems that the circuit realizations of low-dimensional maps are simpler than with high-dimensional continuous systems.

In order to prove hyperchaoticity of map (4.146), we apply Theorem 2.

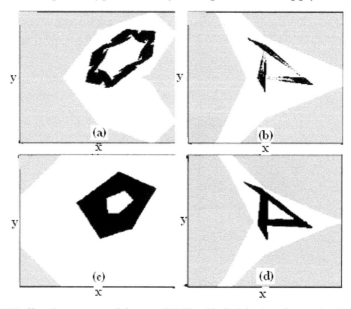

Figure 4.35: Chaotic attractors of the map (4.145) with their basins of attraction (white) for (a) $a = b = 1.1$. (b) $a = 1.2$, $b = -1$. (c) $a = 1.2$, $b = 0.5$. (d) $a = 1.3$, $b = -1.1$., Zeraoulia & Sprott (2010(a)).

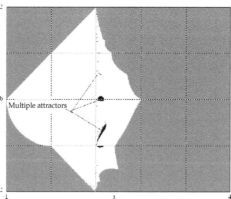

Figure 4.36: The regions of ab-space with multiple attractors (in black) for the map (4.145), Zeraoulia & Sprott (2010(a)).

Hence, some calculations show that:

$$\| f'(x,y) \| = \begin{cases} \dfrac{\sqrt{b^2 + a^2 m_1^2 + \sqrt{2b^2 + b^4 + 2a^2 m_1^2 + a^4 m_1^4 - 2a^2 b^2 m_1^2} + 1} + 1}{2}, & \text{if } |y| \ge 1 \\[4mm] \dfrac{\sqrt{b^2 + a^2 m_0^2 + \sqrt{2b^2 + b^4 + 2a^2 m_0^2 + a^4 m_0^4 - 2a^2 b^2 m_0^2} + 1} + 1}{2}, & \text{if } |y| \le 1 \end{cases} \tag{4.147}$$

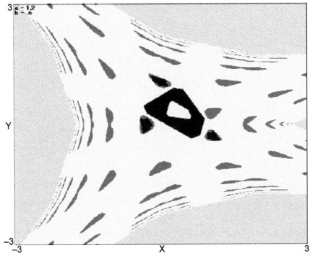

Figure 4.37: Coexisting attractors for the map (4.145) at $a = 1.2$ and $b = -0.6$, where the large chaotic attractor at the center is surrounded by a period-3 chaotic attractor at its periphery, with their basins of attraction, Zeraoulia & Sprott (2010(a)).

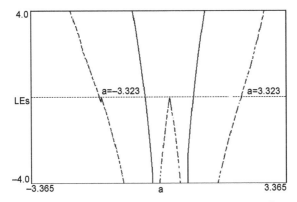

Figure 4.38: Variation of the Lyapunouv exponents of map (4.146) versus the parameter $-3.365 \le a \le 3.365$ with $b = 1.4$, $m_0 = -0.43$, and $m_1 = 0.41$., Zeraoulia & Sprott (2009).

Thus, $\|f'(x, y)\| < +\infty$ and

$$\lambda_{\min}\left(f'^T f'\right)(x,y) = \begin{cases} \dfrac{b^2 + a^2 m_1^2 - dd + 1}{2}, \text{ if}\,|\,y\,|\geq 1 \\ \dfrac{b^2 + a^2 m_0^2 - dd + 1}{2}, \text{if}\,|\,y\,|\leq 1. \\ dd = \sqrt{2b^2 + b^4 + 2a^2 m_1^2 + a^4 m_1^4 - 2a^2 b^2 m_1^2 + 1} \end{cases} \tag{4.148}$$

Hence, if

$$\begin{cases} |\,a\,| > \max\left(\dfrac{1}{|\,m_1\,|}, \dfrac{1}{|\,m_0\,|}\right) \\ |\,b\,| > \max\left(\dfrac{|\,am_1\,|}{\sqrt{a^2 m_1^2 - 1}}, \dfrac{|\,am_0\,|}{\sqrt{a^2 m_0^2 - 1}}\right) \end{cases} \tag{4.149}$$

then both Lyapunov exponents of the map (4.146) are positive for all initial conditions $(x_0, y_0) \in \mathbb{R}^2$, and hence the corresponding attractor is hyperchaotic. Indeed, for $m_0 = -0.43$ and $m_1 = 0.41$, one has that $|\,a\,| > 2.439$, and for $b = 1.4$, one has that $|\,a\,| > 3.323$. Figure 4.38 shows the Lyapunov exponent spectrum for the map (4.146) for $m_0 = -0.43$, $m_1 = 0.41$, $b = 1.4$, and $-3.365 \leq a \leq 3.365$. Thus, the regions of hyperchaos are $-3.365 \leq a \leq -3.323$ and $3.323 \leq a \leq 3.365$. On the other hand, Fig. 4.39 show the discrete hyperchaotic double scroll that results from a stable period-3 orbit transitioning to a fully developed chaotic regime as shown in Fig. 4.40, and it is the only observed scenario. For $b = 1.4$, $m_0 = -0.43$, and $m_1 = 0.41$ and a vary, the map (4.146) exhibits the following dynamical behaviors: For $a < -3.365$, and $a > 3.365$, the map (4.146) does not converge. For $-3.365 \leq a \leq 3.365$, the map (4.146) begins with a *reverse* border-collision bifurcation, leading to a stable period-3 orbit, and then collapses to a point that is *reborn* as a stable period-3 orbit leading to fully developed chaos. Thus, it seems that the map (4.146) behaves in a similar way to the 4-D dynamical system given in [Thamilmaran, et al. (2004)], i.e., both hyperchaotic attractors are obtained by a border-collision bifurcation [Banerjee & Grebogi (1999)].

4.16 Piecewise smooth maps of the plane and robust chaos

Since the Lozi map (3.1) is a special case of the general 2-D piecewise smooth mappings, it is natural to ask about an extension of the results known for the Lozi map (3.1) to this more general form of mappings. Indeed, this general form was the subject of many studies as in [Robert & Robert (2002), Banerjee & Grebogi (1999), Banergee, et al. (1998), Hassouneh, et al. (2002), Banerjee & Verghese (2000), Banerjee, et al. (1999), Kowalczyk (2005)]. The most important results about these maps are about the occurrence of new kind of bifurcations, namely *border collision bifurcations* and *robust chaos*. See [Zeraoulia & Sprott (2011)] for more details about this subject. The last idea comes from the fact that chaotic dynamical systems display two kinds of chaotic attractors: One type has *fragile* chaos which disappears with perturbations of a parameter or coexist with other attractors and the other type has robust chaos defined as follows [Banerjee, et al. (1998)]:

Definition 62 *Robust chaos is defined by the absence of periodic windows and coexisting attractors in some neighborhood in the parameter space.*

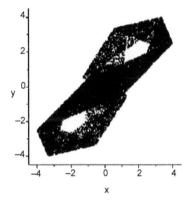

Figure 4.39: The discrete hyperchaotic double scroll attractor obtained from the map (4.146) for $a = 3.36$, $b = 1.4$, $m_0 = -0.43$, and $m_1 = 0.41$ with initial conditions $x = y = 0.1$., Zeraoulia & Sprott (2009).

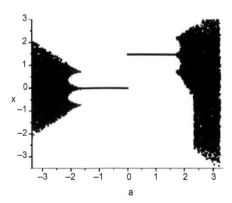

Figure 4.40: The border collision bifurcation route to chaos of map (4.146) versus the parameter $-3.365 \leq a \leq 3.365$ with $b = 1.4$, $m_0 = -0.43$, and $m_1 = 0.41$., Zeraoulia & Sprott (2009).

In other words, the existence of these windows implies that small changes of the parameters would destroy the chaos. A detailed study about robust chaos in dynamical systems can be found in the book [Zeraoulia & Sprott (2011)].

4.16.1 Normal form for 2-D piecewise smooth maps

In order to make it better amenable for analytical treatments the study of bifurcations and chaos in 2-D piecewise smooth maps, a normal form was derived in [Banerjee & Grebogi (1999)]. Indeed, let us consider the following 2-D piecewise smooth system given by:

$$g(x, y; \rho) = \begin{pmatrix} g_1 = \begin{pmatrix} f_1(x,y;\rho) \\ f_2(x,y;\rho) \end{pmatrix}, & \text{if } x < S(y,\rho) \\ g_2 = \begin{pmatrix} f_1(x,y;\rho) \\ f_2(x,y;\rho) \end{pmatrix}, & \text{if } x \geq S(y,\rho) \end{pmatrix} \qquad (4.150)$$

where the smooth curve $x = S(y, \rho)$ divides the phase plane into two regions R_1 and R_2, and the boundary between them Σ as follow:

$$\begin{cases} R_1 = \{(x, y) \in \mathbb{R}^2, x < S(y, \rho)\} \\ R_2 = \{(x, y) \in \mathbb{R}^2, x \geq S(y, \rho)\} \\ \Sigma = \{(x, y) \in \mathbb{R}^2, x = S(y, \rho)\} \end{cases} \qquad (4.151)$$

Assume that the functions g_1 and g_2 are both continuous and have continuous derivatives. Then the map g is continuous, but its derivative is discontinuous at the borderline $x = S(y, \rho)$. It is further assumed that the

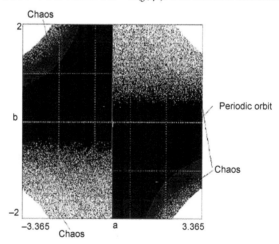

Figure 4.41: Regions of dynamical behaviors in the *ab*-plane for the map (4.146), Zeraoulia & Sprott (2009).

one-sided partial derivatives at the border are finite and in each subregion R_1 and R_2 and the map (4.150) has one fixed point in R_1 and one fixed point in R_2 for a value ρ_* of the parameter ρ. It was shown in [Banerjee & Grebogi (1999)] that the normal form of the map (4.150) is given by:

$$N(x, y) = \begin{cases} \begin{pmatrix} \tau_L & 1 \\ -\delta_L & 0 \end{pmatrix} \begin{pmatrix} x \\ y \end{pmatrix} + \begin{pmatrix} 1 \\ 0 \end{pmatrix} \mu, \text{ if } x < 0 \\ \begin{pmatrix} \tau_R & 1 \\ -\delta_R & 0 \end{pmatrix} \begin{pmatrix} x \\ y \end{pmatrix} + \begin{pmatrix} 1 \\ 0 \end{pmatrix} \mu, \text{ if } x > 0 \end{cases} \tag{4.152}$$

where μ is a parameter and $\tau_{L,R}$, $\delta_{L,R}$ are the traces and determinants of the corresponding matrices of the linearized map in the two subregion R_L and R_R and the boundary between them, Σ, given by:

$$\begin{cases} R_L = \{(x, y) \in \mathbb{R}^2, x \le 0, y \in \mathbb{R}\} \\ R_R = \{(x, y) \in \mathbb{R}^2, x > 0, y \in \mathbb{R}\} \\ \Sigma = \{(x, y) \in \mathbb{R}^2, x = 0, y \in \mathbb{R}\} \end{cases} \tag{4.153}$$

evaluated at the fixed points:

$$\begin{cases} P_L = \left(\dfrac{\mu}{1 - \tau_L + \delta_L}, \dfrac{-\delta_L \mu}{1 - \tau_L + \delta_L} \right) \in R_L \\ P_R = \left(\dfrac{\mu}{1 - \tau_R + \delta_R}, \dfrac{-\delta_R \mu}{1 - \tau_R + \delta_R} \right) \in R_R \end{cases} \tag{4.154}$$

(with eigenvalues $\lambda_{L1,2}$ and $\lambda_{R1,2}$ respectively). The stability of these fixed points is determined by the eigenvalues of the corresponding Jacobian matrix, .i.e., $\lambda = \dfrac{1}{2} \left(\tau \pm \sqrt{\tau^2 - 4\delta} \right)$ in the regions R_L and R_R.

4.16.2 Border collision bifurcations and robust chaos

In this section, we summarize the known sufficient conditions for the possible bifurcations phenomena in the normal form (4.152). First, we note that the study of the dynamics of piecewise smooth maps is due to Feigin through his investigations about border collision bifurcations phenomena [Feigin (1970), Nusse & Yorke (1992)] and later the development is based on the observation of these maps in power electronic circuits studies [Robert & Robert (2002), Banerjee & Grebogi (1999), Banergee, et al. (1998), Hassouneh, et al. (2002), Banerjee & Verghese (2001), Kowalczyk (2005), Banerjee, et al. (1999)]. Non smooth bifurcations are quite common in them [Yuan, et al. 1998]. In [Feigin (1970), di Bernardo, et al. (1999)] the existence of period-1 and period-2 orbits before and after border collision were studied with a classification of border-collision bifurcations in n-dimensional piecewise smooth systems based on the various cases depending on the number of real eigenvalues greater than 1 or less than −1. New methods of classifications (based on the consideration of asymptotically stable orbits (including chaotic orbits) before and after border collision) can be found in [Banerjee & Grebogi (1999), Banerjee, et al. (2000)] for one and two-dimensional maps. Also, several proofs of the existence of various types of border collision bifurcations using essentially the trace and the determinant of the Jacobian matrix on the two sides of the border. These results require that $|\delta_L| < 1$ and $|\delta_R| < 1$, but it was proved that attractors can exist if the determinant in one side is greater than unity in magnitude, if that determinant in the other side is smaller than unity. The situation $|\delta_L| > 1$ and $\delta_R = 0$ (which occurs in some classes in power electronic systems) has been studied in [Parui & Banerjee (2002)].

4.16.3 Fixed points of the normal form map

The possible types of fixed points of the normal form map (4.152) are given by:

(1) For positive determinant

(1-a) For $2\sqrt{\delta} < \tau < (1+\delta)$, the Jacobian matrix has two real eigenvalues $0 < \lambda_{1L}, \lambda_{2L} < 1$, and the fixed point is a regular attractor.

(1-b) For $\tau > (1 + \delta)$, the Jacobian matrix has two real eigenvalues $0 < \lambda_{1L} < 1, \lambda_{2L} > 1$ and the fixed point is a regular saddle.

(1-c) For $-(1 + \delta) < \tau < -2\sqrt{\delta}$, the Jacobian matrix has two real eigenvalues $-1 < \lambda_{1L} < 0, -1 < \lambda_{2L} < 0$ and the fixed point is a flip attractor.

(1-d) For $\tau < -(1+\delta)$, the Jacobian matrix has two real eigenvalues $-1 < \lambda_{1L} < 0, \lambda_{2L} < -1$ and the fixed point is a flip saddle.

(1-e) For $0 < \tau < 2\sqrt{\delta}$, the Jacobian matrix has two complex eigenvalues $|\lambda_{1L}|, |\lambda_{2L}| < 1$ and the fixed point is a clockwise spiral.

(1-g) For $-2\sqrt{\delta} < \tau < 0$, the Jacobian matrix has two complex eigenvalues $|\lambda_{1L}|, |\lambda_{2L}| < 1$ and the fixed point is a counter-clockwise spiral.

(2) For negative determinant

(2-a) For $-(1 + \delta) < \tau < (1 + \delta)$, the Jacobian matrix has two real eigenvalues $-1 < \lambda_{1L} < 0, 0 < \lambda_{2L} < 1$ and the fixed point is a flip attractor.

(2-b) For $\tau > 1 + \delta$, the Jacobian matrix has two real eigenvalues $\lambda_{1L} > 1, -1 < \lambda_{2L} < 0$ and the fixed point is a flip saddle.

(2-c) For $\tau < -(1 + \delta)$, the Jacobian matrix has two real eigenvalues $0 < \lambda_{1L} < 1$, $\lambda_{2L} < -1$ and the fixed point is a flip saddle.

4.16.4 Regions for non robust chaos

In this section, we give a list of the different cases where chaos in map (4.152) is not robust in the sense of Definition 62. Depending on the sign of the system determinants on both sides of the normal form (4.152), robust chaos is not possible in the following cases:

4.16.4.1 The case of positive determinants on both sides of the border

Scenario A: *(Locally unique stable fixed point on both sides of the border).*

If

$$\begin{cases} \delta_L > 0,\ \delta_R > 0 \\ -(1 + \delta_L) < \tau_L < (1 + \delta_L) \\ -(1 + \delta_R) < \tau_R < (1 + \delta_R) \end{cases} \tag{4.155}$$

then a stable fixed point persists as the bifurcation parameter μ is increased (or decreased) through zero. For the parameter range given by (4.155), a stable fixed point yields a stable fixed point after the border crossing with or without *extraneous* periodic orbits emerging from the critical point. These cases are as follow: (1) If $2\sqrt{\delta_L} < \tau_L < (1 + \delta L),- (1 + \delta R) < \tau R < -2\sqrt{\delta_R}$, then a regular attractor yields a flip attractor.

(2) If $2\sqrt{\delta_L} < \tau_L < (1 + \delta L)$, $2\sqrt{\delta_R} < \tau R < (1 + \delta R)$, then a regular attractor yields a regular attractor.

(3) If $-(1 + \delta L) < \tau_L < -2\sqrt{\delta_L}, -(1 + \delta L) < \tau_L < -2\sqrt{\delta_L}$, then a flip attractor yields a regular attractor.

(4) If $-(1 + \delta L) < \tau L < -2\sqrt{\delta_L}, -(1 + \delta R) < \tau R < -2\sqrt{\delta_R}$, then a flip attractor yields a flip attractor.

(5) If $2\sqrt{\delta_L} < \tau_L < (1 + \delta L),- 2\sqrt{\delta_R} < \tau R < 2\sqrt{\delta_R}$, then a regular attractor yields a spiral attractor.

(6) If $-2\sqrt{\delta_L} < \tau_L < 2\sqrt{\delta_L}$, $2\sqrt{\delta_R} < \tau R < (1+\delta R)$, then a spiral attractor yields a regular attractor.

(7) If $-(1 + \delta L) < \tau_L < -2\sqrt{\delta_L}, -2\sqrt{\delta_R} < \tau R < 2\sqrt{\delta_R}$, then a flip attractor yields a spiral attractor.

(8) If $-2\sqrt{\delta_R} < \tau R < 2\sqrt{\delta_R}, -(1 + \delta R) < \tau_R < -2\sqrt{\delta_R}$, then a spiral attractor yields a flip attractor.

(9) If $0 < \tau_L < 2\sqrt{\delta_L}, -2\sqrt{\delta_R} < \tau_R < 0$, then a clockwise spiral attractor yields a anticlockwise spiral attractor.

(10) If $-2\sqrt{\delta_L} < \tau_L < 0, 0 < \tau_R < 2\sqrt{\delta_R}$, then a anticlockwise spiral attractor yields a clockwise spiral attractor.

(11) If $0 < \tau_L < 2\sqrt{\delta_L}, 0 < \tau_R < 2\sqrt{\delta_R}$, then a clockwise spiral attractor yields a clockwise spiral attractor.

(12) If $-2\sqrt{\delta_L} < \tau_L < 0, -2\sqrt{\delta_R} < \tau_R < 0$, then a anticlockwise spiral attractor yields a anticlockwise spiral attractor.

Extraneous periodic orbits (either before, after, or on both sides of the border) occur at bifurcation in the cases (9) and (10) and no known examples or proofs for the cases (5),(6), (11) and (12). While these orbits cannot appear for the cases (1) to (4) because the fixed points on both sides of the border are locally unique and stable and this happens when the fixed point changes from: (1) regular attractor to flip attractor, (2) regular attractor to regular attractor, (3) flip attractor to regular attractor, and (4) flip attractor to flip attractor, as μ is varied through its critical value.

Scenario B: (*Supercritical period doubling border collision bifurcation*). In this case, there are two regions in the parameter space where period doubling border collision bifurcation occurs, i.e., a locally unique stable fixed point leads to an unstable fixed point plus a locally unique attracting period two orbit. This scenario is divided into **Scenario B1** and **Scenario B2** as follows:

Scenario B1: If

$$
\begin{cases}
\delta_L > 0, \delta_R > 0 \\
-(1 + \delta_L) < \tau_L < -2\delta_L \\
\tau_R < -(1 + \delta_R) \\
\tau_R \tau_L < (1 + \delta_R)(1 + \delta_L)
\end{cases}
\tag{4.156}
$$

Scenario B2: If

$$
\begin{cases}
\delta_L > 0, \delta_R > 0 \\
2\delta_L < \tau_L < (1 + \delta_L) \\
\tau_R < -(1 + \delta_R) \\
\tau_R \tau_L > -(1 - \delta_R)(1 - \delta_L)
\end{cases}
\tag{4.157}
$$

4.16.4.2 The case of negative determinants on both sides of the border

In this case, the condition for locally unique stable fixed point on both sides of the border is given in **Scenario C** as follow:

Scenario C: (*Locally unique stable fixed point on both sides of the border*).

If

$$\begin{cases} \delta_L < 0, \delta_R < 0 \\ -(1 + \delta_L) < \tau_L < (1 + \delta_L) \\ -(1 + \delta_R) < \tau_R < (1 + \delta_R) \end{cases} \tag{4.158}$$

then a locally unique stable fixed point leads to a locally unique stable fixed point as μ is increased through zero.

Scenario D: *(Supercritical border collision period doubling).* If

$$\begin{cases} \delta_L < 0, \delta_R < 0 \\ -(1 + \delta_L) < \tau_L < (1 + \delta_L) \\ \tau_R < -(1 + \delta_R) \\ \tau_R \tau_L < (1 + \delta_R)(1 + \delta_L) \\ \tau_R \tau_L > -(1 - \delta_R)(1 - \delta_L) \end{cases} \tag{4.159}$$

then a locally unique stable fixed point to the left of the border for $\mu < 0$ crosses the border and becomes unstable and a locally unique period two orbit is born as μ is increased through zero, i.e., this is a condition for *supercritical* period doubling border collision with no extraneous periodic orbits.

4.16.4.3 The case of negative determinant to the left of the border and positive determinant to the right of the border

In this case, the eigenvalues are real in the left of the border, i.e., $\delta_L < 0$, and they are also real in the right of the border, i.e., $\delta_R > 0$, provided that $\tau^2_R > 4\delta_R$. Thus, a sufficient condition for having a locally unique fixed point leading to a locally unique fixed point as μ is varied through the critical value is given as follows:

Scenario E: *(Locally unique stable fixed point on both sides of the border).*

If

$$\begin{cases} \delta_L < 0, \delta_R > 0 \\ -(1 + \delta_L) < \tau_L < (1 + \delta_L) \\ -(1 + \delta_R) < \tau_R < (1 + \delta_R) \\ \tau^2_R > 4\delta_R \end{cases} \tag{4.160}$$

The conditions (4.160) can be divided into two cases:

Scenario E1: This occurs if

$$\begin{cases} \delta_L < 0, \delta_R > 0 \\ -(1 + \delta_L) < \tau_L < (1 + \delta_L) \\ -(1 + \delta_R) < \tau_R < -2\delta_R \end{cases} \tag{4.161}$$

Scenario E2: This occurs if

$$\begin{cases} \delta_L < 0, \delta_R > 0 \\ -(1 + \delta_L) < \tau_L < (1 + \delta_L) \\ 2\delta_R < \tau_R < (1 + \delta_R) \end{cases} \qquad (4.162)$$

4.16.4.4 The case of positive determinant to the left of the border and negative determinant to the right of the border

In this case, the eigenvalues are real in the right of the border, i.e., $\delta_R < 0$, and they are also real in the left of the border, i.e., $\delta_L > 0$, provided that $\tau^2_L > 4\delta_L$. Hence, a sufficient condition for having a locally unique fixed point leading to a locally unique fixed point as μ is varied through the critical value is given as:

Scenario F: *(Locally unique stable fixed point on both sides of the border).*

If

$$\begin{cases} \delta_L > 0, \delta_R < 0 \\ -(1 + \delta_L) < \tau_L < (1 + \delta_L) \\ \tau^2_L > 4\delta_L \\ -(1 + \delta_R) < \tau_R < (1 + \delta_R) \end{cases} \qquad (4.163)$$

Also, the conditions (4.163) can be divided into two cases:

Scenario F1: This occurs if

$$\begin{cases} \delta_L > 0, \delta_R < 0 \\ -(1 + \delta_L) < \tau_L < -2\delta_L \\ -(1 + \delta_R) < \tau_R < (1 + \delta_R) \end{cases} \qquad (4.164)$$

Scenario F2: This occurs if

$$\begin{cases} \delta_L > 0, \delta_R < 0 \\ 2\delta_L < \tau_L < (1 + \delta_L) \\ -(1 + \delta_R) < \tau_R < (1 + \delta_R) \end{cases} \qquad (4.165)$$

In fact, there are no known conditions for *supercritical* border collision period doubling that occurs without EBOs when the determinants on both sides of the border are of opposite signs. The **Scenarios A-F** are shown in Figs. 4.42 and 4.43 in the space $(\tau_L, \delta_L, \tau_R, \delta_R)$ such that the fixed points are the unique attractors on both sides of the border (in these figures, δ_L and δ_R are fixed whereas τ_L and τ_R are variables).

4.16.4.5 Stable fixed point leading to stable fixed point plus extraneous periodic orbits

From [Banerjee & Grebogi (1999), Banerjee, et al. 2000], we know that there are certain border collision bifurcations that, while not causing a *catastrophic collapse* of the system, may lead to *undesirable* system behavior.

This means that a stable fixed point leading to stable fixed point plus extraneous periodic orbits, that can display multiple attractor bifurcation on either side of the border or both sides of the border in addition to the stable fixed points. This case has no conditions, but the following example shows this situation:

$$l(x,y) = \begin{cases} \begin{pmatrix} 0.50 & 1 \\ -0.90 & 0 \end{pmatrix} \begin{pmatrix} x \\ y \end{pmatrix} + \begin{pmatrix} 1 \\ 0 \end{pmatrix} \mu, \text{if } x < 0 \\ \begin{pmatrix} -1.22 & 1 \\ -0.36 & 0 \end{pmatrix} \begin{pmatrix} x \\ y \end{pmatrix} + \begin{pmatrix} 1 \\ 0 \end{pmatrix} \mu, \text{if } x > 0 \end{cases} \quad (4.166)$$

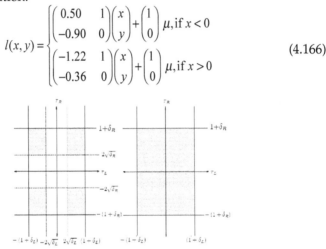

Figure 4.42: No bifurcation occurs as μ is increased (decreased) through zero in the shaded regions for the map (4.152). Only the path of the fixed point changes at $\mu = 0$. (a) $0 < \delta_L < 1$ and $0 < \delta_R < 1$ (b) $-1 < \delta_L < 0$ and $-1 < \delta_R < 0$.

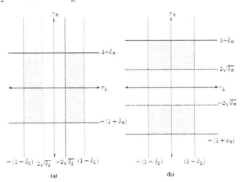

Figure 4.43: No bifurcation occurs as μ is increased (decreased) through zero in the shaded regions for the map (4.152). Only the path of the fixed point changes at $\mu = 0$. (a) $0 < \delta_L < 1$ and $-1 < \delta_R < 0$ (b)$-1 < \delta_L < 0$ and $0 < \delta_R < 1$.

i.e., as shown in Fig. 4.44(a), the case of stable fixed point plus period-4 attractor to stable fixed point plus period-3 attractor as μ is increased through zero. In this case, the fixed point for $\mu < 0$ is spirally attracting and for $\mu > 0$ is a flip attractor. The second example is given by:

$$m(x, y) = \begin{cases} \begin{pmatrix} 1.6 & 1 \\ -0.8 & 0 \end{pmatrix} \begin{pmatrix} x \\ y \end{pmatrix} + \begin{pmatrix} 1 \\ 0 \end{pmatrix} \mu, \text{if } x < 0 \\ \begin{pmatrix} -1.4 & 1 \\ -0.6 & 0 \end{pmatrix} \begin{pmatrix} x \\ y \end{pmatrix} + \begin{pmatrix} 1 \\ 0 \end{pmatrix} \mu, \text{if } x > 0, \end{cases} \quad (4.167)$$

this case is shown in Fig. 4.44(b) and displays a stable fixed point to stable fixed point plus period-7 attractor. In this case, the fixed point for $\mu < 0$ is spirally attracting and for $\mu > 0$ is also spirally attracting with opposite sense of rotation.

4.16.5 Regions for robust chaos

First, we give a definition of the so called dangerous bifurcations:

Definition 63 *The dangerous bifurcations begin with a system operating at a stable fixed point on one side of the border, say the left side.*

The main dangerous bifurcations that can result from border collision are:

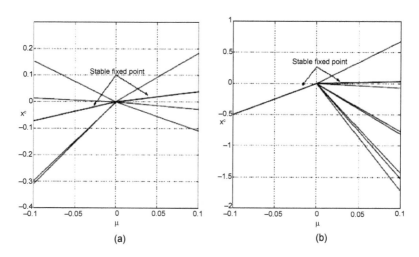

Figure 4.44: (a) Bifurcation diagram of map (4.166) (b) Bifurcation diagram of map (4.167).

(1) *Border collision pair bifurcation:* This is analogous to saddle node bifurcations in smooth maps and occurs if:

$$\begin{cases} -(1+\delta_L) < \tau_L < (1+\delta_L) \\ \tau_R > (1+\delta_R) \end{cases} \tag{4.168}$$

In this case, a stable fixed point and an unstable fixed point merge and disappear as μ is increased through zero. Condition (4.168) implies that the system trajectory diverges for positive values of μ since no local attractors exist.

(2) *Subcritical border collision period doubling:* This bifurcation occurs if:

$$\begin{cases} -(1+\delta_L) < \tau_L < (1+\delta_L) \\ \tau_R < -(1+\delta_R) \\ \tau_R\tau_L > (1+\delta_R)(1+\delta_L) \end{cases} \tag{4.169}$$

In this case, a bifurcation from a stable fixed point and an unstable period-2 orbit to the left of border to an unstable fixed point to the right of the border occurs as μ is increased through zero.

(3) *Supercritical border collision period doubling:* This is not a dangerous bifurcation, but it is undesirable in some applications. As an example of this situation, the one called *cardiac conduction model* proposed in [Sun, et al. (1995)] as a two-dimensional piecewise linear map in which the atrial His interval A is that between cardiac impulse excitation of the lower interatrial septum to the Bundle of His as follow:

$$\begin{pmatrix} A_{n+1} \\ B_{n+1} \end{pmatrix} = f(A_n, R_n, h_n) \tag{4.170}$$

where

$$f(A_n, R_n, h_n) = \begin{cases} \begin{cases} cc + (201 - 0.7A_n)\exp\left(-\dfrac{h_n}{\tau_{rec}}\right), & \text{if } A_n < 130 \\[2ex] cc + (500 - 3.0A_n)\exp\left(-\dfrac{h_n}{\tau_{rec}}\right), & \text{if } A_n > 130 \end{cases} \\[4ex] R_n = \exp\left(-\dfrac{(h_n + A_n)}{\tau_{fat}}\right) + \gamma\exp\left(-\dfrac{h_n}{\tau_{fat}}\right) \\[3ex] cc = A_{min} + R_{n+1} \end{cases} \tag{4.171}$$

The model (4.171) incorporates physiological concepts of recovery, facilitation and fatigue and it confirms and predicts several experiments concerning complex rhythms of nodal conduction. The variable and constants in map (4.171) are as follow: $R_0 = \gamma \exp\left(-\dfrac{h_n}{\tau_{fat}}\right)$, h_0 is the initial h interval, the parameters A_{min}, τ_{fat}, γ and τ_{rec} are positive constants. The variable h_n (usually is the bifurcation parameter) is the interval between bundle of His activation and the subsequent activation (the AV nodal recovery time). The variable R_n (usually is the bifurcation parameter) is a drift in the nodal conduction time. In this case, alternate rhythms with an alternation in conduction time from beat to beat, displays period-doubling bifurcation in the theoretical model in which the map f is piecewise smooth and is continuous at the border $A_b = 130ms$. For $\tau_{rec} = 70ms$, $\tau_{fat} = 30000ms$, $A_{min} = 33ms$, $\gamma = 0.3ms$ a stable period-2 orbit is born after the border collision as shown in Fig. 4.45.

(4) *Stable fixed point leading to chaos:* This situation is also called *instant chaos*. Chaotic behaviors are developed following border collision. As an example of a such situation is the strange bifurcation route to chaos found in a piecewise-linear second-order non autonomous differential equation derived from a simple electronic circuit in [Ohnishi & Inaba (1994)]. In this case, the behavior of this system changes directly to chaos (without period doubling bifurcation or an intermittency) when a limit cycle loses its stability.

4.16.6 Proof of robust chaos

There are several works that devoted to the rigorous proof of robust chaos in the normal form (4.152). We present briefly some recent results. Indeed, it was proved in [Hassouneh, et al. (2002)] that coexisting attractors cannot occur under some conditions on the eigenvalues of the fixed points of the map (4.152):

Theorem 81 *When the eigenvalues at both sides of the border are real, if an attracting orbit exists, it is unique (i.e., coexisting attractors cannot occur).*

It was shown in [Banergee, et al. (1998-1999)] that the resulting chaos from the 2-D map (4.152) is robust in the following cases:

Case 1:

$$\begin{cases} \tau_L > 1 + \delta_L, \text{ and } \tau_R < -(1 + \delta_R) \\ 0 < \delta_L < 1, \text{ and } 0 < \delta_R < 1, \end{cases} \qquad (4.172)$$

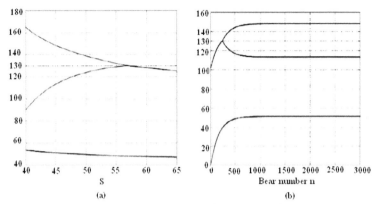

Figure 4.45: (a) Bifurcation diagram for A_n and for R_n for (4.171) with h_n as bifurcation parameter and $\tau_{rec} = 70ms$, $\tau_{fat} = 30000ms$, $A_{min} = 33ms$, and $\gamma = 0.3ms$ (b) Iterations of map (4.171) showing the alternation in A_n as a result of a period doubling bifurcation. The parameter values are the same as in (a) with $h_n = 45ms$.

where the parameter range for boundary crisis is given by:

$$\delta_L \tau_L \lambda_{1L} - \delta_L \lambda_{1L} \lambda_{2L} + \delta_R \lambda_{2L} - \delta_L \tau_R + \delta_L \tau_L - \delta_L - \lambda_{1L} \delta_L > 0 \qquad (4.173)$$

Inequality (4.173) determines the condition for stability of the chaotic attractor. The robust chaotic orbit continues to exist as τ_L is reduced below $1 + \delta_L$.

Case 2: When

$$\begin{cases} \tau_L > 1 + \delta_L, \text{ and } \tau_R < -(1 + \delta_R) \\ \delta_L < 0, \text{ and } -1 < \delta_R < 0 \\ \dfrac{\lambda_{1L} - 1}{\tau_L - 1 - \delta_L} > \dfrac{\lambda_{2R} - 1}{\tau_R - 1 - \delta_R}, \end{cases} \qquad (4.174)$$

The condition for stability of the chaotic attractor is also determined by (4.173). However, if the third condition of (4.174) is not satisfied, then the condition for existence of the chaotic attractor changes to:

$$\frac{\lambda_{2R} - 1}{\tau_R - 1 - \delta_L} < \frac{(\tau_L - \delta_L - \lambda_{2L})}{(\tau_L - 1 - \delta_L)(\lambda_{2L} - \tau_R)} \qquad (4.175)$$

Case 3: The remaining ranges for the quantity $\tau_{L,R}$, $\delta_{L,R}$ can be determined in some cases using the same logic as in the above two cases, or there is no analytic condition for a boundary crisis, and it has to be determined numerically. Namely, the following result was proved in [Banergee, et al. (1998-1999)]:

Theorem 82 *For $1 > \delta_L > 0{,}1 > \delta_R > 0$, the normal form (4.152) exhibits robust chaos in a portion of parameter space bounded by the conditions $\tau_R = -(1 + \delta_R)$, $\tau_L > 2\sqrt{\delta}_L$, and (4.173), as shown in Fig. 4.46. The same logic applies for the cases with negative determinant, i.e., for $-1 < \tau_R < 0$, or $\delta_L < 0$ and $\delta_R < 0$, or $\delta_L < 0$ and the eigenvalues are real for all τ_L.*

4.16.7 Normal form for 2-D non-invertible piecewise smooth maps

Robust chaos in non-invertible piecewise-linear maps was studied in [Kowalczyk (2005)]. This type of maps is the normal form for *grazing–sliding bifurcations* in three-dimensional Filippov type systems. For these maps robust chaos is an essential feature. The main differences with respect to the above analysis for the invertible case are:

(1) In [Banerjee & Grebogi ((1999))] only the case where the determinants of the PWL normal form on both sides of the bifurcation boundary are less then unity, i.e., $|\delta_L| < 1$ and $|\delta_R| < 1$.

(2) If the fixed points are purely real and are of saddle type, then the stable manifold (resp. the unstable manifolds) *folds* at every intersection with the y-axis (at every intersection with the x-axis) and every pre-image of the fold point is a fold for the both cases.

(3) The occurrence of homoclinic intersection between manifolds is not obvious if the map is non-invertible in one of its regions. Compare with [Banerjee & Grebogi (1999), Parui & Banerjee (2002)].

Following [di Bernardo, et al. (2002)], it was shown in [Kowalczyk (2005)] that a normal form map for grazing–sliding bifurcations can be obtained in which on one side of the discontinuity boundary has corank-1:

$$\Pi(x, y, \mu) = \overline{x} = \begin{cases} \begin{pmatrix} \tau_L & 1 \\ -\delta_L & 0 \end{pmatrix} \begin{pmatrix} x \\ y \end{pmatrix} + \begin{pmatrix} 1 \\ 0 \end{pmatrix} \mu, & \text{if } x \leq 0 \\[12pt] \begin{pmatrix} \tau_R & 1 \\ 0 & 0 \end{pmatrix} \begin{pmatrix} x \\ y \end{pmatrix} + \begin{pmatrix} 1 \\ 0 \end{pmatrix} \mu, & \text{if } x > 0, \end{cases} \tag{4.176}$$

Let Π_L, Π_R be the two submappings defined by:

$$\Pi_L = M\overline{x} + C\mu, \text{ if } x \leq 0 \tag{4.177}$$

$$\Pi_R = N\overline{x} + C\mu, \text{ if } x > 0 \tag{4.178}$$

where

$$
\begin{cases}
M = \begin{pmatrix} \tau_L & 1 \\ -\delta_L & 0 \end{pmatrix}, \; N = \begin{pmatrix} \tau_R & 1 \\ 0 & 0 \end{pmatrix} \\
C = \begin{pmatrix} 1 \\ 0 \end{pmatrix}, \; \overline{x} = \begin{pmatrix} x \\ y \end{pmatrix}
\end{cases}
\tag{4.179}
$$

Obviously, $\Pi_L : R_L \to R_L \cup R_R$ is smooth and invertible while $\Pi_R : R_L \to \{x \in \mathbb{R}, y = 0\}$ is smooth but non-invertible. The analysis of map (4.176) is based on the distinction between *admissible* fixed points and *virtual* fixed points of map (4.176) defined in [Kowalczyk (2005)] as follow:

Definition 64 *(a) Admissible fixed points of map (4.176) are fixed points of Π_L and Π_R that lie in the domain of definition of these two submappings. (b) [Banerjee & Grebogi (1999)] Virtual fixed points of map (4.176) are fixed points of Π_L and Π_R but existing outside the corresponding domain of definition.*

Thus, admissible fixed point of Π_L lies in R_L and of Π_R in R_R. The virtual fixed point of Π_L belongs to R_R and of Π_R to RL. In this case, the admissible fixed points of Π_L and Π_R are denoted by the letters A, a and B, b, respectively (see [Feigin (1994)]) and the virtual fixed points of Π_L are denoted by $\overline{A}, \overline{a}$ and of Π_R by $\overline{B}, \overline{b}$:

$$
A, a = \left(\frac{\mu}{1 - \tau_L + \delta_L}, \frac{-\delta_L \mu}{1 - \tau_L + \delta_L} \right) \text{for} \frac{\mu}{1 - \tau_L + \delta_L} \le 0
\tag{4.180}
$$

$$
B, b = \left(\frac{\mu}{1 - \tau_R}, 0 \right), \text{ for } \frac{\mu}{1 - \tau_R} > 0
\tag{4.181}
$$

By definition we have $\dfrac{\mu}{1 - \tau_I + \delta_I} > C$ for a virtual fixed point of Π_L and $\dfrac{\mu}{1 - \tau_R} \le 0$ for a virtual fixed point of Π_R. The corresponding eigenvalues of $A, a, \overline{A}, \overline{a}$ are given by $\lambda_{1,2} = \dfrac{\tau_L \pm \sqrt{\tau_L - 4\delta_L}}{2}$, and for the points $B, b, \overline{B}, \overline{b}$ as $\lambda_{1L} = \tau_R, \lambda_{2L} = 0$. Thus, the border-collision bifurcation can be defined for the map (4.176) as follows [Kowalczyk (2005)]:

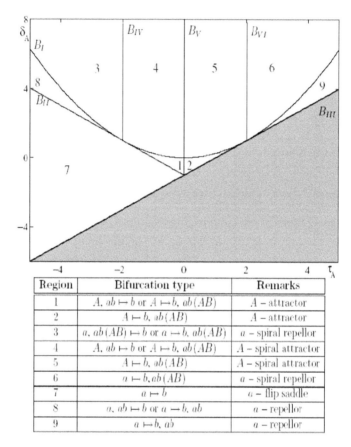

Region	Bifurcation type	Remarks
1	$A,\ ab \mapsto b$ or $A \mapsto b,\ ab\,(AB)$	A – attractor
2	$A \mapsto b,\ ab\,(AB)$	A – attractor
3	$a,\ ab\,(AB) \mapsto b$ or $a \mapsto b,\ ab\,(AB)$	a – spiral repellor
4	$A,\ ab \mapsto b$ or $A \mapsto b,\ ab\,(AB)$	A – spiral attractor
5	$A \mapsto b,\ ab\,(AB)$	A – spiral attractor
6	$a \mapsto b, ab\,(AB)$	a – spiral repellor
7	$a \mapsto b$	a – flip saddle
8	$a,\ ab \mapsto b$ or $a \mapsto b,\ ab$	a – repellor
9	$a \mapsto b,\ ab$	a – repellor

Figure 4.46: Partition of the parameter space τ_L, δ_L into regions characterized by different border collision bifurcation scenarios as μ increases for $\tau_R < -1$ for the map (4.176). The behavior of the simple period-1 and -2 points is given in the table. *Flip saddle* refers to a saddle type fixed point with one eigenvalue less than −1. The six borderlines indicated by B_I to B_{VI} are described in the text. Reused with permission from Kowalczyk, P, Nonlinearity, (2005). Copyright 2005, IOP Publishing Ltd.

Definition 65 *Consider PWL map (4.176), and suppose that the fixed points of this map, which are fixed points of* Π_L *and* Π_R, *depend smoothly on the parameter* μ *in some small neighborhood* $\varepsilon > 0$ *of the origin. Suppose that a fixed point (x, y) = (x*, y*)*

(1) for $-\varepsilon < \mu < 0$ *belongs to* R_L,

(2) for $\mu = 0$ *belongs to the boundary* Σ,

(3) for $0 < \mu < \varepsilon$ *belongs to* R_R.

Then, we say that the fixed point of (4.176) undergoes a border-collision bifurcation and the fixed point is called the border-crossing fixed point.

Definition 65 implies that under the variation of μ in map (4.176) a stable admissible border-crossing fixed point of Π_L become a virtual fixed point of Π_L. For the onset of chaos in map (4.176), four cases were studied in [Kowalczyk (2005)], i.e., (1) border-collision from stable fixed point to flip saddle, (2) two-piece invariant sets, (3) three-piece invariant sets, and (4) N-piece invariant sets. For the first case, the border-collision (from a stable admissible fixed point A existing for $\mu < 0$) is characterized by real eigenvalues, to an admissible saddle, this scenario is possible in the parameter regions:

$$\begin{cases} \delta_L > \tau_L - 1 \\ \tau_R < -1 \end{cases} \tag{4.182}$$

labeled as '1' and '2' in Fig. 4.46. For the second case, some points in the neighborhood of the origin within R_L are mapped into R_R and all the points in R_R are mapped onto the x-axis, a segment of the x-axis might form an Ω-limit set for these points. Thus, the following result was proved in [Kowalczyk (2005)]:

Proposition 4.12 *If the border-collision bifurcation from an admissible fixed point attractor to an admissible flip saddle is exhibited by map (4.176) under the variation of μ and the conditions:*

$$\begin{cases} \tau_L < \dfrac{-1}{1+\tau_R} \\[2mm] \tau_L(1+\tau_R) - \delta_L\left(1+\dfrac{1}{\tau_R}\right) < 0 \\[2mm] \tau_L(1+\tau_R) - \delta_L\left(1+\dfrac{1}{\tau_R}\right) + 1 > 0 \end{cases} \tag{4.183}$$

are satisfied, then there exists an attractor born in the border-collision bifur-cation which must necessarily lie within PWL continuous invariant segment KLC such that:

$$\begin{cases} K = ((\tau_R + 1)\mu, 0) \\ L = (\mu, 0) \\ C = ((\tau_L\tau_R + \tau_L + 1)\mu, -\delta_L(\tau_R + 1)\mu) \end{cases} \tag{4.184}$$

(see Fig. 4.47).

The uniqueness of the attractor in this case is guaranteed by the following result proved in [Kowalczyk (2005)]:

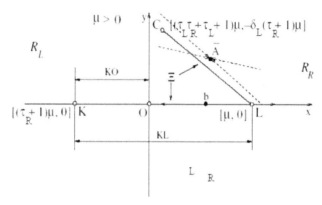

Figure 4.47: An example of a limit set of (4.176) after the border-collision bifurcations from a stable fixed point to a flip saddle. Reused with permission from Kowalczyk, P, Nonlinearity, (2005). Copyright 2005, IOP Publishing Ltd.

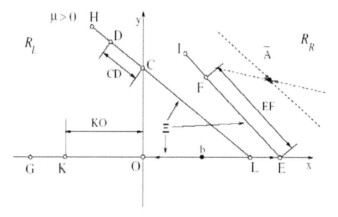

Figure 4.48: An example of a limit set of (4.176) after the border-collision bifurcation consisting of three PWL segments. Reused with permission from Kowalczyk, P, Nonlinearity, (2005). Copyright 2005, IOP Publishing Ltd.

Proposition 4.13 *If the border-collision bifurcation from an admissible fixed point attractor to an admissible flip saddle accompanied by the birth of unstable period-2 points is exhibited by map (4.176) and conditions (4.183) hold, then there is an attractor born in the border-collision bifurcation which is a four-, two- or one-piece chaotic attractor limiting on* Ξ (Ξ *is a well defined set in the proof of Proposition. 4.12 in [Kowalczyk (2005)]).*

For the third case, the image of the point C shown in Fig. 4.47 is not mapped onto the segment *KL* but outside *KL* in which it is impossible to have any stable higher-periodic points within the set Ξ formed as in Fig. 4.48. Hence, the segment *KO* is expanding under forward iteration of (4.176) and the image of *KO* crosses the y-axis and there is a possibility of the existence of period-3 stable periodic points or of a birth of a chaotic attractor living on a higher number of PWL segments. Thus, the following result was proved in [Kowalczyk (2005)]:

Proposition 4.14 *If the border-collision bifurcation from an admissible fixed point attractor to an admissible flip saddle is exhibited by the normal form map (4.176) under the variation of μ and conditions:*

$$\begin{cases} \tau_L > \dfrac{-1}{1+\tau_R} \\[2mm] \tau_L^2(1+\tau_R)+\tau_L+1-\delta_L\left(1+\dfrac{2}{\tau_R}\right)>0 \\[2mm] \tau_L^3(1+\tau_R)+\tau_L^2(1+\tau_L\tau_R)-\delta_L\tau_L\left(1+\dfrac{1}{\tau_R}\right)-\dfrac{\delta_L}{\tau_R}+1>0 \end{cases} \tag{4.185}$$

hold, then there exists an attractor born in the border-collision bifurcation which must necessarily lie within the PWL continuous invariant set built from segments GEI and LH such that:

$$\begin{cases} G=\left(\left(\tau_R\left(\dfrac{\delta_L}{\tau_L}\right)+\tau_R+1\right)\mu,0\right), \\[3mm] E=\left(\left(\left(\dfrac{\delta_L}{\tau_L}\right)+1\right)\mu,0\right), \\[3mm] I=(v,-\delta_L(\tau_R\delta_L+\tau_L\tau_R+\tau_L+1)\mu), \\[3mm] v=(\tau_L(\tau_R\delta_L+\tau_L\tau_R+\tau_L+1)+1)\mu+\left(-\delta_L\left(\left(\dfrac{\tau_R}{\tau_L}\right)\delta_L+\tau_R+1\right)\mu\right), \\[3mm] L=(\mu,0), \\[3mm] H=\left((\tau_R\delta_L+\tau_L\tau_R+\tau_L+1)\mu,-\delta_L\left(\left(\dfrac{\tau_R}{\tau_L}\right)\delta_L+\tau_R+1\right)\mu\right) \end{cases} \tag{4.186}$$

(see Fig. 4.48). This attractor is either a period-3 point or a six-, three- or one-piece chaotic attractor.

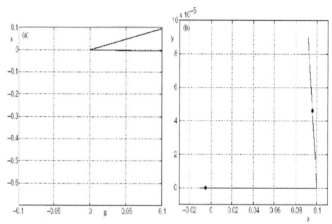

Figure 4.49: (a) Bifurcation diagram presenting period doubling of map (4.176) due to the border-collision bifurcation scenario for $M = \begin{pmatrix} 0.85401 & 1 \\ -0.009 & 0 \end{pmatrix}$, $N = \begin{pmatrix} \tau_R & 1 \\ 0 & 0 \end{pmatrix}$ with $\tau_R = -1.1$ and (b) period-2 points (•) for $\mu = 0.1$. Note that as predicted they lie within Ξ-black PWL line. Reused with permission from Kowalczyk, P, Nonlinearity, (2005). Copyright 2005, IOP Publishing Ltd.

An example of this situation is the one generated by the map defined by $M = \begin{pmatrix} 0.5 & 1 \\ -0.06 & 0 \end{pmatrix}$ and $N = \begin{pmatrix} \tau_R & 1 \\ 0 & 0 \end{pmatrix}$. The fourth case is chaotic attractors with N-piece invariant sets. In this case, the existence of a set Ξ formed by

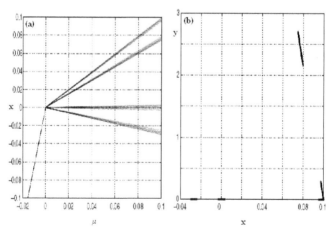

Figure 4.50: Similar to Fig. 4.50 but for $\tau_R = -1.3$. (a) Bifurcation diagram. (b) Ω-limit set. Reused with permission from Kowalczyk, P, Nonlinearity, (2005). Copyright 2005, IOP Publishing Ltd.

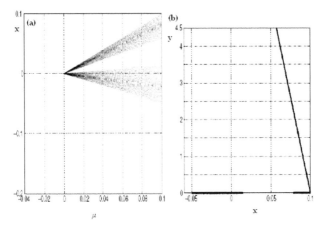

Figure 4.51: Similar to Fig. 4.50 but for $\tau_R = -1.5$. (a) Bifurcation diagram. (b) Ω-limit set. Reused with permission from Kowalczyk, P, Nonlinearity, (2005). Copyright 2005, IOP Publishing Ltd.

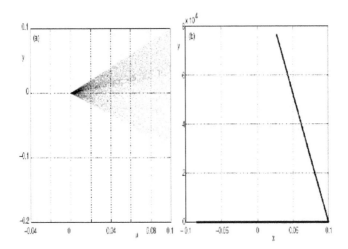

Figure 4.52: Similar to Fig. 4.50 but for $\tau_R = -1.85$. (a) Bifurcation diagram. (b) Ω-limit set. Reused with permission from Kowalczyk, P, Nonlinearity, (2005). Copyright 2005, IOP Publishing Ltd.

a higher number of PWL segments, four, five... etc is possible. For example, if EI shown in Fig. 4.48 is not mapped within GE or if the maximum value of x of the image of OEF is greater than at E, then any bounded set must necessarily contain a higher number of PWL segments than three. Finally, Figs. 4.49 to 4.56 show some examples of the above studied cases.

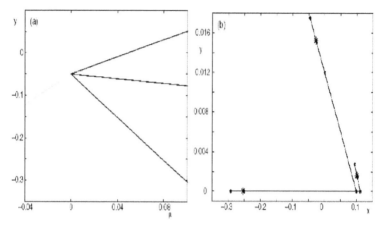

Figure 4.53: (a) Bifurcation diagram representing birth of a stable period-3 point due to the border collision bifurcation scenario for $M = \begin{pmatrix} 0.5 & 1 \\ -0.06 & 0 \end{pmatrix}$, $N = \begin{pmatrix} \tau_R & 1 \\ 0 & 0 \end{pmatrix}$ with $\tau_R = -3.5$ and (b) period-3 points for the map (4.176) for $\mu = 0.1$. Note that as predicted they lie within Ξ, which is denoted in the figure by three black line segments. Small asterisks denote the points K, L and C to I forming the set Ξ. Reused with permission from Kowalczyk, P, Nonlinearity, (2005). Copyright 2005, IOP Publishing Ltd.

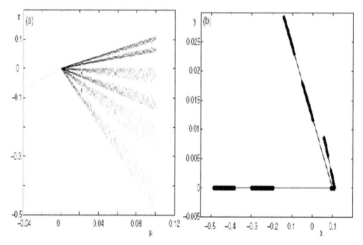

Figure 4.54: Similar to Fig. 4.54 but for $\tau_R = -5.25$. (a) Bifurcation diagram. (b) Ω-limit set. Reused with permission from Kowalczyk, P, Nonlinearity, (2005). Copyright 2005, IOP Publishing Ltd.

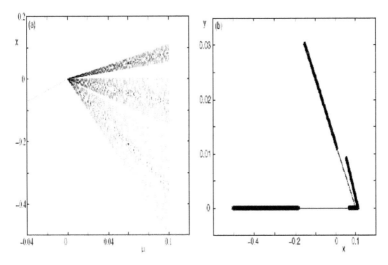

Figure 4.55: Similar to Fig. 4.54 but for $\tau_R = -5.5$. (a) Bifurcation diagram. (b) Ω-limit set. Reused with permission from Kowalczyk, P, Nonlinearity, (2005). Copyright 2005, IOP Publishing Ltd.

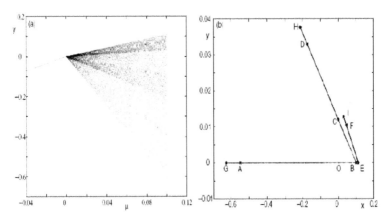

Figure 4.56: Similar to Fig. 4.54 but for $\tau_R = -6.5$. (a) Bifurcation diagram. (b) Ω-limit set. Reused with permission from Kowalczyk, P, Nonlinearity, (2005). Copyright 2005, IOP Publishing Ltd.

5

Real and Mathematical Applications of Lozi Mappings

First, basic application of chaos theory concerns control of irregular behaviors in devices and systems. The list of applications includes but not limited to: engineering, computers, communications, medicine and biology, management and finance, consumer electronics, ...etc. The potential application types of chaos are: control, synthesis, synchronization and information processing. For more details one can see for example [Lozi (2002)]. On the other hand, the Lozi mappings are in some cases the main examples for demonstrating the validity and the correctness of several methods and algorithms related with chaos theory, dynamical systems and their real-world applications. In this chapter, we give only some applications related to the Lozi mappings. Indeed, in Sections 5.1, 5.2 we discuss the utilization of these maps in control and synchronization theories. In Section 5.3 an example of the application of a Lozi map in secure multiple access communications. Section 5.4 shows some relations between game theory and chaotic behaviors. In Section 5.5, we describe the use of Lozi mappings in evolutionary algorithms with one example. In Section 5.6 the Fuzzy modeling of an experimental thermal-vacuum system and its relation with the Lozi map (3.1) is presented and discussed.

Second, this chapter starts with the assumption that the reader is already familiar with basics of control theory. Hence, for this reason, there are many technical terms included without any explanation such as "controlling mixed chaotic signals, proportional-discrete-predictive controller, adaptive control, spatiotemporal chaos, PID controller, AVR and etc."

5.1 The Lozi mappings in control theory

In several works of control theory, the Lozi mappings are the basic examples for demonstrating techniques and methods of this theory. Indeed, it was shown in [Loskutov (1993)] that deterministic noise (chaos) appearing via destruction of the quasi-periodic motion can be easily suppressed by weak parametric perturbation of the system. In [Hern´andez, et al. (2005)] a backpropagation-based neural network architecture was used as a controller to stabilize unsteady periodic orbits. This technique[1] can be considered as a neural network-based method (overcoming the problems that arise with control feedback methods) for transferring the dynamics among attractors. This method leads to more efficient system control. In particular, it was shown that two mixed chaotic signals can be controlled using this method as a filter to separate and control both signals at the same time. One of the advantages of this method is that the control is more effective because it can be applied at any point of the system. This control method remains stable even if the system under consideration have random dynamic noise. In [Graziela (1995)] an adaptive control laws method was applied to the chaotic Lozi map (3.1) (with uncertainties on the parameters) in order to drive its trajectories to certain given reference points. In [Qammar, et al. (1996)] the feedback control of a chaotic polymerization reaction was examined to a steady state using a simple proportional controller, a discrete controller, and a nonlinear model predictive controller. The performance of each method was evaluated from the basin of successful control (has a fractal structure for some values of the control parameter). In this case, model predictive control has the best performance since it yields the most extensive basin. In [Park, et al. (1998)] a generalized predictive control method based on an ARMAX model was proposed for the control of chaos in discrete-time systems. Comparing with the conventional model-referenced adaptive control, this method yields faster settling time, more accurate target tracking, and less initial sensitivity. In [Yang, et al. (1998)] the problem of studying control of chaotic systems with unknown parameters was presented by using a stable adaptive control scheme. This method guarantees (by using a Lyapunov function approach and the center manifold theorem) the convergence of the parameter estimator to stabilizing values such that the controlled chaotic system asymptotically approaches a reference point. This techniques was applied to Chua's circuit (1.50)-(1.51) with cubic nonlinearity and 6 (six) unknown parameters to demonstrate the usefulness of this adaptive control of chaotic systems.

The suppression of spatiotemporal chaos using a feedback technique was presented analytically in [Prashant (1998)] with a suggested improved

---

[1]Which can be applied to every point of the basin.

method for stabilizing periodic solutions and achieving target cluster states in a spatiotemporal system. In [Zhou & Zhang (2002)] a method based on the relativity of the chaotic sequence (considered as an ergodicity random sequence) was used to ensure the controllability of discrete-time chaotic dynamic systems to converge to their equilibrium point or to their multi-periodic orbits. This method was used to support such controllability of the chaotic Hénon and Lozi mappings. In [Starrett (2002)] and based on the OGY method, a an n-step variation, where n is the dimension of the system was proposed and called *time-optimal chaos control by center manifold targeting.* This method sends any initial condition in a controllable region directly to the target orbit instead of its stable subspace and the system becomes completely controlled after n iterations of this method. Some examples such as 2-D piecewise linear and nonlinear maps demonstrate the effectiveness of the control procedure. In [Ueta, et al. (2004)] a simple method for calculating any desired unstable periodic orbits (UPOs) (embedded in a chaotic attractor) and their control in piecewise smooth autonomous systems was described using a simple state feedback control with some examples. In [Feng & Chen (2005)] an adaptive control algorithm was presented based on the so called *T–S model* of discretetime chaotic systems and some conventional adaptive control techniques. One of the advantages of this method is its global stability and robustness as shown by the application to the Hénon and Lozi mappings. In [MingQing (2006)] a direct method for the design of nonlinear discrete-time observers was proposed with an explicit expression of the used change of variables. This method was applied to the chaotic Lozi and Hénon mappings.

A recent result about control of the Lozi mapping (3.1) is the one given in [Coelho (2009)]. This is a tuning method for determining the parameters of PID control for an automatic regulator voltage (AVR) system using a chaotic optimization approach based on Lozi map (3.1) for $a = 1.7$ and $b = 0.5$. **Generally, the PID controller is simple and easy to implement and it is widely used in industry to solve various control problems with many modifications.** The ergodicity and the stochastic property of the Lozi map (3.1) increases the convergence rate and resulting precision for the proposed chaotic optimization. Also, the performance of this chaotic optimization was confirmed numerically for an AVR system for nominal system parameters and step reference voltage input. The general unconstrained optimization problems with continuous variables can be formulated as the following functional optimization problem:

Find X to minimize $f(X)$; $X = (x_1, x_2, ..., x_n)$.

Subject to $x_i \in [L_i, U_i]$; $i = 1, 2, ..., n$.

where f is the objective function, and X is the decision solution vector consisting of n variables $x_i \in R^n$ bounded by lower (L_i) and upper limits (U_i).

If $n = 3$ then $X = [x_1, x_2, x_3] = [K_p, K_i, K_d]$.

Hence, the chaotic search procedure based on Lozi map (3.1) (with $x = y_1$ and $y = y$) is given by:

Inputs:

M_G: maximum number of iterations of chaotic Global search;

M_L: maximum number of iterations of chaotic Local search;

$M_L + M_G$: stopping criterion of chaotic optimization method in iterations;

k: step size in chaotic local search.

Outputs:

X^*: best solution from current run of chaotic search;

f^*: best objective function (minimization problem).

Step 1: Initialization of variables: Set $k = 1$, where k represents the iteration number. Set the initial conditions $y_1(0)$, $y(0)$, $a = 1.7$ and $b = 0.5$ of Lozi map (3.1). Set the initial best objective function $f^* = +\infty$.

Step 2: Algorithm of chaotic global search:

Begin

While $k \le M_G$ do

$x_i (k) = L_i + z_i (k) . (U_i - L_i)$, $i = 1, ..., n$.

If $f (X (k)) < f^*$ then

$X^* = X (k)$ (i.e., $X^* = x_i (k)$, $i = 1, ..., n$)

$f^* = f (X (k))$

End If

$k = k + 1$,

End While

End

Step 3: Algorithm of chaotic local search:

Begin

While $k \le (M_G + M_L)$ do

For $i = 1$ to n

If $r < 0.5$ then (where r is a uniformly distributed random variable with range $[0, 1]$)

$x_i (k) = x^*_i + \lambda._{zi} (k) . | U_i - X^*_i |$ (λ is step size) Else If

$x_i (k) = x^*_i - \lambda._{zi} (k) . | X^*_i - L_i |$ End If

End For
If $f(X(k)) < f^*$ then
$X^* = X(k)$ (i.e., $X^* = x_i(k)$, $i = 1, ..., n$)
$f^* = f(X(k))$
End If
$k = k + 1$,
End While
End

For $Kg = 0.7$ and $sg = 1.5$, we have Minimum (best)= 42.654, Mean= 48.058, Maximum (worst)= 151.333, Median= 42.830, Standard deviation= 20.445, $K_p = 1.174$, $K_i = 0.647$ and $K_d = 0.456$. The result is shown in Fig. 5.1. For $Kg = 1.0$ and $sg = 2.0$, we have Minimum (best)= 43.077, Mean= 47.756, Maximum (worst)= 167.852, Median= 43.287, Standard deviation= 21.243, $K_p = 1.021$, $K_i = 0.456$ and $K_d = 0.418$. The result is shown in Fig. 5.2, 5.3.

5.2 The Lozi mappings in synchronization theory

For the synchronization of discrete-time systems, there are several methods. In this section we discuss only two of them. The first method was proposed in [Nastaran & Johari (2006)] where a fuzzy model-based adaptive approach to synchronize two different discrete-time chaotic systems was presented. The

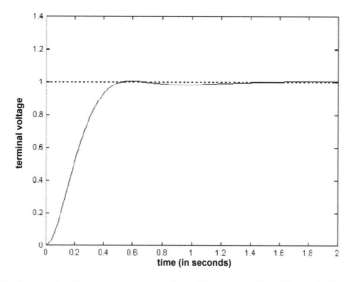

Figure 5.1: Terminal voltage step response of an AVR system ($K_g = 0.7$, $s_g = 1.5$) using PID controller. Reused with permission from Coelho, Chaos, Solitons & Faractals, (2009). Copyright 2009, Elsevier Limited.

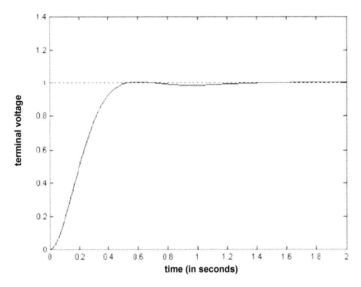

Figure 5.2: Terminal voltage step response of an AVR system (K_g = 1.0, s_g = 2.0) using PID controller. Reused with permission from Coelho, Chaos, Solitons & Faractals, (2009). Copyright 2009, Elsevier Limited.

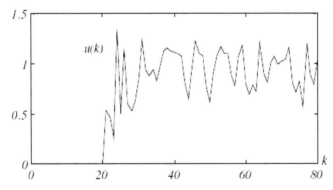

Figure 5.3: Control effort *u* (*k*) for synchronization. Reused with permission from Nastaran, and Johari, Chaos, Solitons & Faractals, (2006). Copyright 2006, Elsevier Limited.

method of analysis uses the Takagi–Sugeno (TS) fuzzy model to represent the chaotic drive with unknown parameters and response systems and an adaptive law to estimate the unknown parameters. Thus a control law was proposed to stabilize the error dynamics by means of two generalized Hénon models with different parameters to demonstrate the effectiveness of the proposed methodology. Also a Hénon model *y* (*k*) given by (3.1) as drive system and Lozi model *x* (*k*) given by (3.1) as response system with a control effort *u* (*k*) for synchronization were given to illustrate asymptotic properties of the proposed method as shown in Figs. 5.3, 5.4. The details can be found in the same paper.

The second example of this situation is given in [Dmitriev, et al. (2001)] where an information theoretic study of the master-slave synchronization of two chaotic systems that interact over a noisy channel was given. The main result of this study is that arbitrarily precise synchronization is possible if the channel capacity is higher than the information production rate of the master system. The application of this result to a specific version of the Lozi map show the validity of this method.

5.3 The Lozi mappings and secure communications

Generally, there has been growing interest in several potential applications of chaotic systems due to the power of easily accessible computer hardware. Many types of these potential applications of chaos theory include secure communications. Some of the Lozi mappings were used to a such applications. For example in [Dmitriev, et al. (2000(a))] the principle of multiple access communications was discussed based on the fine structure of the chaotic attractor displayed by the Lozi map given by:

$$\mathcal{L}_3\left(x, y\right) = \begin{pmatrix} \alpha - 1 - \alpha \left| x \right| + y \\ \beta x \end{pmatrix} \tag{5.1}$$

using control of special chaotic trajectories. This result was also demonstrated experimentally for asynchronous packet data transmission. The principle of this technique is based on the fact that a strange attractor can be treated as a number of countable sets of unstable periodic orbits (UPOs) and transitions between these orbits. In this case, the set of the unstable *skeleton* periodic orbits can be considered as a *reservoir* of potential codes (unlimited) for

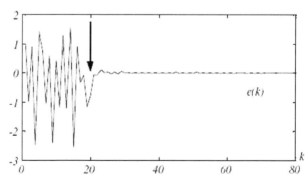

Figure 5.4: Synchronization error output. Reused with permission from Nastaran, and Johari, Chaos, Solitons & Faractals, (2006). Copyright 2006, Elsevier Limited.

multiuser communication systems. In this way, the first pair of users is provided with *own* cycles **X, Y, Z**. The second is provided with another cycle-codes – **A** and **B**. Hence, the total message **AXZABZYXB** contains two individual messages: The first is **XZZYX** and the second is **AABB**, and so on. The processing of the unstable orbits is based essentially on forming of asynchronous data stream and selection of cycle-codes from this stream. Additional information can be found in [Dmitriev, et al. (2000(b))] where a multiple access telecommunication system was provided using chaotic signals. In this case, each transmitter has a chaotic system with one strange attractor, and each receiver has chaotic systems corresponds to those in the transmitters. A set of unstable periodic orbits from a strange attractor is assigned to each pair of users. Also, at the transmitter side, a multiplexer receives data (multiplexed in time produce an asynchronous data stream) from plurality of users. Finally, the data stream is mapped to a sequence of chaotic signals in accordance with the set of trajectories for each transmitter and receiver pair. More details about using dynamical systems (in particular, the Lozi mappings) in secure communications can be found in [Crutchfield (1988), Choi, et al. (1998), Lian, et al. (2001), Tsai, et al. (2005), Prokhorov & Mchedlova (2006)].

5.4 The Lozi mappings in game theory

Generally, game theory is used in the social sciences, especially in economics, biology (in particular, evolutionary biology and ecology), engineering, political science, international relations, computer science, and philosophy and so on. A Parrondo's game has been described as: *A losing strategy that wins*, i.e., *Given two games, each with a higher probability of losing than winning, it is possible to construct a winning strategy by playing the games alternately.* Parrondo's paradox has some applications in engineering, population dynamics, financial risk, etc. Its relation with chaos theory can be seen for example in [Tang, et al. (2004)]. The different effects of chaotic switching on Parrondo's games were investigated, as compared to random and periodic switching. If C is the current capital at discrete-time step n, then it was remarkable that the rate of winning of these games with chaotic switching depends on coefficient(s) defining the chaotic generator (a dynamical system), initial conditions and the proportion of **Game A** played defined as follow:

Game A: consists of a biased coin that has a probability p of winning,

Game B: consists of 2 games, the condition of choosing either one of the games is given as follow: $C \bmod M = 0$ play a biased coin that has probability p_1 of winning, and $C \bmod M \neq 0$ play a biased coin that has probability p_2 of winning[2].

Generally, a Parrondo's games consist of two games, **Game A** and **Game B** defined above such that at discrete-time step n, only one game will be played according to a pattern called *switching strategy* which is either a random or periodic switchings. The parameters for the original Parrondo's games are: $M = 3$, $p = \dfrac{1}{2} - \varepsilon$, $p_1 = \dfrac{1}{10} - \varepsilon$, $p_2 = \dfrac{3}{4} - \varepsilon$, where the biasing parameter, ε (here $\varepsilon = 0.005$) is included in the above equations to control the three probabilities p, p_1 and p_2.

Generally, a chaotic sequence, X is usually generated by nested iteration of some functions defining for example two-dimensional chaotic mappings such as the Lozi system (3.1)[3]. Also, the properties of a chaotic sequence generated from a chaotic system depend on its coefficient(s) and initial conditions.

Now, to play Parrondo's games with chaotic switching, a chosen chaotic generator (such as the Lozi mapping (3.1)) will be used to generate a sequence X used to decide one of the two above games to be played by comparing each value x_n of X with a constant γ, i.e., if $x_n \leq \gamma$, **Game A** will be played (this case is equivalent to the proportion of this game to be played after n discrete-time steps) but if $x_n > \gamma$, **Game B** will be played. The

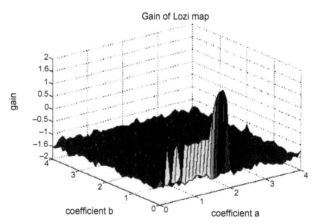

Figure 5.5: Gain of Lozi switching (3.1) after 100 games with different a and b coefficients. Reused with permission from Tang, Allison and Abbott, Proc. SPIE, (2004). Copyright 2004.

---

[2] Here the value p_2 is greater than 0.5 making **Game B** favorable, but the values of C and M are then important to make **Game B** unfavorable on average.

[3] We note that other chaotic dynamical systems could equally be used to study this application.

γ value is a threshold on selection of games to be played on each round. This value makes Parrondo's paradox appear or not. Generally, all the chaotic sequences are normalized to have values in the interval (0, 1).

The effect of the coefficient(s) of chaotic generator on the rate of winning is given by:

$$R(n) = E \mid J_{n+1} - J_n \mid = E \mid J_{n+1} \mid - E \mid J_n \mid = \sum_{j=-\infty}^{\infty} j \left[\pi_j(n + 1) - \pi_j(n) \right] \qquad (5.2)$$

where $\pi_j(n)$ is the *stationary probability* of being in state j at discrete-time step n. In fact, under stable regions, the system shows periodic behaviors in which the maximum rate of winning occurs when the chaotic generator tends toward periodic behavior. However, the rate of winning is much smaller when a chaotic generator behaves truly chaotically. The rate of winning of Parrondo's games for the lozi mapping (3.1) is shown in Fig. 5.5, where the gains after 100 games are plotted with different combinations of a and b values. The maximum gain is obtained when $a = 1.7$ and $b = 0$.

The effect of initial conditions and γ value on the rate of winning can be summarized as follow: The games lose when the initial conditions drive the chaotic sequence towards its attractors, i.e., playing **Game A** or **Game B** individually. However, the same rate of winning can be obtained for the other initial conditions. A plot of the effect of γ on the proportion of **Game A** played is shown in Fig. 5.6 and the capital under different switching regimes for 100 games (averaged over 50.000 trials) is shown in Fig. 5.7.

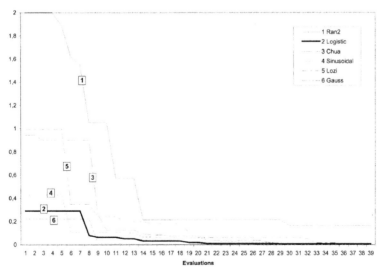

Figure 5.6: Plot of effect of γ on the proportion of **Game A** played. Reused with permission from Tang, Allison and Abbott, Proc. SPIE, (2004). Copyright 2004.

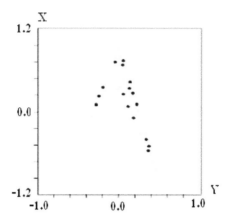

Figure 5.7: Capital under different switching regimes for 100 games (averaged over 50, 000 trials). Reused with permission from Tang, Allison and Abbott, Proc. SPIE, (2004). Copyright 2004.

Finally, initial conditions have no affect on the proportion of **Game A** played, but this proportion is significantly dependent on γ because the γ value is a threshold value and draws the boundary of regions between the two games in the normalized time series plot of the chaotic sequence. In all cases, Parrondo's games with chaotic switching gives higher rate of winning compared to random switching.

5.5 The Lozi mappings and evolutionary algorithms

In the field of evolutionary algorithm (EA) [Holland (1975), Zalzala & Fleming (1997)], the convergence properties are connected to the random sequences applied on variation operators during a run in the sense that different random sequences gives very close but not equal results. Different numbers of generations are required to reach the same optimal values. In this view, chaotic sequences were used instead of random ones for some real-world applications such as secure transmission [Caponetto, et al. (1998)], natural phenomena modeling [Bucolo, et al. (1998)], neural networks [Nozawa (1992), Wang & Smith (1998)], nonlinear circuits [Arena, et al. (2000)], DNA computing procedures [Manganaro & Pineda de Gyvez (1997)] using chaotic time series. In [Caponetto, et al. (2003)] an experimental analysis on the convergence of evolutionary algorithms (EAs) was proposed using chaotic sequences instead of random ones (random number generator

(RNG)) during all the phases of the evolution process. Several chaotic systems were used for this purpose such as the Lozi map (3.1). The main result here are that some chaotic sequences are able to increase the value of some measured algorithm-performance indexes with respect to random sequences and that EAs are extremely sensitive to different RNGs. The method of analysis is based on five functions $(f_i)_{1 \leq i \leq 5}$ (De Jong functions) for optimization test. These functions are:

$$f_1(x) = \sum_{i=0}^{n} x_i^2, -5.12 \leq x_i \leq 5.12,$$

$$f_2(x) = 100 \left(x_0^2 - x_1\right)^2 + (1 - x_0)^2, -2.048 \leq x_i \leq 2.048,$$

$$f_3(x) = 30 + \sum_{i=1}^{5} |x_i|, -5.12 \leq x_i \leq 5.12, \tag{5.3}$$

$$f_4(x) = \sum_{i=1}^{30} ix_i^4 + G(0,1), -1.28 \leq x_i \leq 1.28, G = \text{Gauss map}$$

$$\frac{1}{f_5(x)} = \frac{1}{500} + \sum_{j=1}^{25} \frac{1}{j + \sum_{i=1}^{2} (x_i - a_{ij})^6}, -65.536 \leq x_i \leq 65.536,$$

5.6 The Lozi mappings and fuzzy modeling of an experimental thermal-vacuum system

Particle Swarm Optimization is a population-based swarm algorithm, was developed in [Eberhart & Kennedy (1995), Kennedy & Eberhart (1995)]. An evolutionary algorithm approach (PSO) is an optimization tool based on a population where the position of each member/particle is a potential solution to an analyzed problem, where each particle is associated to a randomized velocity that moves throughout the problem space [Goldberg (1989)]. In [Araujo & Coelho (2008)] a chaotic PSO (CPSO) approach, i.e., a Particle Swarm Optimization (PSO) approach intertwined with a Lozi map chaotic sequences was proposed to obtain the so called *Takagi–Sugeno (TS) fuzzy model* for representing dynamical behaviors. This method was employed for optimizing the premise part of the IF–THEN rules of an TS fuzzy model, where the *least mean squares technique* was used for the consequent part. As a practical example, **this chaotic PSO approach was utilized for a thermal-vacuum system which is employed for space environmental emulation and satellite qualification.** The results show that this method succeeded in eliciting a TS fuzzy model for this nonlinear and time-delay application. A geometric view of the POS algorithm is shown in Fig. 5.8 and a one-step ahead forecasting using TS fuzzy model using PSO approaches and batch mean least squares is shown in Fig. 5.9.

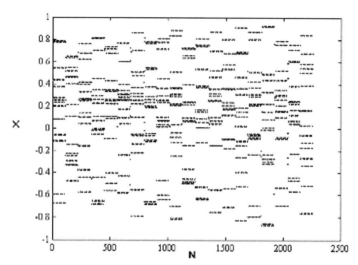

Figure 5.8: Geometric view for PSO algorithm. Reused with permission from Araujo and Coelho, Applied Soft Computing, (2008). Copyright 2008, Elsevier Limited.

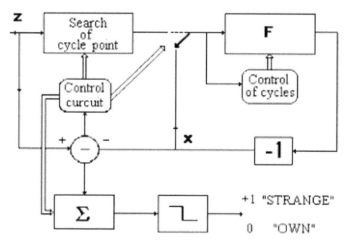

Figure 5.9: One-step ahead forecasting using TS fuzzy model using PSO approaches and batch mean least squares. Reused with permission from Araujo and Coelho, Applied Soft Computing, (2008). Copyright 2008, Elsevier Limited.

As claimed by the authors in [Araujo & Coelho (2008)], the future research in this area includes the hybridization of the PSO and CPSO and other local search methods, such as simulated annealing and quasi-Newton methods. Also, the chaotic PSO approach possibly can give good results for identification of thermal-vacuum system. In the end of this section, we note that it was shown experimentally in [Sucharev, et al. (2004a)] that the *Oxyhydrate gels* of rare metals belong to the group of self-inverse fractal systems. For example the attractors of *yttrium oxyhydrate gels* systems were reconstructed and they are similar each other and have similar fractal geometry, i.e., both light, and mechanical influence, conduct to formation in the gel phase, the same self-inverse structures. However, there are some difference in the geometry of the attractors corresponding to the area of reflexes and to the plateau-area on the full flow curve. Also, it was shown in [Sucharev, et al. (2005)] that discrete electric current occurs in *zirconium oxyhydrate gels* under certain conditions of experiment and displays some chaotic attractors in which the existence of some similar Lozi and Lozi-Hénon strange attractors was established experimentally. The chaotic nature of gel oxyhydrate systems with certain nonchaotic component was confirmed. By processing experimentally received data on kinetics of change of rheological characteristics phase portraits of attractors of yttrium oxyhydrate gel systems were reconstructed in [Sucharev, et al. (2004(b))]. It was shown that the geometry of these attractors have a strange fractal structure. In fact, the geometry of attractors obtained for the *dried up xerogels* is more complex than for just prepared gels. In [Sucharev (2004)] more examinations of nonlinear properties of oxyhydrate gel systems shows the following phenomena: periodic dilatancy, a periodic (pulsatory) electrical conductance accompanied of polarization phenomena, coloring of the gel systems, periodic optical and sorption properties, etc. The attractors of these Oxyhydrate gels systems show *self-inverse Mandelbrot fractal* (Self-inverse fractals are another type of invariant fractal sets. Their definition is as follow: Self-inverse fractals are formed with repeatedly nonlinear inversion operations. The initial points in the image space are transformed into the output fractal object.)

Bibliography

[1] Aarts, J. M. and Fokkink, R. J. 1991. The classification of solenoids. Proc. Amer. Math. Soc. **111**: 1161–1163.

[2] Abraham, R. 1972. *Introduction to Morphology*. Publ. Dept. Math. Lyon. 38–114.

[3] Abraham, R. and Marsden, J. E. 1978. Foundations of Mechanics. Benjamin/ Cummings Publishing, Reading Mass.

[4] Abraham, R. and Robbin, J. 1967. Transversal mappings and flows. Benjamin.

[5] Abraham, R. and Smale, S. 1970. Nongenericity of Ω-stability. Global analysis I, Proc. Symp. Pure Math. AMS. **14**: 5–8.

[6] Abdenur, F., Bonatti, C., and Crovisier, S. 2008. Non-uniform hyperbolicity for C^1-generic diffeomorphisms. Preprint, arXiv0809.3309.

[7] Abdenur, F., C., Bonatti, Crovisier, S., Diaz, L. and Wen, L. 2007. Periodic points and homoclinic classes. Ergod. Th. & Dynam. Sys. **27**: 1–22.

[8] Abu-Saris, R. 1999. On the periodicity of the difference equation $x_{n+1} = \alpha \mid x_n \mid + \beta x_{n-1}$. J. Differ. Equations Appl. **5**(1): 57–69.

[9] Abu-Saris, R., 2000a. Global behavior of rational sequences involving piecewise power function. J. Comput. Anal. Appl. **2**(1): 103–109.

[10] Abu-Saris, R. 2000b. Characterization of rational periodic sequences. J. Differ. Equations Appl. **6**(2): 233–242.

[11] Abu-Saris, R. and Al-Hassan, Q. 2003a. On global periodicity of difference equations. J. Math. Anal. Appl. **283**(2): 468–477.

[12] Abu-Saris, K. and Al-Hassan, Q. 2003b. Characterization and transformation of invariants. The Journal of Difference Equations and Applications. **9**(10): 869–877.

[13] Abu-Saris, R. and Al-Jubouri, N. 2004. Characterization of rational periodic sequences II. J. Differ. Equations Appl. **10**(4): 409–418.

[14] Abu-Saris, R. 2006. A self-invertibility condition for global periodicity of difference equations. Applied Mathematics Letters. **19**: 1078–1082.

[15] Adler, R. and Weiss, B. 1970. Similarity of automorphisms of the torus. Memoirs of the American Mathematical Society. **98**. American Mathematical Society, Providence, R.I.

[16] Adrangi, B. and Chatrath, A. 2003. Non-linear dynamics in futures pricesevidence from the coffee, sugar and cocoa exchange. Applied Financial Economics. **13**(4): 245–256.

[17] Afraimovich, V. S., Bykov, V. V. and Shilnikov, L. P. 1977. On the appearance and structure of Lorenz attractor. DAN SSSR. **234**: 336–339.

[18] Afraimovich, V. S., Bykov, V. V. and Shil'nikov, L. P. 1982. On structurally unstable attracting limit set of the type of Lorenz attractor. Trans. Moscow. Math. Soc. **44**: 153–216.

[19] Afraimovich, V. S. and Shilnikov, L. P. 1983. *Strange attractors and quasi-attractors in* Nonlinear Dynamics and Turbulence eds. by Barenblatt, G. I., Iooss, G. and Joseph, D. D., Pitman, NY, 1–28.

[20] Afraimovich, V. S., Bykov, V. V. and Shilnikov, L. P. 1983. On the structurally unstable attracting limit sets of Lorenz attractor type. Tran. Moscow. Math. Soc. **2**: 153–215.

[21] Afraimovich,V. S. and Shil'nikov, L. P. 1983. Strange attractors and quasiattractors. *In: Dynamics and Turbulence*. New York Pitman.

[22] Afraimovich, V. S., Chernov, N. I. and Sataev, E. A. 1995. Statistical properties of two-dimensional generalized hyperbolic attractors. Chaos. 5(1): 238–252.
[23] Afraimovich, V. and Hsu, S. B. 2003. Lectures on Chaotic Dynamical Systems (Am. Math. Soc., Providence, RI. AMS/IP Studies in Advanced Mathematics, Vol. 28.
[24] Agrachev, A. 2007. The curvature and hyperbolicity of Hamiltonian systems. Proceedings of the Steklov Institute of Mathematics. 256(1): 26–46.
[25] Aharonov, D., Devaney, R. L. and Elias, U. 1997. The dynamics of a piecewise linear map and its smooth approximation. Int. J. of Bif. and Chaos. 7(2): 351–372.
[26] Albers, D. J. and Sprott, J. C. 2006. Structural stability and hyperbolicity violation in high-dimensional dynamical systems. Nonlinearity. 19(8): 1801–1849.
[27] Albers, D. J., Crutchfield, J. P. and Sprott, J. C. 2006. Persistent chaos in high dimensions. Phys. Rev. E 74. 057201.
[28] Alefeld, G. and Herzberger. J. 1983. Introduction to interval computations. Academic Press, New York.
[29] Alekseev. V. M, Jacobson. M. V. 1981. Symbolic dynamics and hyperbolic dynamic systems. Phys. Rep. 75(4): 287–325.
[30] Alexander, J. C., York, J. A., You, Z. P. and Kan, I. 1992. Riddled basins. Int. Bifurc. Chaos. 2: 795–813.
[31] Alligood, K. T. and Sauer, T. 1988. Rotation numbers of periodic orbits in the Hénon map. Comm. Math. Phys. 120(1): 105–119.
[32] Alves, F. and Araujo, V. 2003. Random perturbations of nonuniformly expanding maps. Astérisque. 286: 25–62.
[33] Alves, J. F., Ara´ujo, V., Pacifico, M. J. and Pinheiro, V. 2007. On the volume of singular-hyperbolic sets. Dynamical Systems, An International Journal. 22(3): 249–267.
[34] Andronov, A. and Pontryagin, L. 1937. Syst`emes grossiers. Dokl. Akad. Nauk USSR. 14: 247–251.
[35] Anishchenko, V. 1990. Complex Oscillations in Simple Systems. Nauka, Moscow.
[36] Anishchenko, V. 1995. Dynamical Chaos Models and Experiments. World Scientific, Singapore.
[37] Anishchenko, V. S., Neiman, A. B., Safonova, M. A. and Khovanov, I. A. 1995. Multifrequency Stochastic Resonance. In: Proc. of Euromech Colloquium on Chaos and Nonlinear Mechanics, ed. by Kapitaniak, T. and Brindley, J. (World Scientific, Singapore, p. 41.
[38] Anishchenko, V. S. and Strelkova, G. I. 1997. Attractors of dynamical systems. Control of Oscillations and Chaos. Proc.s., (1997) 1st Inter. Conference., 3: 498–503.
[39] Anishchenko, V. and Strelkova, G. 1998. Irregular attractors. Discrete Dynamics in Nature and Society. 2(1): 53–72.
[40] Anishchenko, V. S., Vadivasova, T. E., Strelkova, G. I. and Kopeikin, A. S. 1998. Chaotic Attractors of Two-dimensional Invertible Maps. Discrete Dynamics in Nature and Society. 2: 249–256.
[41] Anishchenko, V. S., Kopeikin, A. S. and Kurths, J. 2000a. Studying hyperbolicity in chaotic systems. Phys. Lett. A. 270: 301–3.7.
[42] Anishchenko, V. S., Kopeikin, A. S., Vadivasova, T. E., Strelkova, G. I., and Kurths, J. 2000b. Influence of noise on statistical properties of nonhyperbolics attractors. Phy. Rev. E. 62(6): 7886–7893.
[43] Anishchenko, V. S., Tatjana, E. V., Kopeikin, A. S., Kurths, J., and Strelkova, G. I. 2002a. Peculiarities of the relaxation to an invariant probability measure of nonhyperbolic chaotic attractors in the presence of noise. Phys. Rev. E. 65: 036206-1–036206-10.
[44] Anishchenko, V. S., Luchinsky, D. G, McClintock, P. V. E., I. Khovanov, A. and Khovanova, N. A. 2002b. Fluctuational escape from a quasihyperbolic attractor in the Lorenz system. J. Experimental.Theoretical Physics. 94(4): 821–833.
[45] Anishchenko, V. S., Astakhov, V. V., Neiman, A. B., Vadivasova, T. E. and Schimansky-Geier, L. 2002c. Nonlinear Dynamics of Chaotic and Stochastic Systems: Tutorial and Modern Development. Springer, Berlin, Heidelberg.

[46] Anishchenko, V. S., Vadivasova, T. E., Okrokvertskhov, G. A. and Strelkova, G. I. 2003a. Correlation analysis of dynamical chaos. Physica A. **325**: 199–212.

[47] Anishchenko, V. S., Vadivasova, T. E., Kopeikin, A.S., Kurths, J. and Strelkova, G. I. 2003b. Spectral and Correlation Analysis of Spiral Chaos. Fluct. Noise. Lett. **3**: L213–L221.

[48] Anishchenko, V. S., Vadivasova, T. E., Okrokvertskhov, G. A. and Strelkova, G. I. 2003c. Correlation analysis of the regimes of deterministic and noisy chaos. J. Comm. Techn. Electr. **48**: 750–760.

[49] Anishchenko, V. S., Astakhov, V. V., Vadivasova, T. E., Neoeman, A. B., Strelkova, G. I. and Schimansky-Geier, L. 2003. Nonlinear Effects of Chaotic and Stochastic Systems. Inst. Komp'yut. Issled., Moscow (in Russian).

[50] Anishchenko, V. S., Vadivasova, T. E., Strelkova, G. I. and Okrokvertskhov, G. A. 2004.Statistical properties of dynamical chaos. Math. Biosciences. Engineering. **1**(1): 161–184.

[51] Anishchenko, V. S., Neiman, A. B., Vadiavasova, T. E., Astakhov, V. V. and Schimansky-Geier, L. 2007. Nonlinear dynamics of chaotic and stochastic systems. Springer Series in Synergetics, Second Edition.

[52] Andrea, S. A. 1965. On homeomorphisms of the plane, and their embeddings in flows. Bull. AMS. **71**: 381–383.

[53] Andrianov, I. V. and Manevitch, L. I. 2002. Asymptotology: Ideas, Methods, and Applications. Springer.

[54] Anosov, D. V. 1967. Geodesic flows on closed Riemannian manifolds of negative curvature. Proc. Steklov Math. Inst. **90**: 1–235.

[55] Anosov, D. V., Aranson, S. K., Grines, V. Z. and Plykin, R. V. 1995. *Dynamical Systems IX: Dynamical Systems with Hyperbolic Behaviour*. Encyclopaedia Mat. Sci. **9**.

[56] Anosov, D. V. 2001. *Ergodic theory*, in Hazewinkel, Michiel, Encyclopedia of Mathematics, Kluwer Academic Publishers.

[57] Anosov, D. V., Klimenko, A.,V., Kolutsky, G. 2008. On the hyperbolic automorphisms of the 2-torus and their Markov partitions. Preprint of Max-Plank Institute for Mathematics, appeared in.

[58] Ansari. A., Jedidi. K. and Dube. L. 2002. Heterogeneous factor analysis models: A bayesian approach. Psychometrika. **67**(1): 49–77.

[59] Amaricci, A., Bonetto, F. and Falco, P. 2007. Analyticity of the Sinai-Ruelle- Bowen measure for a class of simple Anosov flows. J. Mathematical Physics. **48**(7): 072701-072701-15.

[60] Asaoka, M. 2008. A simple construction of C^1-Newhouse domain for higher dimensions. Preprint.

[61] Ashwin, P. and Rucklidge A. M. 1998. Cycling chaosits creation, persistence and loss of stability in a model of nonlinear magnetoconvection. Physica D. **122**(1): 134–154.

[62] Arai, Z. and Mischaikow, K. 2006. Rigorous Computations of Homoclinic Tangencies. SIAM J. Appl. Dyn. Syst. **5**: 280–292.

[63] Arai. Z. 2007a. On Hyperbolic Plateaus of the Hénon Map. Experiment. Math. **16**(2): 181–188.

[64] Arai. Z. 2007b. On loops in the hyperbolic locus of the complex Hénon map Preprint.

[65] Araujo, V. 1987. Existencia de atratores hiperbolicos para dieomorfismos de superficies, Ph.D. Thesis, IMPA.

[66] Araujo, V., Pacifico, M. J., Pujals, E. R. and Viana, M. 2009. Singularhyperbolic attractors are chaotic. Trans. Amer. Math. Soc. **361**: 2431–2485.

[67] Araujo, V. and Pacifico, M. J. 2007. Three Dimensional Flows. XXV Brazillian Mathematical Colloquium. IMPA, Rio de Janeiro.

[68] Araujo, V. and Bessa, M. 2008. Dominated splitting and zero volume for incompressible three-flows. Nonlinearity. **21**(7): 1637–1653.

[69] Araujo,V. and Pacifico, M. J. 2008. What is new on Lorenz-like attractors. Preprint, arXiv0804.3617.

[70] Araujo, E. and Coelho, L.S. 2008. Particle swarm approaches using Lozi map chaotic sequences to fuzzy modelling of an experimental thermalvacuum system. Applied Soft Computing. **8:** 1354–1364.

[71] Arena, P., Caponetto, R., Fortuna, L. and Rizzo, A. and La Rosa, M. 2000. Self organization in non recurrent complex system, Int. J. Bifurcation and Chaos. **10**(5): 1115–1125.

[72] Arnéodo, A., Coullet, P. and Tresser, C. 1981. Possible new strange attractors with spiral structure. Comm. Math. Phys. **79:** 673–679.

[73] Arnold, V. I. and Avez, A. 1968. Ergodic Problems of Classical Mechanics. New York, Benjamin.

[74] Arnold, V. I. 1988. Geometrical Methods in the Theory of Ordinary Differential Equations. Springer-Verlag, Berlin.

[75] Arnold, V. I. 1992. Ordinary defferential equations. Springer-Verlag.

[76] Aronson, D. G., Chory, M. A., Hall, G. R. and McGehee, R. P. 1982. Bifurcations from an Invariant Circle for Two-Parameter Families of Maps of the planeA Computer-Assisted Study. Commun. Math. Phys. **83:** 303–354.

[77] Arov, D. Z. 1963. Topological similarity of automorphisms and translations of compact commutative groups. Uspekhi Mat. Nauk. **185:** 133–138 (in Russian).

[78] Arrowsmith, D.K. and Plaa, C.M. 1990. An introduction to dynamics systems. Cambridge University Press.

[79] Arroyo, A. and Hertz, F. R. 2003. Homoclinic bifurcations and uniform hyperbolicity for three-dimensional flows. Ann. I. H. Poincaré. **20**(5): 805–841.

[80] Arroyo, A. 2007. Singular hyperbolicity for transitive attractors with singular points of 3-dimensional C^2-flows. Bull. Braz. Math. Soc, New Series. **38**(3): 455–465.

[81] Arroyo, A. and Pujals, E. 2007. Dynamical properties of singular hyperbolic attractors. Discrete an Continuous dynamical systems. **19**(1): 67–87.

[82] Aubry, S. J. 1995. Anti-integrability in dynamical and variational problems. Physica D. **86:** 284–296.

[83] Aulbach, B. and Flockerzi, D. 1989. The past in short hypercycles. J. Math. Biol. **27:** 223–231.

[84] Auerbach, D. 1988. O'Shaughnessy, B., Procaccia, I., Scaling structure of strange attractors. Phys. Rev. A. **37:** 2234–2236.

[85] Auerbach, D. and Procaccia, I. 1990. Grammatical complexity of strange sets. Phys. Rev. A. **41:** 6602–6614.

[86] Avrutin, V. and Schanz. M. 2004. Border-collision period-doubling scenario, Phys. Rev. E. **70**(3): 2, 026222, 11.

[87] Aziz Alaoui. M. A. 2000. Multi-Fold in a Lozi-type map. Preprint.

[88] Aziz Alaoui. M. A., Robert. C., Celso Grebogi. C. 2001. Dynamics of a Hénon-Lozi-type map. Chaos. Solitons & Fractals. **12:** 2323–2341.

[89] Badii, R. and Politi, A. 1997. Complexity-Hierarchical Structures and Scaling in Physics, Cambridge: Cambridge University Press.

[90] Baladi, V. and Young, L. S. 1993. On the spectra of randomly expanding maps. Comm. Math. Phys. **156**(2): 355–385, and its erratum Comm. Math. Phys. **166**(1): (1994) 219–220.

[91] Balint. P. and Gouezel. S. 2006. Limit theorems in the stadium billiard. Comm. Math. Phys. **263:** 461–512.

[92] Ballmann, W. and Wojtkowski, M. P. 1989. An estimate for the measure theoretic entropy of geodesic flows. Ergod. Th. & Dynam. Sys. **9:** 271–279.

[93] Balmforth, N. J. 1995. Solitary waves and homoclinic orbits. Annual review of fluid mechanics. **27:** 335–373 Annual Reviews, Palo Alto, CA.

[94] Bamon, R., Labarca, R., Mané, R. and Pacifico, M. J. 1993. The explosion of singular cycles. Publ. Math. IHES. **78:** 207–232.

[95] Banergee, S., York, J. A. and Grebogi, C. 1998. Robust chaos. Phys. Rev. Lettres. **80**(14): 3049–3052.

[96] Banerjee, S. and Chakrabarty, K. 1998. Nonlinear modeling and bifurcations in the boost converter. IEEE Trans. Power Electron. **13**: 252–260.

[97] Banerjee, S. and Grebogi, C. 1999. Border collision bifurcations in twodimensional piecewise smooth maps. Phy. Rev. E. **59**(4): 4052–4061.

[98] Banerjee, S., Kastha, D., Das, S., Vivek, G. and Grebogi, C. 1999. Robust chaos-the theoretical formulation and experimental evidence. ISCAS **(5)**: 293–296.

[99] Banerjee, S. and Verghese, G. C. 2001. Nonlinear Phenomena in Power Electronics: Attractors Bifurcations, Chaos , and Nonlinear Control. IEEE Press, New York, USA.

[100] Banerjee, S., Parui, S. and Gupta, A. 2004. Dynamical effects of missed switching in current-mode controlled dc-dc converters, IEEE Trans. Circuits & Systems–II **51**: 649–54.

[101] Banks, J. and Dragan, V. 1994. Smale's Horseshoe map via ternary numbers. SIAM Review. **36**(2): 265–271.

[102] Baptista, D. N. 2007. Symbolic dynamics for the Lozi maps. Equadiff.

[103] Baptista, D., Severino, R., Vinagre, S. 2009. The basin of attraction of Lozi mappings. Inter. J. Bifur. Chaos. **19**(3): 1043–1049.

[104] Baraviera, A. T. 2000. Robust nonuniform hyperbolicity for volume preserving maps. PhD thesis, IMPA.

[105] Barge, M. 1987. Homoclinic intersections and indecomposability. Proc. Amer. Math. Soc. **101**: 541–544.

[106] Barge, M. and Kennedy, J. 2007. Continuum Theory and Topological Dynamics. In: Open Problems in Topology II, Elliott Pearl Editor, Elsevier Press.

[107] Barreto, E., Hunt, B., Grebogi, C. and Yorke, J. A. 1997. From high dimensional chaos to stable periodic orbits: The structure of parameter space. Phys. Rev. Lett. **78**: 4561.

[108] Barreira. L., Pesin. Y. and Schmeling. J. 1997. Multifractal spectra and multifractal rigidity for horseshoes. J. Dynam. Control Systems. **3**: 33–49.

[109] Bass, H., Cornell, E. H. andWrigh, D. 1982. The Jacobian conjecture Reduction of degree and formal expansion of inverse. Bull. Amer.Math.Soc. **7**: 287–330.

[110] Bass, H. 1989. Conjecture jacobienne et opérateurs différentiels. Mém. Soc. Math. France. **38**: 39–50.

[111] Bautista, S., Morales, C. and Pacífico, M. J. 2004. There are singular hyperbolic flows without spectral decomposition. Preprint IMPA. Série **A278**.

[112] Bautista, S. 2005. Sobre conjuntos hiperbo´licos-singulares. In Portuguese. Thesis, Universidade Federal do Rio de Janeiro.

[113] Bautista, S., Morales, C. and Pacifico, M.J. 2005. Intersecting invariant manifolds on singular-hyperbolic sets. Preprint.

[114] Bautista, S. and Morales, C. 2006. Existence of periodic orbits for singularhyperbolic sets. Moscow Mathematical Journal. **6**(2): 265–297.

[115] Beardon, A. F., Bullett. S. R. and Rippon. P. J. 1995. Periodic orbits of difference equations. Proc. Roy. Soc. Edinburgh, Ser. A (Math.) **125A**: 657–674.

[116] Becker, T., Weispfenning, V. and Gröbner, B. 1993. Computational Approach to Commutative Algebra. New York, Springer-Verlag.

[117] Bedford, T., Keane, M. and Series, C., eds. 1991. Ergodic theory, symbolic dynamics and hyperbolic spaces. Oxford University Press.

[118] Bedford, E. and Smillie, J. 2006. Real polynomial diffeomorphisms with maximal entropy II. Small Jacobian, Ergod. Th. Dynam. Syst. **26**(5): 1259–1283.

[119] Belykh, V. N. 1995. Chaotic and strange attractors of a two-dimensional map, Sbornik: Mathematics. **186**(3): 311–326, translation from Matematicheskiĭ Sbornik. **186**(3): 3–18.

[120] Belykh, V., Belykh, I. and Mosekilde, E. 2005. Hyperbolic Plykin attractor can exist in neuron models, Inter.J. Bifurcation and Chaos. **15**(11): 3567–3578.

[121] Benedicks, M. and Young, L.S. 1993. SBR-measures for certain Hénon maps. Invent. Math. **112**: 541–576.

[122] Benedicks, M. and Young, L.S. 2000. Markov extensions and decay of correlations for certain Hénon maps. Asterisque. **26:** 113–56.

[123] Benettin, G., Galgani, L., Giorgilli, A. and Strelcyn, J.M. 1980. Lyapunov characteristic exponents for smooth dynamical systems and for Hamiltonian systems; a method for computing all of them, Part 2: numerical applications. Meccanica. **15**(9): 21–30.

[124] Benoist, Y. and Labourie, F. 1993. Sur les difféomorphismes d'Anosov à feuilletages stable et instable différentiables. Invent. Math. **111**(2): 285–308.

[125] Berger, A. 2001. Chaos and Chance: An Introduction to Stochastic. Aspects of Dynamics, Walter de Gruyter.

[126] Berns, D. W., Moiola, J. L. and Chen, G. 2001. quasi-analytical method for period-doubling bifurcation, ISCAS (2001). The (2001) IEEE Inter. Symposium on Circuits and Syst., (Cat. No.01CH37196). IEEE. Part. **2:** 739–742.

[127] Bers, L. and Royden, H. L. 1986. Holomorphic families of injections. Acta Math. **157:** 259–286.

[128] Bertau, M., Mosekilde, E. and Westerhoff, H. V. 2007. Biosimulation in Drug Development. Wiley and son.

[129] Biham, O. and Wenzel,W. 1989. Characterization of unstable periodic orbits in chaotic attractors and repellers. Phys. Rev. Lett. **63:** 819–822.

[130] Bhattacharya, J. and Kanjilal, P. P. 1999. On the detection of determinism in a time series. Physica D. **132**(1): 100–110.

[131] Bedford, E. and Smillie, J. 1991. Polynomial diffeomorphisms of $\mathbb{C}^2$, currents, equilibrium measure and hyperbolicity. Invent. math. **103:** 69–99.

[132] Bedford, E. and Smillie, J. 2006a. The Hénon family. The complex horseshoe locus and real parameter values. Contemp. Math. **396:** 21–36.

[133] Bedford, E. and Smillie, J. 2006b. Real Polynomial diffeomorphisms with maximal entropy: Tangencies, Ergod. Th. Dynam. Syst. **26**(5): 1259–1283.

[134] Béguin, F. and Bonatti, C. 2002. Flots de Smale en dimension 3: présentations finies de voisinages invariants d'ensembles selles (French. English, French summary), (Smale flows in dimension 3finite presentations of invariant neighborhoods of saddle sets). Topology. **41**(1): 119–162.

[135] Belykh, V. 1982. Models of discrete systems of phase locking. In Phase Locking Systems. (L.N. Belyustina and V. V. Shakhgil'dyan, Eds.). Radio Svyaz, Moscow. 161–176 (in Russian).

[136] Belykh, V. 1995. Chaotic and strange attractors of two dimensional map. Math. Sbornik **186**(3) (in Russian).

[137] Benedicks, M. and Carleson, L. 1985. On iterations of $1-ax^2$ on $(-1, 1)$. Ann. Math. **122:** 1–25.

[138] Benedicks, M. and Carleson, L. 1991. The dynamics of the Hénon maps. Ann. Math. **133:** 73–169.

[139] Benedicks, M. 2002. Non Uniformly Hyperbolic Dynamics Hénon Maps and Related Dynamical Systems. *ICM 2002* **III:** 1–3.

[140] Bessa, M. 2007. The Lyapunov exponents of generic zero divergence 3- dimensional vector fields. Ergod. Th. Dynam. Syst. **27**(5): 1445–1472.

[141] Bessa, M. and Duarte, P. 2007. Abundance of elliptic dynamics on conservative 3-flows. Preprint arXiv0709.0700v.

[142] Birkhoff, G. D. 1931. Proof of the ergodic theorem. Proc. Natl. Acad. Sci. USA. **17:** 656–660.

[143] Birkhoff, G. D. 1942. What is the ergodic theorem?. Amer. Math. Monthly. **49**(4): 222–226.

[144] Bloch, W. L. 1995. Extending flows from isolated invariant sets. Ergodic Theory Dynam. Systems **15** (6): 1031–1043.

[145] Bochi, J. 2002. Genericity of zero Lyapunov exponents. Ergodic Theory Dynam. Systems. **22**(6): 1667–1696.

[146] Bonani, F. and Gilli, M. 1999a. A harmonic balance approach to bifurcation analysis of limit cycles. ISCAS'99. Proc. (1999) IEEE Inter. Symp. Circuits. Syst. VLSI (Cat. No.99CH36349). IEEE. Part, **6**: 298–301.

[147] Bonani, F. and Gilli, M. 1999b. A harmonic-balance based method for computing Floquet's multipliers in Lur'e systems. Proc. of the 7th Inter. Specialist Workshop on Nonlinear Dynamics of Electronic Systems., Tech. Univ. 13–16.

[148] Bonatti, C. and Diaz, L. J. 1996a. Persistent nonhyperbolic transitive diffeomorphisms. Ann. of Math. **143**: 357–396.

[149] Bonatti, C. 1996b. A local mechanism for robust transitivity. Conference at IMPA, Seminar of Dynamical systems, August.

[150] Bonatti, C. and Diaz, L. J. 1999. Connexions hétérocliniques et généricite d'une infinité de puits ou de sources, Annales Scientifiques de l'école Normal Suprieure de Paris. **32**(4): 135–150.

[151] Bonatti, C. and Viana, M. 2000. SRB measures for partially hyperbolic systems whose central direction is mostly contracting. Israel J. Math. **115**: 157–193.

[152] Bonatti, C., Diaz, L. J. and Pujals, E. R. 2003. A C^1-generic dichotomy for diffeomorphisms: Weak forms of hyperbolicity or infinitely many sinks or sources. Annals of Mathematics. **158**: 355–418.

[153] Bonatti, C., Boyle, M. and Downarowicz,T. 2004. The entropy theory of symbolic extensions. Invent. Math. **156**: 119–161.

[154] Bonatti, C., Diaz, L. and Viana, M. 2005. Dynamics beyond uniform hyper-bolicity. A Global Geometric and Probabilistic Perspective. Encyclopaedia Mat. Sci. **102**. Mathematical Physics, III (BerlinSpringer-Verlag).

[155] Bonatti, C., Gan, S. and Wen, L. 2007. On the existence of non-trivial homoclinic classes. Ergod. Th. & Dynam. Sys. **27**: 1473–1508.

[156] Bonatti, C., Crovisier, S. and Wilkinson, A. 2008. Centralizers of C^1 generic diffeomorphisms. Preprint.

[157] Bonatti, C., Diaz, L. and Fisher, T. 2008. Supergrowth of the number of periodic orbits for non-hyperbolic homoclinic classes. Preprint.

[158] Bonetto, F., Falco, P. and Giuliani, A. 2004. Analyticity of the SRB measure of a lattice of coupled Anosov diffeomorphisms of the torus. J. Math. Phys. **45**: 3282–3300.

[159] Bonetto, F., Kupiainen, A. and Lebowitz, J. 2005. Absolute continuity of projected SRB measures of coupled Arnold cat map lattices. Ergod. Th. Dynam. Sys. **25**: 59–88.

[160] Bonasera, A., Bucolo,M., Fortuna, L. and Rizzo, A. 2000. The d/sub infinity/parameter to characterise chaotic dynamics. Proc. the IEEE-INNS-ENNS Inter. Joint Conference on Neural Networks. IJCNN (2000). Neural ComputingNew Challenges and Perspectives for the New Millennium. IEEE Comput. Soc. Part., **5**: 565–570.

[161] Borges. E. P., Tirnakli. U. 2004. Two-dimensional dissipative maps at chaos threshold: Sensitivity to initial conditions and relaxation dynamics, Physica A. **340**(1-3): 227–233.

[162] Bothe, H. G. 2001. Strange attractors with topologically simple basins. Topology Appl. **114**(1): 1–25.

[163] Bourbaki, N., Topologie generale Actualités. Sci. Ind., 2nd ed., (Hermann, Paris) 1942–1947 nos. 916 1029; Russian transl. of Topologie generale, (Gos. Izdat. Fiz. Mat. Lit., Moscow) Chaps. IV, VI, VII, 1959.

[164] Bourbaki, M. 1969. Elements de Mathematiques, Integration chapter IX. Paris: Hermann.

[165] Bowen, R. 1970a. Topological entropy and Axiom A. Proc. Sympos. Pure Math., Vol. **XIV**, Berkeley, Calif., (1968) 23–41 Amer. Math. Soc., Providence, R.I.

[166] Bowen, R. 1970b. Markov partitions and minimal sets for Axiom A diffeomorphisms. Amer. J. Math. **92**: 907–918.

[167] Bowen, R. 1971. Periodic points and measures for axiom A diffeomorphisms. Trans. Amer. Math. Soc. **15**(4): 377–397.

[168] Bowen, R. 1973. Symbolic dynamics for hyperbolic flows. Amer. J. Math. **95**: 429–460.

[169] Bowen. R. 1975a. Equilibrium states and the ergodic theory of Anosov diffeo-morphisms. Lect. Notes Math. 470 Springer-Verlag, Berlin.

[170] Bowen, R. 1975b. ω-limit sets for Axiom A diffeomorphisms. J. Differential Equations. **18**(2): 333–339.

[171] Bowen, R. and Ruelle, D. 1975. The ergodic theory of Axiom A flows. Invent. Math. **29**: 181–202.

[172] Box, G. E. P. and Jenkins, G. 1976. Time Series Analysis: Forecasting and Control. Holden-Day.

[173] Boyle, M., Fiebig, D. and Fiebig, U. 2002. Residual entropy, conditional entropy and subshift covers. Forum Math. **14**: 713–757.

[174] Boyle, M. and Sullivan, M. 2005. Equivariant flow equivalence for shifts of finite type. Proceedings of the London Mathematical Society. **91**(1): 184–214 .

[175] Breiman, L. 1968. *Probability*. Original edition published by Addison-Wesley. Reprinted by Society for Industrial and Applied Mathematics (1992).

[176] Briden. W. J., Ladas. G. and Nesemann. T. 2000. On the nonautonomous difference equation $x_{n+1} = \dfrac{\max\{x, A_n\}}{x_n x_{n-1}}$, in "Proceedings of the 4th International Conference on Difference Equations," Gordon and Breach, New York.

[177] Brin, M. and Pesin, Y. 1974. Partially hyperbolic dynamical systems. Proc. Sov. Acad. Sci, Ser. Math., (Izvestia) **38**: 170–212.

[178] Brin, M. 1981. Bernoulli diffeomorphisms with nonzero exponents. Ergod. Th. and Dyn. Syst. 1: 1–7.

[179] Brown, R. 1993a. From the Chua circuit to the generalized Chua map. J. Circuits Systems & Computers. **3**(1): 11–32.

[180] Brown, R. 1995. Horseshoes in the measure–preserving Hénon map. Ergodic Theory Dyn. Syst. **15**: 1045–1059.

[181] Brucks, K.M., Misiurewicz, M. and Tresser, C. 1991a. Monotonicity properties of the family of trapezoidals maps. Comm. Math. Phys. **137**: 1–12.

[182] Brucks. K. M., Diamond. B., Otero-Espinar. M. V. and Tresser. C. 1991b. Dense orbits of critical points for the tent map. Contemp. Math. **117**: 57–61.

[183] Brucks, K. and Misiurewicz, M. 1996. Trajectory of the turning point is dense for almost all tent maps. Ergodic. Theory. Dynam. Systems **16**: 1173–1183.

[184] Brucks, K. and Buczolich, Z. 2000. Trajectory of the turning point is dense for a co-σ-poruous set of tent maps. Fund. Math. **165**: 95–123.

[185] Brucks, K. and Bruin, H. 2004. Topics from one-dimensional dynamics, London Mathematical Society Student Texts, Cambridge Univer. Pr.

[186] Bruin, H., Keller, G., Nowicki, T. and van Strien, S. 1996. Wild Cantor attractors exist. Ann. of Math. **143**: 97–130.

[187] Bruin, H. 1998. For almost every tent map, the turning point is typical. Fund. Math. **155**: 215–235.

[188] Bunimovich, L.A. and Sinai, Y. 1980. In Nonlinear Waves, edited by Gaponov-Grekhov, A.V, Nauka, Moscow, P. 212 (in Russian).

[189] Bucolo, M., Caponetto, R., Fortuna, L. and Xibilia, M. G. 1998. How the Chua circuit allows to model population dynamics. presented at the Proc. NOLTA' 98, La Regent, Crans-Montana, Switzerland, Sept. 14–17.

[190] Bunimovich, L.A. 1983. Statistical Properties of Lorenz Attractors. In Non-linear Dynamics and Turbulence. (ed. Barenblatt, G. I.) Boston ets., Pitman, 71–92.

[191] Bunimovich, L.A., Sinai, Y. and Chernov, N. I. 1990. Markov partitions for two dimensional hyperbolic Billiards. Russ. Math. Surv. **45**(3): 105–152.

[192] Bunimovich, L. A., Sinai, Y. and Chernov, N. I. 1991. Decay of Correlations and the Central Limit Theorem for two dimensional Billiards. Russ. Math. Surv. **46**(4): 47–106.

[193] Bunimovich, L. A. 2000. Dynamical Systems, Ergodic Theory and Applications. Encyclopedia of Mathematical Sciences Vol. **100** Springer, New York.

[194] Burns K., Pugh, C., Shub, M. and Wilkinson, A. 2001. Recent results about stable ergodicity, in Proceedings of Symposia in Pure Mathematics Vol **69** "Smooth Ergodic Theory and Its Applications" (Katok, A., de la Llave, R., Pesin, Y., Weiss, H., Eds), AMS, Providence, R.I. 327–366.

[195] Burns, K., Dolgopyat, D. and Pesin, Y. 2002. Partial hyperbolicity, Lyapunouv exponents and stable ergodicity. J. Statistical Physics. **108** (5-6): 927–942.

[196] Butler, L. T. and Gelfreich, V. 2008. Positive-entropy geodesic flows on nilmanifolds. Nonlinearity. **21**(7): 1423–1434.

[197] Buzzi, J. 2009. Maximal entropy measures for piecewise affine surface homeomorphisms. Ergodic Theory and Dynamical Systems. **29**(06): 1723–1763.

[198] Caladrini, G., Berns, D., Paolini, E. and Moiola, J. 2000. On cyclic fold bifurcations in nonlinear systems. (2000) IEEE Inter. Symposium on Circuits and Systems. Emerging Technologies for the 21st Century. Proc.s (IEEE Cat No.00CH36353).Presses Polytech. Univ. Romandes. Part. **2**: 485–488.

[199] Campbell, D. K., Galeeva, R., Tresser, C. and Uherka, D.J. 1996. Piecewise linear models for the quasiperiodic transition to chaos. Chaos. **6**(2): 121–154.

[200] Cao, Y. and Liu, Z. 1998a. Strange attractors in the orientation-preserving Lozi map. Chaos, Soliton, Fractals. **9**: 1857–1863.

[201] Cao, Y and Liu, Z. 1998b. The geometric structure of strange attractors in the Lozi map. Commun. Nonlin. Sci. Numer. Simul. **3**(2): 119–123.

[202] Cao, Y. and Kiriki, S. 2000a. The basin of the strange attractors of some Hénon maps. Chaos, Solitons & Fractals. **11**(5): 729–734.

[203] Cao, Y. and Mao, M. J. 2000b.The non-wandering set of some Hénon maps. Chaos, Solitons & Fractals. **11**(13): 2045–2053.

[204] Cao, Y. 2004. A note about Milnor attractor and riddled basin. Chaos, Solitons and Fractals. **19**: 759–764.

[205] Cao, Y., Luzzatto, S. and Rios, I. 2008. The boundary of hyperbolicity for Hénon-like families. Ergod. Th. Dynam. Syst. **28**: 1049–1080.

[206] Carballo, C. M., Morales, C., and Pacifico, M. J. 2000. Maximal transitive sets with singularities for generic C^1 vector fields. Bol. Soc. Brasil.Mat. (N.S.) **31**(3): 287–303.

[207] Carballo, C. M., Morales, C. A. and Pacifico, M. J. 2007. Homoclinic classes for generic C^1 vector fields. Ergodic Theory Dynam. Systems, to appear.

[208] Casselman, B. 2005. Picturing the Horseshoe Map. Notices.Amer. Math. Soc. **52**(5): 518–519.

[209] Caponetto, R., Criscione, M., Fortuna, L., Occhipinti, D. and Occhipinti, L. 1998. Synthesis of a programmable chaos generator, based on CNN architectures, with applications in chaotic communication, in Proc. CNNA '98, London, UK., Apr. 14–17, 124–129.

[210] Caponetto, R., Fortuna, L., Fazzino, S. and Xibilia, M. G. 2003. Chaotic sequences to improve the performance of evolutionary algorithms, IEEE Trans. Evolutionary Computaion. **7**(3): 289–304.

[211] Cessac, B. 2007. does the complex susceptibility of the Hénon map have a pole in the upper-half plane? A numerical investigation. Nonlinearity. **20**(12): 2883–2895.

[212] Changming, D. 2005. The omega limit sets of subsets in a metric space. Czechoslovak Mathematical Journal. **55**(1): 87–96.

[213] Chazottes, J. R. 2005. Collet, P. and Schmitt, B., Statistical consequences of Devroye inequality for processes: Applications to a class of nonuniformly hyperbolic dynamical systems, Nonlinearity. **18**(5): 2341–2364.

[214] Chen, G. and Dong, X. 1998. From chaos to order: Methodologies, Perspectives and Applications. World Scientific, Singapore.

[215] Chernov, N. I. 1992. Ergodic and statistical properties of piecewise linear hyperbolic automorphisms of the 2-torus J. Stat. Phys. **69**(1-2): 111–134.

[216] Chernov, N. I. 1999. Decay of correlations and dispersing billiards. J. Stat. Phys. **94**: 513–556.

[217] Choi, C., Eom, T. -D., Hong, S. -G. and Lee, J. -J. 1998. Information processing using chaos with application to mobile robot navigation problems. Artificial Life and Robotics. **2**(2).

[218] Chomsky, N. 1959. On certain formal properties of grammars, Inf. Control. **2**: 137–167.

[219] Chow, S. N. and Palmer, K. J. 1992. On the numerical computation of orbits of dynamical systemsthe higher dimensional case. J. Complexity. **8**: 398–423.

[220] Christiansen, F. and Rugh, H. H. 1997. Computing Lyapunov spectra with continuous Gram–Schmidt orthonormalization. Nonlinearity. **10**(5): 1063–1073.

[221] Christy, J. 1993. Branched surfaces and attractors. I. Dynamic branched surfaces.Trans. Amer. Math. Soc. **336**: 759–784.

[222] Chua, L. O., Komuro, M. and Matsumoto, T. 1986. The double scroll family. IEEE Trans. Circuits Syst. **CAS-33**(11): 1073–1118.

[223] Chua, L. O. and Lin, G-. N. 1990. Canonical realization of Chua's circuit family. IEEE Trans. Circuits Syst. **377**: 885–902.

[224] Chua, L. O. and Tichonicky, I. 1991. 1-D map for the double scroll. IEEE Trans. Circuits Syst.-IFund. Th. Appl. **38**(3): 233–243.

[225] Chua, L. O. 1992. The genesis of Chua's circuit. Archiv fur Elektronik und Uebertragungstechnik. **46**(4): 250–257.

[226] Chua, L. O. and Huynh, L.T. 1992. Bifurcation analysis of Chua's circuit. Proc. 35th Midwest Symposium on Circuits. Syst., (Cat. No.92CH3099-9). IEEE, **1**: 746–751.

[227] Chua, L. O., Wu, C. W., Huang, A. and Zhong, G.-Q. 1993a. A universal system for studying and generating chaos – Part IRoutes to chaos. IEEE Trans. Circuits Syst.-IFund. Th. Appl. **40**: 732–744.

[228] Chua, L. O., Wu, C. W., Huang, A. and Zhong, G-Q. 1993b. A universal circuit for studying and generating chaos. II. Strange attractors. IEEE Trans. Circuits Syst.-IFund. Th. Appl. **40**(10): 745–761.

[229] Chua, L. O. 1993. Global unfolding of Chua's circuit. IEICE Trans. Fund. Electronics Comm. Comput. Sc. **E76-A**(5): 704–734.

[230] Chua, L. O. 1994. Chua's circuitten years later. IEICE Trans. Fund. Electronics Comm. Comput. Sc. **E77-A**(11): 1811–1822.

[231] Chua, L. O., Pivka, L. and Wu, C-W. 1995. A universal circuit for studying chaotic phenomena. Ph. Trans. Royal Soc. London. **353**(1701): 65–84.

[232] Chunyan, Z. and Wang, X. 2007. Attractors and quasi-attractors of a flow. J. Applied Mathematics and Computing. **23**(1-2): 411–417.

[233] Cleveland, C. 1999. Rotation for attractors in the Lozi family. Ph. D. University of California, Los Angeles.

[234] Coelho, L. 2009. Tuning of PID controller for an automatic regulator voltage system using chaotic optimization approach. Chaos, Solitons & Fractals. **39**(4): 1504–1514.

[235] Cohen, S. D. and Hindmarsh, A. C. 1996. CVODE, A Stiff/Nonstiff ODE Solver in C. Comput. Phys. **10**: 138–141.

[236] Colmenarez, W. 2002. Algumas propriedades de atratores para fluxos em dimensao trés. Thesis Universidade Federal do Rio de Janeiro (in Portuguese).

[237] Collet. P. and Eckmann. J. P. 1980. Iterated maps on the interval as dynamical systems. Birkhatiser. Stuttgart.

[238] Collet, P. and Levy, Y. 1984. Ergodic properties of the Lozi mappings. Comm. Math. Phys. **93**: 461–482.

[239] Cornfeld, I., Fomin, S. V. and Sinai, Y. 1982. Ergodic Theory. Springer-Verlag, Berlin.

[240] Cook, H., Ingram, W. T., Kuperberg, K. and Lelek, A. 1995. Continua. Lecture notes in pure and applied mathematics.

[241] Coomes, B. A., Kocak, H. and Palmer, K. J. 1997. Computation of long periodic orbits in chaotic dynamical systems. The Australian Math. Soc.Gazette. **24**: 183–190.

[242] Coomes, B. A., Ko‚cak, H. and Palmer, K. J. 2005. Homoclinic shadowing. J. Dynamics and Differential Equations. **17**(1): 175–215.

[243] Coven, E. M., Kan, I. and Yorke, J. A. 1988. Pseudo-orbit shadowing in the family of tent maps. Transactions of the American Mathematical Society. **308**: 227–241.

[244] Crovisier, S. 2006. Birth of homoclinic intersectionsa model for the central dynamics of partially hyperbolic systems. Arxiv preprint math.DS/0605387.

[245] Crovisier, S. 2008. Partial hyperbolicity far from homoclinic bifurcations. Preprint (2008).

[246] Crutchfield, J. P. 1988. Spatio-Temporal Complexity in Nonlinear Image Processing. IEEE Transactions on Circuits and Systems. **35**(7): 770–780.

[247] Crutchfield, J. P, and Young, K. 1990. Computation at the onset of chaos. In Complexity, Entropy, and Physics of Information, ed Zurek, W., New York: Addison-Wesley. 223–69.

[248] Cvitanović, P., Gunaratne, G. and Procaccia. I. 1988. Topological and metric properties of Hénon-type strange attractors. Phys. Rev. A. **38**: 1503–1520.

[249] Cvitanovic. P. 1991. Periodic orbits as the skeleton of classical and quantum chaos. Physica. D. **5**: 138–151 .

[250] Daido, H. 1985. Coupling sensitivity of Chaos and the Lyapunov dimension: The case of coupled two-dimensional maps. Phys. Lett. **110A**: 5–9.

[251] D'Alessandro. G., Gransberger. P., Isola. S. and Politi. A. 1990. On the topology of the Hénon map. J. Phys. A. **23**(22): 5285–5295.

[252] D'aissanddro. D., Isola. S. and Poloti, A. 1991. Geometric properties of the pruning front. Prog. Theor. Phys. **86**(6): 1149–1157.

[253] Dan, T. 2006. Turing instability leads oscillatory systems to spatiotemporal chaos. Prog. theor. phys., Suppl. **61**: 119–126.

[254] Danca, M-F. and Codreanu, S. 2002. On a possible approximation of discontinuous dynamical systems. Chaos Solitons & Fractals. **13**(4): 681–691.

[255] Davidchack, R., Lai, Y. C., Klebanoff, A. and Bollt, E. 2001. Towards complete detection of unstable periodic orbits in chaotic systems. Phys. Lett. A. **287**: 99–104.

[256] Davis, M. J., MacKay, R. S. and Sannami, A. 1991. Markov shifts in the Hénon family. Physica.D. **52**(2-3): 171–178.

[257] Davies, H. G. and Rangavajhula, K. 2002. Noisy parametric sweep through a period-doubling bifurcation of the Henon map. Chaos, Solitons & Fractals. **14**(2): 293–299.

[258] Dawson, S., Grebogi, C., Sauer, T. and Yorke, J. A. 1994. Obstructions to shadowing when a Lyapunov exponent fluctuates about zero. Phys. Rev. Lett. **73**(14): 1927–1930.

[259] de Carvalho, A. 1999. Pruning fronts and the formation of horseshoes. Ergod. Theor. Dynam. Syst. **19**: 851–94.

[260] de Carvalho, A., Hall, T. 2002. How to prune a horseshoe. Nonlinearity. **15**: R19–68.

[261] Dedieu, H. and Ogorzalek, M. J. 1997. Identifiability and identification of chaotic systems based on adaptive synchronization. IEEE Trans. Circuits Syst.-IFund. Th. Appl. **44**(10): 948–962.

[262] Dedieu, J. P. and Shub, M. 2003. On random and mean exponents for unitarily invariant probability measures on $GL(n,\mathbb{C})$ in "Geometric methods in dynamical systems (II)-Volume in Honor of Jacob Palis". Asterisque. **287**: 1–18. Soc. Math. De France.

[263] de Faria, E. and de Melo, W. 2000.Rigidity of critical circle mappings. II. J. A. M. S. **13**: 343–370.

[264] Dellnitz, M. and Junge, O. 2002. Set oriented numerical methods for dynamical systems. Handbook of dynamical systems. **2**: 221–264 North-Holland, Amsterdam.

[265] Denker M. 1989. The central limit theorem for dynamical systems. Dyn. Syst. Ergod. Th. Banach. Center Publ. **23**. Warsaw PWN–Polish Sci. Publ.

[266] de Melo, W. 1973. Structural stability of diffeomorphisms on two-manifolds. Invent. Math. **21**: 233–246.

[267] de Melo, W. and Palis, J. 1982. Geometric theory of dynamical systems. An introduction. Springer-Verlag, New York–Berlin.

[268] Derbyshire, J. 2004. Prime obsession Bernhard Riemann and the greatest unsolved problem in mathematics. New York Penguin.

[269] de Oliveira, K. A., Vannucci, A. and da Silva, E. C. 2000. Using artificial neural networks to forecast chaotic time series. Phys. A. **284**(1): 393–404.

[270] Dettmann, C. P., Cohen, E. G. D. and van Beijeren, H. 1999. Microscopic chaos from Brownian motion?. Nature. **401**: 875–875.

[271] Dettmann, C. P. and Cohen, E. G. D. 2000. Microscopic chaos and diffusion. J. Stat. Phys. **101**: 775–817

[272] Devaney, R.L. and Nitecki, Z. 1979. Shif automorphism in the Hénon mapping. Connun. Math. Phys., **67**: 137–146.

[273] Devaney, R. L. 1984a. A piecewise linear model for the zones of instability of an area-preserving map. Physica. D. **10**: 387–393.

[274] Devaney, R.L. 1984b. Homoclinic bifurcations and the area conserving Hénon mapping. J. Diff. Equations. **51**(2): 254–266.

[275] Devaney, R.L. 1988. Reversibility, Homoclinic Points, and the Hénon Map. In Dynamical Systems, Approaches to Nonlinear Problems in Systems and Circuits. Philadelphia SIAM. 3–14.

[276] Devaney, R. L. 1989. An Introduction to Chaotic Dynamical Systems. Addison-Wesley, New York.

[277] Diamond, P., Kloeden, P. E., Kozyakin, V. S. and Pokrovskii, A. V. 1995. Semi-hyperbolic mappings. J. Nonlinear Sci. **5**: 419–431.

[278] Diamond, P., Kloeden, P. E., Kozyakin, V. S. and Pokrovskii, A. V. 2008. Semi-hyperbolic mappings. Manuscript.

[279] Diaz, L. J., Pujals, E. and Ures, R. 1999. Partial hyperbolicity and robust transitivity. Acta Math. **183**: 1–43.

[280] di Bernardo, M., Feigin, M. I., Hogan, S. J. and Homer, M. E. 1999. Local analysis of C-bifurcations in *n*-dimensional piecewise smooth dynamical systems. Chaos, Solitons & Fractals. **10**(11): 1881–1908.

[281] di Bernardo. M., Kowalczyk. P. and Nordmark. A. B. 2002. Bifurcations of dynamical systems with sliding: derivation of normal form mappings. Physica. D. **170**: 175–205.

[282] di Bernardo. M., Kowalczyk. P. and Nordmark. A. B. 2003. Sliding bifurcations: a novel mechanism for the sudden onset of chaos in dry-friction oscillators. Int. J. Bifurc. Chaos. **13**: 2935–2948.

[283] Ding, M. and Yang, W. 1997. Stability of synchronous chaos and on-off intermittency in coupled map lattices. Phys. Rev. E. **56**: 4009–4016.

[284] Ding, C. 2004. On the intertwined basins of attraction for planar flows. Applied Mathematics and Computation. **148**(3): 801–805.

[285] Ditza, A. 1990. Scaling of periodic orbits in to-dimensional Chaotic systems. Phys. Rev. A. **41**(12): 6692–6701.

[286] Djellit, I. and Boukemara, I. 2007. Dynamics of a three parameters family of piecewise maps. Facta universitatis. Nis, Ser. Elec. Energ. **20**(1): 85–92.

[287] Dmitriev, A.S., Andrey Panas, I. and Starkov, S. O. 2000a. Multiple access communication based on control of special chaotic trajectories, Proceedings of International Conference on Control of Oscillations and Chaos (COC-2000), St. Petersburg, Russia, July 5–7, **3**: 518–522.

[288] Dmitriev, A. S., Starkov, S. O., Shirokov, O. M., Panas, A. I., Yong, L., Wen, T. and Wang, R. 2000b. Software Patent: A multiple access communication system using chaotic signals and method for generating and extracting chaotic signals, EP1183842, Nortel networks ltd (CA).

[289] Dmitriev, A. S, Hasler, M., Kassian, G. A., Khilinsky, A. D. 2001. Chaotic Synchronization of 2-D maps via information transmission, Proceedings of 2001 International Symposium on Nonlinear Theory and its Applications, Miagi, Japan, October 28–November 1, **1**: 79–82.

[290] Dobrynskiy, V., The Lozi-like non-expansive endomorphism of plane having a 2-Dimensional chaotic attractor. No referenced paper.

[291] Dobrynskiy, V. A. 1999. On attractors of piecewise linear 2-endomorphisms. Nonlinear Analysis. **36**(4): 423–455.

[292] Dobrynskiy, V. A. 2005. On the Structure of generalized hyperbolic attractors of mappings that are not one-to-one, Differential Equations. **41**(6): 780–790.

[293] Dobrynskiy, V. 2008. On the structure of invariant attracting sets of systems of finite-difference equations generating Hénon-Lozi mappings, Differential Equations. **44**(9).

[294] Doering, C. I. 1987. Persistently transitive vector fields on three-dimensional manifolds. Dynamical Systems and Bifurcation Theory, Pitman Research Notes in Mathematics Series **160:** 59–89.

[295] Doerner, R., Hubinger, B. and Martienssen, W. 1994. Advanced chaos forecasting. Phys. Rev. E. **50**(1): R12–15.

[296] Dolgopyat, D. 1998. Prevalence of rapid mixing in hyperbolic flows. Ergodic Theory Dynam. Systems **18**: 1097–1114.

[297] Dolgopyat, D. 2000. On dynamics of mostly contracting diffeomorphisms. Comm. Math. Phys. **213**: 181–201.

[298] Dolgopyat, D. 2004. On differentiability of SRB states for partially hyperbolic systems. Invent. Math. **155**: 389–449.

[299] Dolgopyat, D. 2002. On mixing properties of compact group extensions of hyperbolic systems. Israel. J. Math. **130**: 157–205.

[300] Dolgopyat, D. and Pesin, Y. 2002. On the existence of Bernoulli diffeomorphisms with nonzero Lyapunov exponents on compact smooth manifolds. To appear in Ergod. Th. and Dyn. Syst.

[301] Dolgopyat, D., Hu, H. and Pesin, Y. 2001. An example of a smooth hyperbolic measure with countably many ergodic components. in Smooth ergodic theory and its applications by Katok, A. and de la Llave, R., Proceedings of Symposia in Pure Mathematics. **69**: 95–106.

[302] Dolgopyat, D. and Pesin, Y. 2002. Every compact manifold carries a completely hyperbolic diffeomorphism. Ergod. Th. Dynam. Syst. **22**(2): 409–435.

[303] Driebe, D. J. 1999. Fully chaotic maps and broken time symmetry. Kluwer Academic Publishers.

[304] Duchesne, L. 1993. Using characteristic multiplier loci to predict bifurcation phenomena and chaos-a tutorial. IEEE Trans. Circuits Syst.-IFund. Th. Appl. **40**(10): 683–688.

[305] Dullin, H .R., Sterling, D. and Meiss, J. D. 2000. Self-rotation number using the turning angle. Physica. D. **145**: 25–46.

[306] Duarte, P. 1994. Plenty of elliptic islands for the standard family of area preserving maps. Ann. Inst. H. Poincaré Anal. Non. Linéaire. **11**: 359–409.

[307] Easton, R. W. 1988. Geometric methods for discrete dynamical systems. Oxford university press.

[308] Eberhart, R. C. and Kennedy, J. 1995. A new optimizer using particle swarm theory. In: Proceedings of the International Symposium on Micro Machine and Human Science, Nagoya, Japan. 39–43.

[309] Eckmann, J. P. and Ruelle, D. 1985. Ergodic theory of chaos and strange attractors. Rev. Mod. Phys. **57:** 617–656.

[310] Eckmann, J. P. and Procaccia, L. 1986. Fluctuations of dynamical scaling indices in nonlinear systems. Phys. Rev. A. **34:** 659–661.

[311] El Hamouly, H. and Mira, C. 1981. Lien entre les propriétés d'un endomorphisme de dimension unet celles d'un diff'eomorphisme de dimension deux. C. R. Acad. Sci. Paris Sér. I Math. **293:** 525–528.

[312] Endler, A. and Gallas, J. A. C. 2001. Period four stability and multistability domains for the Hénon map. Phys. A. **295**(1): 285–290.

[313] Eric, S. Van Vleck. 1995. Numerical shadowing near hyperbolic trajectories. SIAM J. Sci. Comp. **16** (5): 1177–1189.

[314] Ermentrout, G. B., XPPAUT, http//www.pitt.edu/;phase.

[315] Farmer. J. D. 1982. Information and the probabilistic structure of chaos, Z. Naturforsh. **37a**: 1304–1325.

[316] Farmer, J. D., Ott, E. and Yorke, J. A. 1983. The dimension of chaotic attractors. Physica. D. **7**: 153–180.

[317] Farrell, F. T. and Jones, L. E. 1978. Anosov diffeomorphisms constructed from π_1 (*Diff* (S_n)). Topology. **17**(3): 273 282.

[318] Feigenbaum, M. J. 1984. Universal behavior in nonlinear systems. Los Alamos Sci. **1** (1980) 4–27 (reprinted in Universality in Chaos (P. Cvitanovid, Ed.). Adam Hilger,.

[319] Feigin, M. I. 1970. Doubling of the oscillation period with C-bifurcations in piecewise continuous systems. Prikladnaya Matematika i Mechanika. **34**: 861–869.

[320] Feigin, M. I. 1994. Forced oscillations in systems with discontinuous nonlinearities (Moscow: Nauka) (in Russian).

[321] Feit, S. 1978. Characteristic exponents and strange attractors. Commun. Math. Phys. **61**: 249–260.

[322] Feely, O. and Chua, L. O. 1992. Nonlinear dynamics of a class of analog-to-digital converters. Inter. J. Bifur.Chaos. **22**: 325–340.

[323] Feldman, J. and Katok, A. 1981. Bernoulli diffeomorphisms and group extensions of dynamical systems with nonzero characteristic exponents. Ann. Math. **113**: 159–179.

[324] Feng, B. Y. 1998. The heteroclinic cycle in the model of competition between *n*-species and its stability. Acta. Math. Appl. Sinica. **14**: 404–413.

[325] Feng, G. and Chen, G. 2005. Adaptative control of discrete-time chaotic systems: a fuzzy control approach. Chaos, Solitons and Fractals. **23**(2): 459–467.

[326] Feuer. J. and Janowski, E. 2000. Global behavior of solutions of $x_{n+1} = \dfrac{\max\{x_n,\, A\}}{x_n x_{n-1}}$. J. Comput. Anal. Appl. **2**(3).

[327] Fisher, T. 2006a. Hyperbolic sets that are not locally maximal. Ergod. Th. and Dynam. Sys. **26**(05): 1491–1509.

[328] Fisher, T. 2006b. Hyperbolic sets with nonempty interior. Discrete and Contin. Dynam. Systems. **15**(2): 433–446.

[329] Fisher, T. 2006c. The topology of hyperbolic attractors on compact surfaces. Ergod. Th. and Dynam. Sys. **26**(05): 1511–1520.

[330] Fisher, T. and Rodriguez-Hertz, J. 2008. Quasi-Anosov diffeomorphisms of 3-manifolds. Preprint.

[331] Fisher, T. 2009. Hyperbolic chain recurrent classes for commuting diffeomorphisms. Preprint.

[332] Flaminio, L. and Katok, A. 1991. Rigidity of symplectic Anosov diffeomorphisms on low-dimensional tori. Ergodic Theory Dynam. Systems. **11**(3): 427–441.

[333] Fokkink, R. J. 1991. *The structure of trajectories*. PhD Thesis, (Technical University of Delft).

[334] Fomin, S. V. and Gelfand, I. M. 1952. Geodesic flows on manifolds of constant negative curvature. Uspehi. Mat. Nauk. **7**(1): 118–137.

[335] Fontich, E. 1990. Transversal homoclinic points of a class of conservative diffeomorphisms. J. Differ. Equations. **87**: 1–27.

[336] Fomin, S. W., Kornfeld, I. P. and Sinaj, J. G. 1980. Ergodic theory. Nauka Moskwa, (in Russian).

[337] Fornæss, J. E. and Gavosto, E. A. 1992. Existence of generic homoclinic tangencies for Hénon mappings. J. Geom. Anal. **2**: 429–444.

[338] Formanek, E. 1994. Observations About the Jacobian Conjecture. Houston J. Math. **20**: 369–380.

[339] Fornæss, J. E. and Gavosto, E. A. 1999. Tangencies for real and complex Hénon maps: An analytic method. Experiment. Math. **8**: 253–260.

[340] Forni, G., Lyubich, M., Pugh, C. and Shub, M. 2007. Partially Hyperbolic Dynamics, Laminations, and Teichmuller flow. Fields Institute Communications.

[341] Franks, J. 1970. Anosov diffeomorphisms. Proc. Sympos. Pure Math. **14**: 61–93.

[342] Franks, J. 1971. Necessary conditions for the stability of diffeomorphisms. Trans. A. M. S. **158**: 301–308.

[343] Franks, J. and Williams, R. 1980. Anomalous Anosov flows. Global theory of dynamical systems (Proc. Internat. Conf., Northwestern Univ., Evanston, Ill., (1979)) 158-174 Lecture Notes in Math. **819** Springer, Berlin.

[344] Franks, J. 1985. Period doubling and the Lefschetz formula. Trans. Amer. Math. Soc. **287**(1): 275–283.

[345] Freire. E., Rodriguez-Luis. A.J., Gamero. E. and Ponce. E. 1993. A case study for homoclinic chaos in an autonomous electronic circuit: A trip from Takens-Bogdanov to Hopf-Shil'nikov. Physica D. **62**: 230–253.

[346] Freeman, W. J. 1995. *Societies of Brains.* Lawrence Erlbaum Associates, Mahwah.

[347] Freiling, C. 1986. Axioms of symmetry: throwing darts at the real number line. J. Symbolic Logic. **51**(1): 190–200.

[348] Friedland, S. and Milnor, J. 1989. Dynamical properties of plane polynomial automorphisms. Erg. th. and dyn. syst. **9**: 67–99.

[349] Franceschini. V., Giberti, C. and Zheng, Z. 1993. Characterization of the Lorenz attractor by unstable periodic orbits. Nonlinearity. **6**: 251–258.

[350] Fujii, H., Aihara, K. and Tsuda, I. 2007a. Corticopetal acetylcholine: A role in attentional state transitions and the genesis of quasi-attractors during perception. Advances in Cognitive Neurodynamics ICCN.

[351] Fujii, H., Aihara, K. and Tsuda, I. 2007b. Corticopetal acetylcholine: Possible scenarios on the role for dynamic organization of quasi-attractors. Lecture Notes in Computer Science, Neural Information Processing14th International Conference, ICONIP (2007) Kitakyushu, Japan, November 13–16 (2007) Revised Selected Papers, Part I , 170–178.

[352] Fujisaka, H. 1983. Statistical dynamics generated by fluctuations of local Lyapunov exponents, Prog. Theor. Phys. **70**: 1264–1275.

[353] Fujisaka, H. and Sato. C. 1997. Computing the number, location and stability of fixed points of Poincaré maps. Circuits and Systems IFundamental Theory and Applications, IEEE Transactions on (see also Circuits and Systems IRegular Papers, IEEE Transactions on)., **44**(4): 303–311.

[354] Gallas, A. and Jason, C. 1993. Structure of the parameter space of the Hénon map. Phys. Rev. Letters. **70**(18): 2714–2717.

[355] Gallavotti, G., Bonetto, F. and Gentile, G. 2004. Aspects of ergodic, qualitative and statistical theory of motion. Springer.

[356] Galias, Z. 1997a. Numerical studies of the Hénon map. In: Proc. Int. Symposium on Scientific Computing, Computed Arithmetic and Validated Numerics, SCAN'97., **XIV5-6** Lyon.

[357] Galias, Z. 1998a. Rigorous Numerical Studies of the Existence of Periodic Orbits for the HénonMap. J. Universal Computer Science. **4**(2): 114–125.

[358] Galias, Z. 1998b. Existence and uniqueness of low-period cycles and estimation of topological entropy for the Hénon map. In: Proc. Int. Symposium on Nonlinear Theory and its Applications, NOLTA'98. **1**: 187–190 Crans-Montana.

[359] Galias, Z. 1999. All periodic orbits with period n ≤ 26 for the Hénon map. In: Proc. European Conference on Circuit Theory and Design, EC- CTD'99. **1**: 361–364 Stresa.

[360] Galias. Z. and Zgliczynski, P. 2001. Abundance of homoclinic and heteroclinic orbits and rigorous bounds for the topological entropy for the Hénon map. Nonlinearity. **14**(5): 909–932.

[361] Galias, Z. 2001. Interval methods for rigorous investigation of periodic orbits. Inter.J. Bifurcation & Chaos. **11**(9): 2427–2450.

[362] Galias, Z. 2003. Mean value form for evaluation of Poincaré map in piecewise linear systems. In: Proc. European Conference on Circuit Theory and Design, **ECCTD'03., I:** 283–286 Kraków.

[363] Galias, Z. 2004. Towards full characterization of continuous systems in terms of periodic orbits. In: Proc. IEEE Int. Symposium on Circuits and Systems., **ISCAS'04 IV**, 716–719 Vancouver, Canada.

[364] Galias. Z. and Zgliczynski. P. 1998. Computer assisted proof of chaos in the Lorenz equations. Physica. D. **115:** 165–188.

[365] Gambaudo, J.M. and Tresser, C. 1985. Dynamique régulière ou chaotique. Applications du cercle ou l'intervalle ayant une discontinuité. C. R. Acad. Soc. Paris, Sér. **I 300**(10): 311–313.

[366] Gan, S., Yang, D. and Wen, L. 2007. Minimal non-hyperbolicity and index completeness. Preprint Beijing University.

[367] Gardini, L., Abraham, R., Record, R. J. and Fournier-Prunaret, D. 1994. A double logistic map. Int. J. Bifurcation Chaos. **4**(1): 145–176.

[368] Gaspard, P. 2005. Maps. Encyclopedia of Nonlinear Science. Routledge, New York, 548–553.

[369] Gavrilov, N. K. and Shil'nikov, L. P. 1973. On three dimensional dynamical system close to systems with a structuraly satble homoclinic curve. Math. USSR Sb. **19:** 139–156.

[370] Gelfreich, V. G. 1991. Separatrices splitting for polynomial area-preserving maps. In: M. Sh. Birman., editor, Topics in Math. Phys. **13:** 108–116. Leningrad State university.

[371] Gelfreich, V. G. and Sauzin, D. 2001. Borel summation and splitting of separatrices for the Hénon map. Ann. Inst. Fourier. **51:** 513–567.

[372] Ghrist, R. W. and Zhirov, A. Yu. 2004. Combinatorics of one-dimensional hyperbolic attractors of diffeomorphisms of surfaces, (Russian. Russian summary) Trudy Matematicheskogo Instituta Imeni V. A. Steklova. Rossi kaya Akademiya Nauk, **244** Din. Sist. i Smezhnye Vopr. Geom. 143–215.

[373] Ghys, E. 1988. Codimension one Anosov flows and suspensions. Dynamical systems, Val-paraiso (1986) 59–72 Lecture Notes in Math., **1331** Springer, Berlin.

[374] Ghys, E. 1995. Holomorphic Anosov systems. Invent. Math. **119:** 585–614.

[375] Glendinning, P. and Sparrow, C. 1984. Local and global behavior near homoclinic orbits. J. Stat. Phys. **35:** 645–697.

[376] Glendinning, P. and Sparrow, C. 1993. Prime and renormalizable kneading invariants and the dynamics of expanding Lorenz maps. Physica. D. **62:** 22–50.

[377] Glendinning, P. 1994. Stability, Instability and Chaos. Cambridge Texts in Applied Mathematic.

[378] Goldberg, D. E. 1989. Genetic Algorithms in Search, Optimization and Machine Learning, Addison Wesley, Reading, MA.

[379] Golub, G., VanLoan, C. F. 1989. Matrix Computations. Johns Hopkins University Press, Baltimore.

[380] Golumb. M. 1992. Periodic recursive sequences. Amer. Math. Monthly. **99:** 882–883.

[381] Gomez, G. and Sim´o, C. 1983. Homoclinic and heteroclinic points in the Hénon map. Lect. Notes in Physics. **179:** 245–247.

[382] Gomez, A. and Meiss, J. D. 2003. Reversible polynomial automorphisms of the planethe involutory case. Phys. Lett. A. **312**(1): 49–58.

[383] Gomez, A. and Meiss, J. D. 2004. Reversors and symmetries for polynomial automorphisms of the plane. Nonlinearity. **17**(3): 975–1000.

[384] Gonchenko, S. V., Shilnikov, L. P., Turaev, D. V. 1992. On models with a structurally unstable homoclinic Poincaré curve. Sov. Math. Dokl. **44**(2): 422–426.

[385] Gonchenko, S. V., Shil'nikov, L. P. and Turaev, D. V. 1993. On models with nonrough Poincaré homoclinic curves. Physica. D. **62:** 1–14.

[386] Gonchenko, S. V., Turaev, D. V. and Shilnikov, L. P. 1993. Dynamical phenomena in multi-dimensional systems with a structurally unstable homoclinic Poincaré curve. Russian Acad. Sci. Dokl. Math. **47**(3): 410–415.

[387] Gonchenko, S. V., Ovsyannikov, I. I. and Simo, C. 2005. Three dimensional Hénon-like maps and wild Lorenz-like attractors. Int. J. Bifur. Chaos. **15:** 3493–3508.

[388] Gonchenko, S. V., Turaev, D. V .and Shil'nikov, L. P. 2005. On dynamic properties of diffeomorphisms with homoclinic tangency. J. Math. Sci. **126**(4): 1317–1343.

[389] Gorodnik, A. 2007. Open problems in dynamics and related fields. J. Modern Dynamics. **1**(1) (2007) 1–35.

[390] Gorodetski A. S. and YIlyashenko, U. S. 1996. Minimal and strange attractors. Internat. J. Bifur. Chaos. **6**: 1177–1183.

[391] Gorodetski, A. and Kaloshin, V. 2007. How often surface diffeomorphisms have infinitely many sinks and hyperbolicity of periodic points near a homoclinic tangency. Advances in Mathematics. **208**: 710–797.

[392] Gourmelon, N. 2007. Generation of homoclinic tangencies by C^1- perturbations. Preprint Université de Bourgogne.

[393] Gramberger, P. and Kantz. H. 1985. Generating partitions for the dissipative Hénon map. Phys. Lett.A. **113**: 235–238.

[394] Grassberger, P. 1983. Fractal dimension of the strange attractor in a piecewise linear two-dimensional map, Phys. Lett. A. **97**: 219–223.

[395] Grassberger, P. and Procaccia, I. 1983. Measuring the Strangeness of Strange Attractors. Physica. D. **9**: 189–208.

[396] Grassberger, P. 1985. Generalizations of the Hausdorff dimension of fractal measures. Phys. Lett. A. **107**(3): 101–105.

[397] Grassberger, P. 1986. Toward a quantitative theory of self-generated complexity. Int. J. Theor. Phys. **25**: 907–938.

[398] Grassberger, P. 1989. On Lyapunov and dimension spectra of 2D attractors, with an application to the Lozi map. J. Phys. A: Math. Gen. **22**: 585–589.

[399] Grassberger, P., Kantz, H. and Moenig, U. 1989. On the symbolic dynamics of the Hénon map. J. Phys. AMath. Gen. **22**: 5217–5230.

[400] Grassberger, P., Nadler, W., and Yang, L. 2002. Heat conduction and entropy production in a one-dimensional hard-particle gas. Phys. Rev. Lett. **89**: 180601.

[401] Graziela, G. A. 1995. Controlling chaos of an uncertain Lozi system via adaptive techniques. Int. J. of Bifur. Chaos. **5**(2): 559–562.

[402] Graczy, J. and Swiatek, G. 1997. Generic hyperbolicity in the logistic family. Ann. Math. **146**: 1–52.

[403] Grebogi, C., Ott, E. and Yorke. J. A. 1983. Crises, sudden changes in chaotic attrators, and transient chaos. Physica. D. **70**: 191–200.

[404] Grebogi, C., Ott, E. and Yorke, J. A. 1987a. Basin Boundary Metamorphoses: Changes in Accessible Boundary Orbits. Physica. D. **24**: 243.

[405] Grebogi, C., Ott, E. and Yorke, J. A. 1987b. Critical exponent of chaotic transients in nonlinear dynamical systems. Phys. Rev. Lett. **57**(11): 1284–1287.

[406] Grebogi, C., Ott, E., Romeiras, F. and Yorke, J. A. 1987. Critical exponeats for crisis-induced intermittency. Phy. Rev. A. **36**: 5365–5380.

[407] Grebogi, C., Ott, E. and Yorke. J. A. 1988a. Unstable periodic orbits and the dimension of multihctd chaotic amactors. Phys. Rev. A. **37**: 1711–1724.

[408] Grebogi, C., Hammel, S. and Yorke, J. A. 1988b. Numerical orbits of chaotic processes represent true orbits . Bull. Am. Math. Soc. **19**: 465–469.

[409] Grebogi, C., Hammel, S. M., Yorke, J. A. and Sauer, T. 1990. Shadowing of Physical trajectories in chaotic dynamics containment and refinement. Phys. Rev. Lett. **65**: 1527–1530.

[410] Greene, J. M. 1983. In Long-Time prediction in dynamics. edited by Horton, W., Reichl, L. and Szebehely, V., Wilev. New York.

[411] Grove, E. A, Kent. G, and Ladas. G. 1998. Boundedness and persistance of nonautonomous Lyness and max Equations. J. Diff. Equa. Appl. **3**: 241–258.

[412] Gu, Y. 1987. Most stable manifolds and destruction of tori in dissipative dynamical systems. Phys. Lett. A. **124**: 340–344.

[413] Guckenheimer, J. 1976. A strange attractor. The Hopf bifurcation and its applications. Applied Mathematical Series. **19**: 368–381.

[414] Guckenheimer, J. and Williams, R. F. 1979. Structural Stability of Lorenz Attractors. Publ. Math. IHES. **50:** 307–320.

[415] Guckenheimer, J. and Holmes, P. 1983. Nonlinear Oscillations, Dynamical Systems, and Bifurcations of Vector Fields. New York, Springer Verlag.

[416] Guckenheimer, J. and Holmes, P. 1997. Nonlinear Oscillations, Dynamical Systems, and Bifurcations of Vector Fields. New York, Springer Verlag. Corrected fifth printing.

[417] Gumowski, I. and Mira, C. 1980. Dynamique Chaotique. Toulouse Cepadues Editions.

[418] Hahn, W. 1967. Stability of Motion. Springer-Verlag, Berlin.

[419] Halanay, A. 1966. Differential equations: Stability, Oscillations, Time Lags. Academic Press, New York.

[420] Halsey, T. C., Jensen, M. H., Kadanoff, P. L., Procaccia, I. and Shraiman, B.I. 1986. Fractal measures and their singularities: The characterization of strange sets. Phys. Rev. A. **33:** 1141–1151.

[421] Hammel, S.M, Jones, C. K. R. T. andMoloney, J. V. 1985. Global dynamical behavior of the optical field in a ring cavity. J. Opt. Soc. Am. B. **2:** 552–564.

[422] Hammel, S. M, Yorke, J. A and Grebogi, C. 1987. Do numerical orbits of chaotic dynamical processes represent true orbits?. Complexity. **3:** 136–145.

[423] Hammel, S. M. 1990. A noise reduction method for chaotic systems. Phys. Lett. A. **148:** 421–428.

[424] Han, P. 2007. Perturbed basins of attraction. Mathematische Annalen. **337**(1): 1–13.

[425] Hansen, K.T. and Cvitanovic, P. 1998. Bifurcation structures in maps of Hénon type. Nonlinearity. 11(5): 1233–1261.

[426] Hao, B. L. 1989. Elementary symbolic dynamics. World Scientific, Singapore.

[427] Hao, B. L. 1991. Symbolic dynamics and characterization of complexity. Physica. D. **51:** 161–176.

[428] Hasselblatt, B. 2002. Hyperbolic dynamical systems. Handbook of Dynamical Systems **1A:** 239–319, Elsevier North Holland.

[429] Hasselblatt, B. and Pesin,Y. 2005. Partially Hyperbolic Dynamical Systems. Handbook of Dynamical Systems 1B, 1-55 Elsevier North Holland.

[430] Hasselblatt, B. 2007. Dynamics, ergodic theory and geometry. Mathematical Sciences Research Institute Publications.

[431] Hassouneh, M. A. and Abed, E. H. 2002. Feedback control of border collision bifurcations in piecewise smooth systems. ISR Technical research report, TR. 26.

[432] Hassouneh, M. A., Abed, E. H. and Banerjee, S. 2002. Feedback control of border collision bifurcations in two-dimensional discrete-time systems. ISR Technical research report.

[433] Hata, H., Morita, T., Tomita, K. and Mori, H. 1987. Spectra of singularities for the Lozi and H`enon maps. Prog. Theor. Phys. **78**(4): 721–726.

[434] Hata, H., Horita, T., Mozi, H., Morita, T. and Tomita, K. I. 1988. Characterization of local structures of chaotic attractors in terms of coarsegrained local expension rates. Prog. Theoret. Phys. **80:** 809–826.

[435] Hayashi, S. 1992. Diffeomorphisms in C^1 (M) satisfy Axiom A. Ergod. Th. and Dynam. Sys. **12:** 233–253.

[436] Hayashi, S. 1997. Connecting invariant manifolds and the solution of the C^1 stability and 2126-stability conjectures for flows. Ann. of Math. **145:** 81–137.

[437] Hegger, R., Kantz, H. and Schreiber, T. 1999. Practical implementation of nonlinear time series methods: The TISEAN package. Chaos. **9:** 413–424.

[438] Helleman, R. H. 1980. Fundamental problems in statistical mechanics, (5 North Holland Publ, Amsterdam) 165.

[439] Hempel, J. 2004. 3-Manifolds. Annals of Mathematics Studies. 86 (Paperback) AMS Chelsea Publishing.

[440] Herman, M. 1986. Sur les Courbes Invariantes par les Difféomorphismes de l'Anneau. Astérisque. **144**(2) Soc. Math. de France: Paris , **VIII.2.4.**

[441] Hern´andez, C., Castellanos, J., Gonzalo, R., Palencia, V. 2005. Neural control of chaos and applications. Intern. J. Information Theories & Applications. **12**: 103–109.

[442] Hénon, M. 1969. Numerical study of quadratic area preserving mappings. Q. Appl. Math. **27**: 291–312.

[443] Hénon, M. 1976. A two dimensional mapping with a strange attractor. Commun. Math. Phys. **50**: 69–77.

[444] Hénon, M. 1982. On the numerical computation of Poincaré maps. Physica. D. **5**(2-3): 412–414.

[445] Hirsch, M. and Pugh, C. 1968. Stable manifolds and hyperbolic sets. Proc. Sympos. Pure Math., Vol. XIV, Berkeley, Calif., 133–163 Amer. Math. Soc., Providence, R.I., (1970).

[446] Hirsch, M. 1970. On invariant subsets of hyperbolic sets. Essays on topology and related topics, Mémoires dédiés `a Georges de Rham 126–135.

[447] Hirsch, M. and Pugh, C. 1970. Stable manifolds and hyperbolic sets. Proc. of Symposium in Pure Math., Amer. Math. Soc. **14**: 133–165.

[448] Hirsch, M., Palis, J., Pugh, C. and Shub, M., Neighborhoods of hyperbolic sets. Invent. Math. **9** (1969)/(1970) 121–134.

[449] Hirsh, M. and Smale, S. 1974. Differential Equations, and Linear Algebra. Dynamical Systems. New York, Academic Press.

[450] Hirsch, M. 1976. Differentiable Topology. Graduate Texts in Mathematics. **33** Springer-Verlag.

[451] Hirsh, M., Pigh, C. and Shub, M. 1977. Invariant manifolds. Lecture Notes in Math. Springer-Verlag.

[452] Hirsch, M., Smale, S. and Devaney, R. L. 2004. Differential Equations Dynamical Systems and an Introduction to Chaos, Elsevier.

[453] Hitzl, D. H and Zele, F. 1985. An Exploration of the Hénon quadratic map. Physica. D. **14**: 305–326.

[454] Hochster, M. 2004. Lectures on Jacobian Conjecture. Sci. Math. Research post forwarded by I. Algol. Nov. 11.

[455] Hoensch, U. A. 2008. Some hyperbolicity results for Hénon-like diffeomorphisms. Nonlinearity. **21**: 587–611.

[456] Hofbauer, F. and Keller, G. 1982. Ergodic properties of invariant measures for piecewise monotonic transformations. Math. Z. **180**: 119–140.

[457] Holland, J. H. 1975. Adaptation in Natural and Artificial System. Ann Arbor, MI: Univ. Michigan Press.

[458] Holmes, P. J. and Whitley, D. C. 1984. Bifurcation of one- and two-dimensional maps. Phil. Trans. Roy. Lond. **A311**: 43–102.

[459] Horita, V. and Tahzibi, A. 2006. Partial hyperbolicity for symplectic diffeomorphisms. Ann. I. H. Poincaré – AN. **23**: 641–661.

[460] Hruska, S. L. 2006a. A numerical method for constructing the hyperbolic structure of complex Hénon mappings. Foundations of Computational Mathematics. **6**: 427–455.

[461] Hruska, S. L. 2006b. Rigorous numerical studies of the dynamics of polynomial skew products of $\mathbb{C}^2$. Contemp. Math. **396**: 85–100.

[462] Hsu, G., Ott, E. and Grebogi, C. 1988. Strange saddles and the dimension of their invariant manifolds, Phys. Lett. A. **127**: 199–204.

[463] Hu, H. and Young, L.S. 1995. Nonexistence of SBR measure for some diffeomorphisms that are "almost Anosov". Ergodic Theory Dynamical Systems. **15**: 67–76.

[464] Hubbard, J. H. and Sparrow, C. 1990. The classification of topologically expansive Lorenz maps. Comm. Pure Appl. Math. **XLIII**: 431–443.

[465] Hudson. J. L., Rôssler. Ô. E. 1984. A piecewise-linear invertible noodle map. Physica. D. **11**: 239–242.

[466] Hunt, T. J. 2000. Low dimensional dynamics bifurcations of cantori and realisations of uniform hyperbolicity, Phd thesis, Univ. of Cambridge.

[467] Hunt, T. J. and MacKay, R. S. 2003. Transcritical bifurcation with $O(3)$ symmetry. Nonlinearity. **16**(4): 1499–1473.

[468] Hurewicz, W., Wallman, H. 1984. *Dimension Theory*. Princeton University Press, Princeton, NJ.

[469] Hurley, M. 1982. Attractors: Persistence and density of their basins. Trans. Amer. Math. Sot. **269**: 247–271.

[470] Isaeva, O.V., Jalnine, A.Yu., Kuznetsov, S. P. 2006. Arnold's cat map dynamics in a system of coupled non-autonomous van der Pol oscillators. Phys. Rev. E. **74**: 046207.

[471] Ishii, Y. 1997a. Towards a kneading theory for Lozi mappings. I: A solution of the pruning front conjecture and the first tangency problem. Nonlinearity. **10**: 731–747.

[472] Ishii, Y. 1997b. Towards a kneading theory for Lozi mapping. II: Monotonicity of the topological entropy and Housdorff dimension of attractors. Comm. Math. Phys. **190**: 375–394.

[473] Ishii, Y. and Sands, D. 1998. Monotonicity of the Lozi family near the tentmaps. Comm. Math. Phys. **198**: 397–406.

[474] Ishii, Y. and Sands, D. 2007. Lap number entropy formula for piecewise affine and projective maps in several dimensions. Nonlinearity. **20**: 2755–2007.

[475] Ivanov, A., Liz, E. and TroBmchuk, S. 2002. Halanay inequality, Yorke 3/2 stability criterion, and differential equations with maxima. Tohoku Math. J. **54**: 277–295.

[476] Jackson, E. A. 1991. Perspectives of nonlinear dynamics. Cambridge, **1**.

[477] Jakobson, M. V. 1971. Smooth mappings of the circle into itself. Mat. Sb. **85**(127): 163–188.

[478] Jakobson, M. V. 1981. Absolutely continues invariant measures for oneparameter families of one-dimensional maps. Comm. Math. Phys. **81**: 39–88.

[479] Jakobson, M. V., Newhouse. S. E. 1996. A two-dimensional version of the folklore theorem. Am. Math. Soc. Trans., Ser. 2 **171**: 89–105.

[480] Jakobson, M. V., Newhouse. S. E. 2000. Asymptotic measures for hyperbolic piecewise smooth mappings of a rectangle. Astérisque. **261**: 103–159.

[481] Jafarizadeh, M. A. and Behnia, S. 2002. Hierarchy of Chaotic maps with an invariant measure and their compositions. J. Nonlinear. Math. Phy. **9**(1): 26–41.

[482] Janowski, E. J., Kocic, V. L., Ladas, G, Schultz, S. W. 1995. Global behavior of solutions of $x_{n+1} = \dfrac{\max\{x_n, A\}}{x_{n-1}}$, in: S. Elaydi, J. Greaf, G. Ladas, A. Peterson (Eds.), Proc. of the First ICDEA. 273–282.

[483] Jarvenpaa, E. and Jarvenpaa, M. 2001. On the definition of SRB measures for coupled map lattices. Comm. Math. Phys. **220**: 109–143.

[484] Jiang, M. 2003. SRB measures for lattice dynamical systems. J. Statistical Physics. **111** (3-4): 863–902.

[485] Jing-ling, S., Hua-Wei, Y., Jian-Hua, D. and Hong-Jun, Z. 1996. Riddled basin of laser cooled-Ions in a Paul trap. Chinese Phys. Lett. **13**: 81–84.

[486] Kalinin, B. and Sadovskaya, V. 2003. On local and global rigidity of quasiconformal Anosov diffeomorphisms. J. Inst. Math. Jussieu. **2**(4): 567–582.

[487] Kapitaniak,T., Maistrenko, Y. and Grebogi, C. 2003. Bubbling and riddling of higher-dimensional attractors. Chaos, Solitons and Fractals. **17**(1): 61–66.

[488] Kaplen, J. L. and Yorke. J. A. 1979. Chaotic behavior of multi-dimensional difference equations, Springer. Lect. Notes Math. **730**: 204–227.

[489] Kan, I., Kocak, H. and Yorke, J. A. 1995. Persistent homoclinic tangencies in the Hénon family. Physica. D. **83**(4): 313–325.

[490] Kaplan, J. and Yorke, J. A. 1987. Chaotic behavior of multidimensional difference equations. In Peitgen, H. O. and Walther, H. O., editors, Functional Differential Equations and Approximation of Fixed Points. Springer, New York.

[491] Kathryn, E. L., Lomel, H. E. and Meiss, J. D. 1998. Quadratic volume preserving maps: An extension of a result of Moser. Regular and Chaotic Dynamics. **33**: 122–131.

[492] Katok, A. 1979. Bernoulli Diffeomorphism on Surfaces. Ann. Math. **110**: 529–547.

[493] Katok, A. 1980. Lyapunov exponents, entropy and periodic orbits for diffeomorphisms. Inst. Hautes Etudes Sci. Publ. Math. **51**: 137–173.

[494] Katok, A. and Strelcyn, J. M. 1980. Invariant manifolds for smooth maps with singularities, Part I: Existence. Preprint.

[495] Katok, A., Strelcyn, J., Ledrappier, F., Przytycki, F. 1986. Invariant man-ifolds, entropy and billiards; smooth maps with singularities. Lecture Notes in Mathematics, 1???. Springer-Verlag, Berlin.

[496] Katok, A. and Hasselblatt, B. 1995. Introduction to the Modern Theory of Dynamical Systems, Cambridge University Press.

[497] Kennedy, J. and Eberhart, R. C. 1995. Particle swarm optimization. In: Proceedings of the IEEE International Conference on Neural Networks, Perth, Australia. 1942–1948.

[498] Kennedy, J., Kocak, S. and Yorke, J. A. 2001. A chaos lemma. Amer. Math. Monthly. **108**: 411–423.

[499] Kennedy, J. and York, J. A. 2001. Topological horseshoes. Trans. Amer. Math. Soc. **353**: 2513–2530.

[500] Kennel, M. B. and Isabelle, S. 1992. Method to distinguish possible chaos from colored noise and to determine embedding parameters. Phys. Rev. A. **46**: 3111–3118.

[501] Kevorkian, P. 1993. Snapshots of dynamical evolution of attractors from Chua's oscillator. IEEE Trans. Circuits Syst.-IFund. Th. Appl. 40(10): 762–780.

[502] Keynes, H., B., Sears, M. 1981. Real-expansive flows and topological dimension. Ergodic Theory Dynamical Systems **1**: 179–195.

[503] Khan, A. M., Mar, D. J. and Westervelt, R. M. 1992. Spatial measurements near the instability threshold in ultrapure G_e. Phys. Rev. B. **45**: 8342–8347.

[504] Kifer, Y. 1974. On small random perturbations of some smooth dynamical systems. Math. USSR Izvestija. **8**: 1083–110.

[505] Kifer, Y. 1988. Random perturbations of dynamical systems. Boston, MA: Birkhauser.

[506] Kifer, Y. 1986. General random perturbations of hyperbolic and expanding transformations. Journal D'Analyse Mathematique. **47**: 111–150.

[507] Kirchgraber, U. and Stoffer, D. 2006. Transversal homoclinic points of the Hénon map. Annali di Matematica Pura ed Applicata. **18**(5): 187–204.

[508] Kiriki, S. 2000. The turning orbit is dense in the attractor for almost all Lozi families. New developments in dynamical systems (Japanese) (Kyoto 2000), Sūrikaisekikenkyūsho Kōkyūroku. **1179**: 6–12.

[509] Kiriki, S. 2004. Forward limit sets singularities for the Lozi family. Hokkaido Mathematical Journal. **33**: 491–510.

[510] Kiriki, S. and Soma, T. 2007. Parameter-shifted shadowing property of Lozi maps. Dynamical Systems: An International Journal. **22**(3): 351–363.

[511] Kiriki, S., Li, M. C. and Soma, T. 2008. Coexistence of homoclinic sets with/without SRB measures in Hénon maps. Preprint.

[512] Kiriki, S. 2009. The dense singularity in the Lozi attractors. Preprint.

[513] Klinshpont, N. E., Sataev, E. A. and Plykin, R.V. 2005. Geometrical and dynamical properties of Lorenz type system. Journal of Physics, Conference Series. **23**: 96–104.

[514] Klinshpont, N. E. 2006. On the problem of topological classification of Lorenz-type attractors. Math. Sbornik. **197**(4): 75–122.

[515] Kodama, H., Sato, S. and Honda, K. 1991. Renormalization-group theory on intermittent chaos in relation to its universality. Prog. Theor. Phys. **86**: 309–314.

[516] Koiran, P. 2001. The topological entropy of iterated piecewise affine maps is uncomputable. Discrete Mathematics and Theoretical Computer Science. **4**: 351–356.

[517] Komuro, M., Tokunaga, R., Matsumoto, T., Chua, L. O. and Hotta, A. 1991. Global bifurcations analysis of the double scroll circuit. Inter. J. Bifur. Chaos. **1**: 139–182.

[518] Kotani, S. 1990. Jacobi matrices with random potential taking finitely many values. Reviews in Math. Phys. **1**: 129–133., VIII.2.4.

[519] Kowalczyk, P. 2005. Robust chaos and border-collision bifurcations in noninvertible piecewise-linear maps. Nonlinearity. **18**: 485–504.

[520] Kozlovskii, O. S. 2003. Axiom A maps are dense in the space of unimodal maps in the C^k topology. Ann. of Mathematics. **157**: 1–44.

[521] Krishchenko, A. 1997. Estimations of domain with cycles. Comput. Math. Appl. **34**(2-4): 325–332.

[522] Komuro, M. 1984. Expansive properties of Lorenz attractors. In The theory of dynamical systems and its applications to nonlinear problems, 4–26. World Sci. Publishing, Kyoto.

[523] Kuang, Y. 1993. *Delay differential equations with applications in population dynamics*, Academic Press, New York, Section 4.5.

[524] Kubo, G. T., Viana, R. L., Lopes, S. R. and Grebogi, C. 2008. Crisis-induced unstable dimension variability in a dynamical system. Phys. Lett. A. **372**: 5569–5574.

[525] Kulenovic, M., Ladas, G. 2002. Dynamics of second order rational difference equations, Chapman & Hall/CRC, p. 50.

[526] Kupka, I. 1963 & 1964. Contribution à la théorie des champs génériques, Contributions to differential equations. **2**: 457–484 and **3**: 411–420.

[527] Kuptsov, P. V., Kuznetsov, S. P. and Sataev, I. R. 2008. Hyperbolic attractor of Smale-Williams type in a system of two coupled non-autonomous amplitude equations. Preprint.

[528] Kuramitsu, M. 1995. A classfication of the 3rd order oscillators with respect to chaos. Procs of (1995) internatiojal symposium on Nonlinear theory and its applications. **1**: 599–602.

[529] Kurshan, R. and Gobinath, B. 1974. Recursively generated periodic sequences. Canad. J. Math. **XXVI**(6): 1356–1371.

[530] Kuznetsov, S. P. 2001. Dynamical Chaos. Fizmatlit, Moscow, (in Russian).

[531] Kuznetsov, Y. A. 2004. Elements of Applied Bifurcation Theory. Springer, 3rd edition.

[532] Kuznetsov, S. P. 2005. Example of a Physical System with a hyperbolic attractor of the Smale-Williams type. Phys. Rev. Lett. **95**: 144101.

[533] Kuznetsov, S. and Seleznev, E. 2006. A strange attractor of the Smale-Williams type in the chaotic dynamics of a Physical system. J. Exper. Theor. *Physics*. **102**(2): 355–364.

[534] Kuznetsov, S. P. and Sataev, I. R. 2007. Hyperbolic attractor in a system of coupled non-autonomous van der Pol oscillators Numerical test for expanding and contracting cones. Phys. Lett. A. **365**: 97–104.

[535] Kuznetsov, S. P. and Pikovsky, A. 2007. Autonomous coupled oscillators with hyperbolic strange attractors. Physica. D. **232**: 87–102.

[536] Kuznetsov, S. P. 2008. On the Feasibility of a parametric generator of hyperbolic chaos. J. Experimental and Theorectical Physics. **106**(2): 380–387.

[537] Kuznetsov, S. P. and Pikovsky, A. 2008. Hyperbolic chaos in the phase dynamics of a Q-switched oscillator with delayed nonlinear feedbacks. Eur. Phys. Lett. **84**: 10013.

[538] Kuznetsov, S. P. and Ponomarenko, V. I. 2008. Realization of a strange attractor of the Smale–Williams type in a radiotechnical delay-feedback oscillator. Technical Physics Letters. **34**(9): 771–773.

[539] Labarca, R. and Pacifico, M. J. 1986. Stability of singular horsshoes. Toplogy. **25**(3): 337–352.

[540] Labarca, R. and Moreira, C. G. 2001. Bifurcations of the essential dynamics of Lorenz maps and the application to Lorenz like flowscontributions to the study of the expanding case. Bol. Soc. Bras. Mat. (N.S.) **32**(2): 107–144.

[541] Labarca, R. and Moreira, C. G. 2001. Bifurcations of the essential dynamics of Lorenz maps and the application to Lorenz like flows: contributions to the study of the expanding case. Bulletin of the Brazilian Mathematical Society. **32**(2): 107–144.

[542] Labarca, R. and Moreira, C. G. 2003. Bifurcations of the essential dynamics of Lorenz maps and the application to Lorenz like flows: contributions to the study of the contracting case. Preprint.

[543] Labarca, R. and Moreira, C. G. 2006. Essential dynamics for Lorenz maps on the real line and the Lexicographical World. Ann. I. H. Poincaré –AN. **23**: 683–694.

[544] Lai. Y. 1985a. Bowen-Ruelle measures for certain piecewise hyperbolic maps. Transactions of the AmericanMathematical Society. **287**(1).

[545] Lai, Y., Grebogi, C. and Yorke, J. 1993. How often are chaotic saddles nonhyperbolic?. Nonlinearity. **6**: 779–797.

[546] Lai, Y. C., Grebogi, C., Yorke, J. A. and Venkataramani, S. C. 1996. Riddling bifurcation in chaotic dynamical systems. Phys. Rev. Lett. **77**(1): 55–58.

[547] Lagarias, J. C and Rains, E. 2005a. Dynamics of a family of piecewise-linear area-preserving plane maps I. Rational rotation numbers. J. Difference Eqns. Appl. **11**(12): 1089–1108.

[548] Lagarias, J. C, Rains, E. 2005. Dynamics of a family of piecewise-linear area-preserving plane maps II. J. Difference Equ. Appl. **11**(13): 1137–1163.

[549] Lagarias, J. C. and Eric, R. 2005c. Dynamics of a family of piecewise-linear areapreserving plane maps III. Cantor Set Spectra. J. Difference Eqns. Appl. **11**(14) (2005(c)): 1205–1224.

[550] Lakdawala, P. 1996. Computational complexity of symbolic dynamics at the onset of chaos. Phys. Rev. E. **53**: 4477–4485.

[551] Lasota, A. and Yorke. J. A. 1973. On the existence of invariant measures for piecewise monotonic transformations. Trans. Amer. Math. Soc. **186**: 481–488.

[552] Ledrappier, F and Strelcyn, J. M. 1982. A proof of the estimation from below in Pesin entropy formula. J. Ergodic Theory and Dynam. Syst. **2**: 203–219.

[553] Ledrappier, F. 1983. Quelques proprietes des exposants caracteristiques. Ecole d'ete de probabilites de St. Flour 1982. In: Lecture Notes in Mathematics. Berlin, Heidelberg, New York: Springer.

[554] Ledrappier, F., Shub, M., Simo, C. and Wilkinson, A. 2003. Random versus deterministic exponents in a rich family of diffeomorphisms. Journal of Statistical Physics. **113**: 85–149.

[555] Lefschetz, S. 1926. Intersections and transformations of complexes and manifolds. Trans. Amer. Math. Soc. **28**: 1–49.

[556] Lehto, O. and Virtanen, K. I. 1973. Quasiconformal Mappings in the Plane. Springer-Verlag, New York.

[557] Levy. Y. 2007. Ergodic properties of the Lozi map, in Stochastic aspects of classical and quantum systems. 103–116.

[558] Li, T. Y. and Yorke, J. A. 1975. Periodic three implies chaos. Am. Math. Monthly. **82**: 985–989.

[559] Li, C. and Chen, G. 2004. Estimating the Lyapunov exponents of discrete systems. Chaos. **14**(2): 343–346.

[560] Lian, K. Y., Chiang, T. S., Chiu, C. S. and Liu, P. 2001. Synthesis of fuzzy model-based designs to synchronization and secure communications for chaotic systems. IEEE. Trans. Syst. Man. Cybern. B. Cybern. **31**(1): 66–83.

[561] Liao, S. T. 1980. On the stability conjecture. Chinese Ann. of Math. **1**: 9–30.

[562] Liao, S. T. 1983. Hyperbolicity properties of the non-wandering sets of certain 3-dimensional systems. Acta Math. Sci. **3**: 361–368.

[563] Lichtenberg, A. and Lieberman, M. 1983. Regular and Stochastic Motion. Springer-Verlag.

[564] Liu, Z. and Cao, Y. 1991. Discussion on the geometric of strange attractor. Chinese Physics Letter. **8**(10): 503–506.

[565] Liu, Z., Zhou, Z., Xie, H. and Lu, Q. 1992a. The strange attractor of the Lozi mapping. Int. J. Bifurcations and Chaos. **2**: 831–839.

[566] Liu, Z., Qin, W. and Xie, H. 1992b. The structure of the Lauwerier attractor and the dynamical behavior on it. Kexue Tongbao. **37**(14): 1269.

[567] Liu, Z., Qin,W., Xie, H. and Cao, Y. 1993. The structure of strange attractor of a kind of two dimensionel map and dynamical properties on it. Sci. China. **23**: 702-.

[568] Liverani, C. 1995. Decay of correlations. Annals Math. **142**: 239–301.

[569] Livsic, A. N. 1972. The homology of dynamical systems. Uspehi Mat. Nauk. **273**(165): 203–204.

[570] Liz. E. and TroBmchuk, S. 2000. Existence and stability of almost periodic solutions for quasilinear delay systems and the Halanay inequality, J. Math. Anal. Appl. **248**: 625–644.

[571] Liz. E. and Ferreiro, J. B. 2002. A note on the global stability of generalized difference equations, Appl. Math. Lett. **15**: 655–659.

[572] Liz, E., Ivanov, A. and Ferreiro, J. B. 2003. Discrete Halanay-type inequalities and applications. Nonlinear Analysis. **55**: 669–678.

[573] Lizana, C. and Mora, L. 2008. Lower bounds for the Hausdorff dimension of the geometric Lorenz attractor: the homoclinic case. Discrete and Continuous Dynamical Systems. **22** (3).

[574] Lohner, R. 1992. Computation of guaranteed enclosures for the solutions of ordinary initial and boundary value problems. *Computational ordinary differential equations.*, Cash, J. R. and Gladwell, I. (eds). Clarendon Press, Oxford.

[575] Lorenz, E. N. 1963. Deterministic Non-periodic Flow. J. Atmos. Sci. **20**: 130–141.

[576] Lorenz, E. N. 2008. Compound windows of the Hénon-map. Physica. D. **237**: 1689–1704.

[577] Loskutov, A. Y. 1993. Dynamics control of chaotic systems by parametric destochastization. J. Phys. A: Math. Gen. **26**: 4581–4594.

[578] Lozi, R. 1978. Un attracteur étrange (?) du type attracteur de Hénon. J. Phys. (Paris) **39** Colloq. C5: 9–10.

[579] Lozi, R. 1982. Dimensional bifurcation between thread and sheet strange attractors. Colloque International du C.N.R.S., n°332, Editions du C.N.R.S., pp. 145–152.

[580] Lozi, R. 2002. The importance of strange attractors for industrial mathematics, Trends in Industrial and Applied Mathematics, Proceedings of the 1st International Conference on Industrial and Applied Mathematics of the Indian Subcontinent, Siddiqi, A. H. and Ko˘cvara, M. (eds.), Kluwer Academic Publishers. 275–303.

[581] Lu, Y-Y., Xue, L-P., Zhu,M-C. and Qiu, S-S. 2003. Frequency band estimate and change for chaos systems. J. Shenzhen. University. Sc. Eng. **20**(2): 35–41.

[582] Luchinski, D. G. and Khovanov, I. A. 1999. Fluctuation-induced escape from the basin of attraction of a quasiattractor. JETP Letters. **69**(11): 825–830.

[583] Luzzatto, S. and Viana, M. 2000. Positive Lyapunouv exponents for Lorenzlike families with criticalities. Astérisque. **261**: 201–237.

[584] Luzzatto, S., Melbourne, I. and Paccaut, F. 2005. The Lorenz attractor is mixing. Comm. Math. Phys., **260**(2): 393–401.

[585] Lyubich, M. 1994. Combinatorics, geometry and attractors of quasiquadratic maps. Ann. of Math. **140**: 347–404.

[586] Lyubich, M. 1997. Dynamics of quadratic polynomials I-II. Acta Math. 178: 185–297.

[587] Lyubich, M. 2000. Dynamics of quadratic polynomials, III. Parapuzzle and SBR measure. Astérisque. **261**: 173–200.

[588] Lyubich, M. 2002. Almost every real quadratic map is either regular or stochastic. Ann. Math. **156**: 1–78.

[589] Maistrenko, Y., Maistrenko, V. L. and Popovich, S. I. 1998. On unimodalbimodalbifurcation in a family of piecewise linear maps, **NDES '98**. Proc. 6th Inter. Specialist Workshop on Nonlinear Dynamics of Electronic Systems. Tech. Univ. Budapest. 329–332.

[590] Maistrenko, Y. L., Maistrenko,V. L., Popovich, A. and Mosekilde, E. 1998. Transverse instability and riddled basins in a system of two coupled logistic maps. Phys. Rev. E. **57**: 2713–2724.

[591] Mallet-Paret, J. and Yorke, J. A. 1982. Snakes: oriented families of periodic orbits, their sources, sinks, and continuation. J. Differential Equations. **43**(3): 419–450.

[592] Malykhin, V. I. 2001. Connected space. In Hazewinkel, Michiel, Encyclopaedia of Mathematics, Kluwer Academic Publishers.

[593] MacKay, R. S. and van Zeijts, J. B. J. 1988. Period doubling for bimodal maps: A horseshoe for a renormalization operator. Nonlinearity. **1**: 253–277.

[594] MacKay, R. S. and Meiss, J. D. 1992. Cantori for symplectic maps near the anti-integrable limit. Nonlinearity. **5**: 49–160.

[595] Maggio, G. M., di Bernardo, M. and Kennedy, M. P. 2000. Nonsmooth bifurcations in a piecewise-linear model of the colpitts oscillator. IEEE Trans. Circuits & Systems. **I 8**: 1160–1177.

[596] Mané, R. 1978. Contributions to the C^1-stability conjecture. Topology. 17: 386–396.

[597] Mané, R. 1979. Expansive homeomorphisms and topological dimension. Trans. Amer. Math. Soc. **252**: 313–319.

[598] Mañé, R. 1982. An ergodic closing lemma. Ann. of Math. **116**: 503–540.

[599] Mané, R. 1984. Oseledec's Theorem from the generic viewpoint. Proceedings of the International Congress of Mathematicians, **12** (Warsaw, (1983)) 1269–1276.

[600] Mané, R., Sad, P. and Sullivan, D. 1983. On the dynamics of rational maps. Ann. Sci. Ecole Norm. Sup. **16**: 193–217.

[601] Mané, R. 1985. Hyperbolicity, sinks and measure in one dimensional dynamics. Commun. Math. Phys. **100**: 495–524.

[602] Mané, R. 1985. Hyperbolicity, sinks and measure in one-dimensional dynamics. Comm. Math. Phys. **100**(4): 495–524.

[603] Mané, R. 1987. Ergodic theory and differentiable dynamics. Springer Verlag.

[604] Mañé, R. 1988. A proof of the C1 stability conjecture. Publ. Math. IHES. **66**: 161–210.

[605] Mané, R. 1996. The Lyapunov exponents of generic area preserving diffeomorphisms. International Conference on Dynamical Systems (Montevideo (1995)) Pitman Res. Notes Math. Ser., Longman, Harlow. **362**: 110–119.

[606] Mané, R. and Res, P. 1996. Notes Math. Ser. In International Conference on Dynamical Systems (Montevideo (1995)) volume 362 chapter: The Lyapunov exponents of generic area preserving diffeomorphisms. Longman, Harlow pp. 110–119.

[607] Manganaro, G. and Pineda de Gyvez, J. 1997. DNA computing based on chaos. In Proc. 1997 IEEE International Conference on Evolutionary Computation. Piscataway, NJ: IEEE Press. 255–260.

[608] Manning, A. 1974. There are no new Anosov diffeomorphisms on tori. Amer. J. Math. **96**: 422–429.

[609] Markarian, R. 2004. Billiards with polynomial decay of correlations. Ergod. Th. Dynam. Syst. **24**: 177–197.

[610] Marotto, F. R. 1978. Snap-back repellers imply chaos in $\mathbb{R}^n$. J. Math. Anal. Appl. **3**: 199–223.

[611] Marotto, F. R. 1979a. Perturbation of stable and chaotic difference equation. J.Math. Anal. Appl. **72**(2): 716–729.

[612] Marotto, F. R. 1979b. Chaotic behavior in the Hénon mapping. Commun. Math. Phys. **68**: 187–194.

[613] Martens, M. 1990. Interval dynamics. Ph.D. thesis, Delft.

[614] Martınez, V. J, Domınguez-Tenreiro, R. and Roy, L. J. 1993. Hausdorff dimension from the minimal spanning tree. Phys. Rev. E. **47**(1): 735–738.

[615] Marsden, J. and McCracken, M. 1976. The Hopf Bifurcation and its Applications. Appl. Math. Sciences. **19** Springer-Verlag.

[616] Mautner, F. I. 1957. Geodesic flows on symmetric Riemann spaces. Ann. of Math. **65**: 416–431.

[617] May, R 1974. Biological populations with nonoverlapping generations: Stable points, Stable cycles, and Chaos. Science. **186**: 645–647.

[618] Mazure, M. 2008. On some useful conditions for hyperbolicity. 2008 International Workshop on Dynamical Systems and Related Topics. Trends in Mathematics - New Series. **10**(2): 57–64.

[619] Mazur, M., Tabor, J. and Koscielniak, P. 2008. Semi-hyperbolicity and hyperbolicity. Discrete Contin. Dynam. Syst. **20**: 1029–1038.

[620] Mazur, M. and Tabor, J. 2009. Computational hyperbolicity. Preprint.

[621] McDonald, S.W., Grebogi, C., Ott, E., and Yorke, J.A. 1985. Fractal Basin Boundaries. Physica. D. **17**: 125–135.

[622] Mcdonough, P., Noonan, J. P. and Hall, G. R. 1995. A new chaos detector, Computers & Electrical Engineering. **21**(6): 417–431.

[623] Medvedev, V. and Zhuzhoma, E. 2004. There are no structurally stable diffeomorphisms of odd-dimensional manifolds with codimension one nonorientable expanding attractors. arXiv math. DS/0404416.

[624] Meiss, J. D. 1997. Average exit times in volume preserving maps. Chaos. **7**: 139–147.

[625] Mestel. B. 2003. On globally periodic solutions of the difference equation $x_{n+1} = \dfrac{f(x_n)}{x_{n-1}}$. J. Differ. Equations Appl. 9(2): 201–209.

[626] Michelitsch, M. and Rössler, O. E. 1998. A new feature in Hénon's map. Comput. & Graphics. **13** (1989) 263–275. Reprinted in Chaos and Fractals, A Computer Graphical Journey: Ten Year Compilation of Advanced Research (Ed. Pickover, C. A). Amsterdam, Netherlands Elsevier. 69–71.

[627] Miller, D. A. and Grassi, G. 2001. A discrete generalized hyperchaotic Hénon map circuit. Circuits and Systems. MWSCAS (2001). Proc.s of the 44th IEEE (2001) Midwest Symposium on. 1: 328–331.

[628] Milnor, J. 1965. Topology from the differentiable viewpoint. Based on notes by David W. Weaver. The University Press of Virginia, Charlottesville, Va.

[629] Milnor, J. and Thurston, R. 1977. On iterated maps of the interval I ans II. Unpublished notes. Prinseton University Press, Princeton.

[630] Milnor, J. 1985a. On the Concept of Attractor. Commun. Math. Phys. **99:** 177–195.

[631] Milnor, J. 1985b. On the Concept of Attractor: Correction and Remarks. Commun. Math. Phys. **102:** 517–519.

[632] MingQing, X. 2006. A direct method for the construction of nonlinear discrete-time observer with linearizable error dynamics, IEEE Trans. Automat. Control. **51**(1): 128–135.

[633] Mischaikow, K. 2002. Topological techniques for efficient rigorous computations in dynamics. Acta Numerica. **11:** 435–477.

[634] Misiurewicz, M. and Szewc, B. 1980. Existence of a homoclinic point for the Hénon map. Comm. Math. Phys. **75**(3): 285–291.

[635] Misiurewicz, M. 1980. Strange attractor for the Lozi mapping. Ann.N.Y. Acad. Sci. **357:** 348–358.

[636] Mira, C. 1997. Chua's circuit and the qualitative theory of dynamical systems. Inter. J. Bifur. Chaos. 7(9): 1911–1916.

[637] Mischaikow, K. and Mrozek, M. 1995. Chaos in the Lorenz equations: A Computer-assisted proof. Bull. Amer. Math. Soc. **32:** 66–72.

[638] Mischaikow, K. and Mrozek, M. 1998. Chaos in the Lorenz equations: A computer assisted proof, Part II, Detail. Math. Comp. **67**(223): 1023–1046.

[639] Moerdijk, I. and Mrčun, J. 2003. Introduction to Foliations and Lie groupoids. Cambridge University Press.

[640] Mohamad, S. and Gopalsamy, K. 2000. Continuous and discrete Halanaytype inequalities. Bull. Aust. Math. Soc. **61:** 371–385.

[641] Moore, C. C. 1966. Ergodicity of flows on homogeneous spaces. Amer. J. Math. **88:** 154–178.

[642] Moore, R. E. 1966. Interval Analysis. Prentice Hall, Englewood Cliffs, NJ.

[643] Moore, R. E. 1979. Methods and applications of interval analysis. SIAM, Philadelphia, PA.

[644] Mosekilde, E., Zhusubaliyev, Z. T., Rudakov, V. N. and Soukhterin, E. A. 2000. Bifurcation analysis of the Hénon map. Discrete Dynamics in Nature and Society. **53:** 203–221.

[645] Morita, T., Hata, H., Mori, H., Horita, T. and Tomita, K. 1987. On partial dimensions and spectra of singularities of strange attractors. Prog. Theor. Phys. **78:** 511–515.

[646] Morita, T., Hata, H., Mori, H., Horita, T. and Tomita, K. 1988. Spatial and temporal scaling properties of strange attractors and their representations by unstable periodic orbits. Prog. Theoret. Phys. **78**: 296–312.

[647] Morosawa, S., Nishimura, Y., Taniguchi, M. and Ueda T. 2000. Holomorphic Dynamics. Cambridge University Press.

[648] Moser, J. 1960. On the integrability of area preserving Cremona mappings near an elliptic fixed point. Bol. Soc. Mat. Mexicana. **2**(5): 176–180.

[649] Moser, J. K. 1962. On invariant curves of area-preserving mappings of an annulus. Nachr. Akad.Wiss. Göttingen II. Math. Phys. Kl. 1–20.

[650] Moser, J. 1969. On a theorem of Anosov. Differential Equations. **5**: 411–440.

[651] Moser, J. 1973. Stable and random motions in dynamical systems. Annals of Math. Studies. Princeton University Press.

[652] Moser, J. K. 1994. On quadratic symplectic mappings. Math.Zeitschrift. 216–417.

[653] Mora, L. and Vianna, M. 1993. Abundance of strange attractors. Acta. Math. **17**(1): 1–71.

[654] Morales, C. and Pujals, E. 1997. Singular strange attractors on the boundary of Morse-Smale systems. Ann. Sci. école Norm. Sup. **30**: 693–717.

[655] Morales, C., M. Pacifico, J. and Pujals, E. 1998. On C1 robust singular transitive sets for three-dimensional flows. C. R. Acad. Sci. Paris, Série I. **326**: 81–86.

[656] Morales, C., Pacifico, M. J. and Pujals, E. R. 1999. Singular hyperbolic systems. Proc. Amer. Math. Soc. **127**(11): 3393–3401.

[657] Morales, C. A., Pacifico, M. J., Attractors and singularities robustly accumulated by periodic orbits. International Conference on Differential Equations. 1, 2 (Berlin, (1999)) World Sci. Publishing, 64–67.

[658] Morales, C., Pacifico, M. J. and Pujals, E. 2000. Strange attractors across the boundary of hyperbolic systems. Comm.Math. Phys. **211**(3): 527–558.

[659] Morales, C. and Pacifico, M. J. 2001. Mixing attractors for 3-flows. Nonlinearity. **14**: 359–378.

[660] Morales, C. A. 2003. Singular-hyperbolic sets and topological dimension. Dynamical Systems. **18**(2): 181–189.

[661] Morales, C. and Pacifico, M. J. 2003a. A dichotomy for three-dimensional vector fields. Ergodic Theory Dynam. Systems **23**: 1575–1600.

[662] Morales, C. and Pacifico, M. J. 2003b. Transitivity and homoclinic classes for singular-hyperbolic systems. Preprint Série A 208/2003.

[663] Morales, C. 2004. The explosion of singular hyperbolic attractors. Ergod. Th. Dynam. Syst. **24**(2): 577–592.

[664] Morales, C. A., Pacifico, M. J. and Pujalls, E. R. 2004. Robust transitive singular sets for 3-flows are partially hyperbolic attractors or repellers. Annals of mathematics. **160**(2): 375–432.

[665] Morales, C. A. and Pacifico, M. J. 2004. Sufficient conditions for robustness of attractors. Pacific. J. Mathematics. **216**(2): 327–342.

[666] Morales, C. 2004. A note on periodic orbits for singular-hyperbolic flows. Discrete Contin. Dyn. Syst. **11**(2-3): 615–619.

[667] Morales, C. A., Pacifico, M. J. and San Martin, B. 2005. Expanding Lorenz attractors through resonant double homoclinic loops. SIAM. J Math. Anal. **36**(6): 1836–1861.

[668] Morales, C. 2006. Poincaré–Hopf index and singular-hyperbolic sets on 3-balls. Preprint.

[669] Morales, C. A., Pacifico, M. J. P. and San Martin, B. 2006. Contracting Lorenz attractors through resonant double homoclinic loops. SIAM. J. Mathematical Analysis. 38: 309–332.

[670] Morales, C. 2007. Singular-hyperbolic attractors with handlebody basins. J. Dynamical and Control Systems. **13**(1): 15–24.

[671] Morales, C. 2008. Topological dimension of singular-hyperbolic attractors. Preprint published at IMPA.

[672] Morales, C. 2008. Poincaré-Hopf index and partial hyperbolicity. Ann. Fac. Sci. Toulouse Math. **XVII**(1): 193–206.

[673] Morse, M. 1921. A One-to-One representation of Geodesics on a surface of negative curvature. Amer. J. Math. **43**(1): 33–51.

[674] Murakami, C., Murakami, W. and Hirose, K. 2002. Sequence of global period doubling bifurcation in the Hénon maps. Chaos, Solitons & Fractals. **14**(1): 1–17.

[675] Myrberg, P. 1962. Sur l'itération des polynomes réels quadratiques. J. Math. Pures. Appl. **9**(41): 339–351.

[676] Nastaran, V. and Johari, M. V. 2006. Adaptative fuzzy synchronization of discrete-time chaotic systems, Chaos Solitons Faractals. **28**(4): 1029–1036.

[677] Naudot, V. 1996. Strange attractor in the unfolding of an inclination-flip homoclinic orbit. Ergodic Theory, Dynamical systems. **16** (5): 1071–1086.

[678] Neimark, Y. and Landa, P. 1989. Stochastic and Chaotic Oscillations. Nauka, Moscow.

[679] Nemytskii, V. V. and Stepanov, V. V. 1960. Qualitative theory of differential equations. Princeton University Press, Princeton, NJ.

[680] Nepomuceno, E. G., Takahashi, R. H. C., Amaral, G. F. V. and Aguirre, L. A. 2003. Nonlinear identification using prior knowledge of fixed points: A multiobjective approach. Inter. J. Bifur.Chaos. **13**(5): 1229–1246.

[681] Neunh¨auserer, J. 2000. A Douady-Osterlé type estimate for the Hausdorff dimension of invariant sets of piecewise smooth maps. Conference "Differential Equations and Application", Russia 2000, University Dresden.

[682] Neumaier, A. 1990. Interval methods for systems of equations. Cambrigde University Press.

[683] Newcomb, R. W. and Sathyan, S. 1993. An RC op amp chaos generator. IEEE trans, Circuits & Systems. **CAS-30**: 54–56.

[684] Newhouse, S. 1970a. On codimension one Anosov diffeomorphisms. Amer. J. Math. **92**: 761–770.

[685] Newhouse, S. 1970b. Non-density of Axiom A(a) on S². Proc. A. M. S. Symp. Pure. Math. **14**: 191–202, 335–347.

[686] Newhouse, S. and Palis, J. 1973. Bifurcations of Morse-Smale dynamical systems. In M. M. Peixoto, editor, Dynamical SystemsProc. Symp. Bahia, Brazil, July 26-Aug. **14** (1971) 303–366. Academic Press.

[687] Newhouse, S. E. 1972a. Hyperbolic limit sets. Trans. Amer. Math. Soc. **167**: 125–150.

[688] Newhouse, S. E. 1972b. The abundence of wild hyperbolic sets and non-smooth stable sets for diffeomorphisms. Publ. Math. IHES. **50**: 101–151.

[689] Newhouse, S. 1974. Diffeomorphisms with infinitely many sinks. Topology. **13**: 9–18.

[690] Newhouse, S. 1975. On simple arcs between structurally stable flows. In Dynamical systems—Warwick (1974) (Proc. Sympos. Appl. Topology and Dynamical Systems, Univ. Warwick, Coventry, (1973)/(1974))Lect. Notes Math. **468**: 209–233. Springer-Verlag, Berlin.

[691] Newhouse, S. 1977. Quasi-elliptic periodic points in conservative dynamical systems. Amer. J. Math. **99**(5): 1061–1087.

[692] Newhouse, S., Ruelle, D. and Takens, F. 1978. Occurrence of strange axiom A attractors near quasi periodic flows on $\mathbb{T}^m$, $m \geq 3$. Comm. Math. Phys. **64**: 35–40.

[693] Newhouse, S. 1979. The abundance of wild hyperbolic sell and non-smooth stable sets for diffeomorphisms. Publ. Mnrh. IHES. **50**: 101–51.

[694] Newhouse, S. 1980. Asymptotic behavior and homoclinic points in nonlinear systems. Ann. of N.Y. Acad. Sci. **357**: 292–299.

[695] Newhouse, S. 2004a. New Phenomena associated with homoclinic tangencies. Ergodic Theory, Dynamical systems. **24**(5): 1725–1738.

[696] Newhouse, S. 2004b. Cone-fields, domination, and hyperbolicity, in Modern dynamical systems and. applications, 419–432, Cambridge University press.

[697] Newhouse, S., Ruelle, D. and Takens, F. 1978. Occurrence of strange axiom A attractors near quasi periodic flows on $\mathbb{T}^m$, $m \geq 3$. Comm. Math. Phys. **64**: 35–40.

[698] Newhouse, S., Berz, M., Grote, J., Makino, K. 2008. On the Estimation of Topological Entropy on Surfaces, Contemporary Mathematics. **469**: 243–270.

[699] Nikodym, O. 1930. Sur une généralisation des intégrales de M. J. Radon (in French). Fundamenta Mathematicae. **15**: 131–179.

[700] Nikolaev, I. 2001. Foliations on Surfaces. Ergebnisse der Mathematik und ihrer Grenzgebiete. 3. Folge/A Series of Modern Surveys in Mathematics.

[701] Nossek, J. A. 1995. Experimental verification of horseshoes from electronic circuits. Ph. Trans.Royal Soc.London. **353** (1701): 59–64.

[702] Novikov, S. P. 1965. The topology of foliations. Trudy Moskov.Mat. Obshch. **14** (1965) 248–278, English transl. Trans.MoscowMath. Soc. **14**: 268–304.

[703] Nozawa, H. 1992. A neural network model as globally coupled map and application based on chaos. Chaos. **2**: 377–386.

[704] Nusse, H. E. and Yorke, J. A. 1988. Is every approximate trajectory of some process near an exact trajectory of a near process?. Communications in Mathematical Physics. **114**: 363–379.

[705] Nusse, H. E. and Yorke, J. A. 1989. A procedure for finding numerical trajectories on chaotic saddles. Physica. D. **36**: 137–56.

[706] Nusse, H. E. and Tedeschini-Lalli, L. 1992. Wild Hyperbolic Sets, Yet no chance for the coexistence of infinitely many KLUS-simple newhouse attracting sets. Commun. Math. Phys. **144**: 429–442.

[707] Nusse, H. E. and Yorke, J. A. (First Edition (1994)). Dynamics: Numerical Explorations, Applied Mathematical Sciences 101, Springer-Verlag, New York, Second Edition (1997).

[708] Nusse, H. E. and Yorke, J. A. 1995. Border-collision bifurcations for piecewise smooth one-dimensional maps. Inter. J. Bifur. Chaos. **5**: 189–207.

[709] Nusse, H. E. and Yorke, J. A. 1996. Basins of Attraction. Science. **27**(1): 1376–1380.

[710] Nunez, P. L. 1981. Electric Fields of the Brain. Oxford University Press, New York.

[711] Nunez, P. L. 2000. Toward a quantitative description of large-scale neocortical dynamic function and EEG. Behav. Brain Sci. **23**(3): 371–437.

[712] Ohnishi, M. and Inaba, N. 1994. A singular bifurcation into instant chaos in piecewise-linear circuit. IEEE Transactions on Circuits and Systems I Communications and Computer Sciences. **41**(6): 433–442.

[713] Ott, E. 1993. Chaos in Dynamical Systems. Cambridge Univ. Press, Cambridge.

[714] Ottino, J. M. 1989. The kinematics of mixingstretching, chaos, and transport. Cambridge Cambridge University Press.

[715] Ottino, J. M., Muzzion, F. J., Tjahjadi, M., Franjione, J. G., Jana, S. C. and Kusch, H. A. 1992. Chaos, symmetry, and self-similarity, exploring order and disorder in mixing processes. Science. **257**: 754–760.

[716] Ortega, R. and Anchez, L. A. S. 2000. Abstract competitive systems and orbital stability in $\mathbb{R}^3$. Proc. Amer. Math. Soc. **128**: 2911–2919.

[717] Ovsyannikov, I.M. and Shilnikov, L. P. 1987. On systems with a saddle-focus homoclinic curve, Mat. Sbornik, **58** (1986) 557–574; English translation in Math. USSR Sb. **58**: 557–574.

[718] Ovsyannikov, I. M. and Shilnikov, L. P. 1992. Systems with a homoclinic curve of multidimensional saddle-focus type, and spiral chaos Mat. Sb. 182 (1991)1043–1073; English translation in Math. USSR Sb. **73**: 415–443.

[719] Paar, V. and Pavin, N. 1998. Intermingled fractal arnold tongues. Phy. Rev. E. **57**(2): 1544–1549.

[720] Pacifico, M. J., Pujals, E. R. and Viana, M. 2002. Sensitiveness and SRB measure for singular hyperbolic attractors. Preprint.

[721] Palacios, A. 2002. Cycling chaos in one-dimensional coupled iterated maps. Inter. J. Bifur.Chaos. **12**(8): 1859–1868.

[722] Palis, J. 1969. On Morse-Smale dynamical systems. Topology. **8**: 385–405.

[723] Palis, J. 1968. On the structure of hyperbolic points in Banach spaces. Anais. Acad. Bras. Ciecias. **40**.

[724] Palis, J. and Smale, S. 1970. Structural stability theorems, in *Global Analysis*, Berkeley (1968) in Proc. Sympos. Pure Math., vol. **XIV**, Amer. Math. Soc. 223–232.

[725] Palis, J. and Takens, F. 1987. Hyperbolicity and the creation of homoclinic orbits. Annals of Mathematics. **125**: 337–374.

[726] Palis, J. and Takens, F. 1993. Hyperbolicity and sensitive chaotic dynamic at homoclinic bifurcation. Cambridge University Press.

[727] Palis, J. and Viana, M. 1994. High dimension diffeomorphisms displaying infinitely sinks. Ann. Math. **140**: 1–71.

[728] Palis, J. 2000. A global view of dynamics and a conjecture on the denseness of finitude of attractors. Asterisque. **261**: 339–351.

[729] Palis, J. 2005. A global perspective for non-conservative dynamics. Ann. I. H. Poincaré -AN. **22**: 485–507.

[730] Palis, J. 2008. Open questions leading to a global perspective in dynamics. Nonlinearity. **21**: T37–T43.

[731] Palmer, K. J. 1988. Exponential dichotomies, the shadowing lemma and transversal homoclinic points. In: U. Kirchgraber and H. O. Walther, editors, *Dynamics Reported*, 1. Wiley and Teubner.

[732] Papaschinopoulos, G. and Schinas, C. J. 1999. Invariant boundedness and persistence of nonautonomous difference equations of rational form. Comm. Appl. Nonlinear Anal. **6**: 71–88.

[733] Papaschinopoulos, G. and Hatzifilippidis, V. 2001. On a max difference equation. Journal of Mathematical Analysis and Applications. **258**: 258–268.

[734] Panti, G. 2008. Multidimensional continued fractions and a Minkowski function. Monatsh. Math. **154**: 247–264.

[735] Paradıs, J., Viader, P. and Bibiloni, L. 2001. The derivative of Minkowski $s?(x)$ function. J. Math. Anal. Appl. **253**(1): 107–125.

[736] Park, K-S., Jin-Bae Park, J-B., Choi, Y-H., Yoon, T-S. and Chen, G. 1998. Generalized predictive control of discrete-time chaotic systems, Int. J. of Bifurcation and Chaos. **8**(7): 1591–1597.

[737] Parker, T. and Chua, L. O. 1989. Pracfical numerical algorithms for chaotic systems. New York Springer-Verlag.

[738] Parry, W. 1979. The Lorenz attractor and a related population model. Springer-Verlag. Lecture Notes in Math. **729**: 169–187.

[739] Parui, S. and Banerjee, S. 2002. Border collision bifurcations at the change of state-space dimension. Chaos. **12**: 1054–1069.

[740] Pastor-Satorras, R. and Riedi, R. H. 1996. Numerical estimates of the generalized dimensions of the Hénon attractor for negative q. J. Physics A: Mathematical and General. **29**(15): L391–L398.

[741] Pei-Min, X. and Bang-Chun, W. 2004. A new type of global bifurcation in Hénon map. Chinese Phys. **13**(5): 618–624.

[742] Peixoto, M. 1962. Structural stability on two-dimensional manifolds. Topology. **1**: 101–120.

[743] Pellegrini, L., Tablino, C., Albertoni, S. and Biardi, G. 1993. Different scenarios in a controlled tubular reactor with a countercurrent coolant. Chaos, Solitons & Fractals. **3**(5): 3537–3549.

[744] Perov, A. I. and Egle, I. Yu. 1972. On the Poincaré-Denjoy theory of multidimensional differential equations, Differentsial'nye Uravneniya. 8 (1972) 801–810, English transl. Differential Equations. **8**: 608–615.

[745] Pereira, R. F., Pinto, S. E., Viana, R. L., Lopes, S. R. and Grebogi, C. 2007. Periodic orbit analysis at the onset of the unstable dimension variability and at the blowout bifurcation. Chaos **17**: 023131.

[746] Pesin, Y. 1977. Characteristic Lyapunov exponents and smooth ergodic theory. Russian Math. Surveys. **32**(4): 55–114.

[747] Pesin, Y., and Sinai, Y. 1982. Gibbs measures for partially hyperbolic attractors. Erg. Th. & Dynam. Sys. **2**: 417–438.

[748] Pesin, Y. 1992. Dynamical systems with generalized hyperbolic attractors: Hyperbolic, ergodic and topological properties. Ergodic Theory Dynam. Systems. **12**: 123–151.

[749] Pesin, Y. and Weiss, H. 2001 (eds.). Smooth ergodic theory and its applications. Amer. Math. Soc.

[750] Pesin, Y. 2004. Lectures on Partial Hyperbolicity and Stable Ergodicity. Zürich Lectures in Advanced Mathematics, EMS.

[751] Peter, G. 2007. On the determination of the basin of attraction of discrete dynamical systems. J. Difference Equations and Applications. **13**(6): 523–546.

[752] Peter, G. and Heiko,W. 2007. Lyapunov function and the basin of attraction for a single-joint muscle-skeletal model. J. Mathematical Biology. **54**(4): 453–464.

[753] Petersen, K. 1990. Ergodic Theory. Cambridge Studies in Advanced Mathematics. Cambridge Cambridge University Press.

[754] Petrisor, E. 2003. Entry and exit sets in the dynamics of area preserving Hénon map. Chaos Solitons & Fractals. **17**(4): 651–658.

[755] Pilyugin, S. Y. 1999. Shadowing in Dynamical Systems. Lect. Notes Math., 1706. Berlin, Heidelberg, New YorkSpringer.

[756] Pingel, D. 1999. Schmelcher, P. and Diakonos, F. K., Theory and examples of the inverse Frobenious-Perron problem for complete chaotic maps. Chaos. **9**: 357–366.

[757] Pinto, A. A., Rand, D. A. and Ferreira, F. 2008. Fine structures of hyperbolic diffeomorphisms. Springer Monographs in Mathematics.

[758] Pivka, L., Wu, C. -W. and Huang, A. 1994. Chua's oscillator: a compendium of chaotic phenomena. J. Franklin Instit. **331B**(6): 705–741.

[759] Pivka, L., Wu, C-W. and Anshan, H. 1996. Lorenz equation and Chua's equation. Inter. J. Bifur.Chaos. **6**(12B): 2443–2489.

[760] Plykin, R. V. 1974. Sources and sinks for A-diffeomorphisms of surfaces. Math. USSR Sb. **23**: 233–253.

[761] Plykin, R. V. 1984. On geometry of hyperbolic attractors of smooth cascades. Uspekhi Matem. Nauk. **39**(6): 75–113.

[762] Plykin, R. V. 1977. The existence of attracting (repelling) periodic points of axiom A diffeomorphisms of the projective plane and of the Klein bottle. Uspekhi Mat. Nauk. **323**: 179 (in Russian).

[763] Plykin, R. V. 1980. Hyperbolic attractors of diffeomorphisms. Internat. Topology. Conf., (Moscow State Univ., Moscow) (1979); Uspekhi Mat. Nauk. 353 (1980) 94–104 English transl. Russian Math. Surveys. **353**: 109–121.

[764] Plykin, R. V. 1984. On the geometry of hyperbolic attractors of smooth cascades. Uspekhi Mat. Nauk. **396**: 75–113, English transl. Russian Math. Surveys. 396 (1984) 85–131.

[765] Plykin, R. V. and Zhirov, A. Y. 1993. Some problems of attractors of dynamical systems. Topology. Appl. **54**: 19–46.

[766] Plykin, R. V., Sataev, E. A. and Shlyachkov, S. V. 1995. Strange attractors, Itogi Nauki i Tekhniki Sovremennye Problemy Mat.Fundamental'nye Napravleniya. 66 (VINITI, Moscow) (1991) 100–147 English transl. Dynamical systems *IX*. Encyclopaedia Math. Sci. **66**, Springer-Verlag, Berlin. 93–139.

[767] Plykin, R. V. 2002. On the problem of topological classification of strange attractors of dynamical systems. Russ. Math. Surv. **576**: 1163–1205.

[768] Poincaré, H. 1890. Sur le probl`eme des trois corps et les équations de la dynamique. Acta. Math. **13**: 1–270.

[769] Pollicott, M. 1993. Lectures on ergodic theory and Pesin theory on compact manifolds. London Mathematical Society Lecture Note Series.

[770] Pomeau, Y. and Manneville, P. 1980. Intermittent transition to turbulence in dissipative dynamical systems. Commun. Math. Phys. **74**: 189–97.

[771] Prashant, M. G 1998. Feedback control in coupled map lattices. Phy. Rev. E. **57**(6): 7309–7312.

[772] Press, W. H., Teukolsky, S. A., Vettering, W. T. and Flannery, B. P. 1992. Numerical recipes in C. Cambridge University Press.

[773] Prokhorov, A. A. and Mchedlova, E. S. 2006. Complex dynamics of a generator with a piecewise-linear current-voltage characteristic subjected to an external periodic multifrequency signal. Journal of Communications Technology and Electronics. **51**(4): 419–423.

[774] Przytycki, F. 1980. Construction of invariant sets for Anosov diffeomorphisms and hyperbolic attractors. Studia Math. **68**: 199–213.

[775] Pugh, C. and Shub, M. 2000. Stable ergodicity and Juliene quasiconformality. J. European Mathematical. Society. **2**: 1–52.

[776] Pujals, E. R., Sambarino, M. 2000a. Homoclinic tangencies and hyperbolicity for surface diffeomorphisms. Annals. Math. **151**: 961–1023.

[777] Pujals, E. R. and Sambarino, M. 2000b. On homoclinic tangencies, hyperbolicity, creation of homoclinic orbits and variation of entropy. Nonlinearity. **13**: 921–926.

[778] Pujals, E. R. 2002. Tangent bundles dynamics and its consequences. ICM (2002). III 1–3.

[779] Pujals, E. R. 2006. On the density of hyperbolicity and homoclinic bifurcations for 3D-diffeomorphisms in attracting regions. Discrete and Continuous dynamical systems. **16**(1): 179–226.

[780] Pujals, E. R., Robert, L. and Shub, M. 2006. Expanding maps of the circle rerevisited: Positive Lyapunov exponents in a rich family. Ergod. Th. Dynam. Syst. **26**: 1931–1937.

[781] Pujals, E. and Sambarino, M. 2007. Integrability on codimension one dominated splitting. Bull. Braz. Math. Soc. **38**: 1–19.

[782] Pujals. E. R. 2008. Density of hyperbolicity and homoclinic bifurcations for topologically hyperbolic sets. Discrete and Continuous Dynamical System. **20**(2): 337–408.

[783] Qammar, H., Seshadhri, K. R., Gomatam, R. and Venkatesan, A. 1996. Control of a chaotic polymerization reaction using linear and nonlinear controllers. The Chemical Engineering Journal. **64**: 141–148.

[784] Rabinovich, M. and Trubetskov, D. 1984. The Introduction to the Theory of Oscillations and Waves. Nauka, Moscow.

[785] Rajaraman, R., Dobson, I. and Jalali, S. 1996. Nonlinear dynamics and switching time bifurcations of a thyristor controlled reactor circuit. IEEE Trans. Circuits & Systems. I **43**: 1001–1006.

[786] Ramdani, S., Chua, L.O., Lozi, R. and Rossetto, B. 1999. A qualitative study comparing Chua and Lorenz systems. Proc.s of the 7th Inter. Specialist Workshop on Nonlinear Dynamics of Electronic Systems. Tech. Univ. Denmark. 205–208.

[787] Rand, D. 1978. The topological classification of Lorenz attractors. Math. Proc.Camb. Phil. Soc. **83**: 451–460.

[788] Rand, D., The topological classification of Lorenz attractors. Proc. Cambridge Philos. Soc. **83** (1978) 451–460 Russian transl. Strange Attractors, (Mir, Moscow) (1981) 239–251.

[789] Robert, B. and Robert, C. 2002. Border collision bifurcations in a onedimensional piecewise smooth map for a PWM currentprogrammed H-bridge inverter. Int J. Control. **75**(16–17): 1356–1367.

[790] Robbin, J. 1971. A structural stability theorem. Ann. of Math. **94**: 447–493.

[791] Robinson, C. and Verjovsky, A. 1971. Stability of Anosov diffeomorphisms, "Seminario de Sistemas Dinamicos" edited by Palis, J., Monografias de Matematica. 4. IMPA Rio de Janeiro. Brazil, Chapter 9.

[792] Robinson, C. 1975. Structural stability of vector fields. Ann. of Math. 99 (1974) 154–175; Errata in Robinson, C., Ann. of Math. **101**: 368.

[793] Robinson, C. 1983. Bifurcation to infinitely many sinks. Comm. Math Phys. **90**: 433–459.

[794] Robinson, C. 1984. Transitivity and invariant measures for the geometric model of the Lorenz attractor. Ergod Th Dynam Sys. **4**: 605–611.

[795] Robinson, C. 1989. Homoclinic Bifurcation to a Transitive Attractor of Lorenz Type. Nonlinearity. **2**: 495–518.

[796] Robinson, C. 2000. Nonsymmetric Lorenz attractor from a homoclinic bifurcation. SIAM. J. Math Anal. **32**(1): 119–141.

[797] Robinson, C. 2004. Dynamical systemsstability, symbolic dynamics, and chaos. CRC Press.

[798] Rodriguez Hertz, F. 2005. Stable ergodicity of certain linear automorphisms of the torus. Annals of mathematics. **162**(1): 65–107.

[799] Rohlin, V. A. 1949. On the fundamental ideas of measure theory. Math. Sbornik. **25**(67): 107–150.

[800] Rolfsen, D. 1976. Knots and Links. Wilmington, DEPublish or Perish Press. 287–288.

[801] Rosen, R. 1970. Dynamical system theory in biology, Wiley, New York.

[802] Rosenblatt, J. M. and Weirdl, M. 1995. Pointwise ergodic theorems via harmonic analysis. (1993). Appearing in Ergodic Theory and its Connections with Harmonic Analysis, Proceedings of the (1993) Alexandria Conference. Petersen, K. E. and Salama, I. A. eds., Cambridge University Press.

[803] Rosenthal, J. S. 2000. A first look at rigorous probability theory, World Scientific.

[804] Rossler, Ô. E. 1976. An equation for continuous chaos. Phys. Lett. A. **57**: 397–98.

[805] R^ossler, Ô. E. 1979. Continuous chaos-Four prototype equations. Ann. N.Y. Acad. Sci. **31**: 376–392.

[806] Rossler, Ô. E. 1980. Chaos and bijections across dimensions. In: New approaches to nonlinear problems in dynamics, Holmes, P., ed. (SIAM, Philadelphia) 477–486.

[807] Rôssler, Ô. E., Hudson, J. L. and Farmer, J. D. 1983. Noodle-map chaos: A simple example, in: Stochastic phenomena and chaotic behavior in complex systems, P. Schuster, ed, UNESCO Cocference, Flattnitz, Austria, June 6-10.1983 (Springer, New York).

[808] Rovella, A. 1993. The dynamics of perturbations of the contarcting lorenz attractor. Bol. Soc. Bras. Mat. **24**(2): 233–259.

[809] Russell, D. A., Hanson, J. D. and Ott, E. 1980. Dimension of Strange Attractors. Phys. Rev. Let. **45**: 1175–1178.

[810] Ruelle, D. and Takens, F. 1971. On the nature of turbulence. Comm. Math. Phys. **20**: 167–192.

[811] Ruelle, D. 1976. A measure associated with Axiom A attractors. Am. J. Math. **98**: 619–654.

[812] Ruelle, D. 1978. An inequality for the entropy of differentiable maps. Bol. Soc. Bras. Math. **9**: 83–87.

[813] Rychlik, M. 1983a. Bounded variation and invariant measures. Studia Math. **LXXVI**: 69–80.

[814] Rychlik, M. 1983b. Invariant measures and the variation principle for Lozi mappings. PhD dissertation. University of California, Berkeley.

[815] Rychlik, M. 1983c. Mesures invariantes et Principe variationel pour les applications de Lozi. C. R. Acad. Sc. Paris. **296**. Serie I, 19–22.

[816] Ryouichi, H. and Akira, S. 2004a. An algorithm to prune the area-preserving Hénon map. J. Phys. A: Math. Gen. **37**: 10521–10543.

[817] Ryouichi, H. and Akira, S. 2004b. Grammatical complexity for twodimensional maps. J. Phys. A: Math. Gen. **37**: 10545–10559.

[818] Sanchez-Salas, F. J. 2001. Sinai-Ruelle-Bowen measures for piecewise hyperbolic transformations. Divulgaciones Matematicas. **9**(1): 35–54.

[819] Sander, E. and Yorke, J. A. 2011. Period-doubling cascades galore. Ergodic Theory and Dynamical Systems. **31**(4): 1249–1267.

[820] Sander, E. and Yorke, J. A. 2009. Period-doubling cascades for large perturbations of Hénon families, Journal of Fixed Point Theory and Applications. **6**(1): 153–163.

[821] Sands, D. 1995. Topological conditions for positive Lyapunov exponent in unimodal maps. Preprint 95–59, Université de Paris-XI (Orsay).

[822] Sannami, A. 1983. The stability theorems for discrete dynamical systems on two-dimensional manifolds. Nagoya Math. J. **90**: 1–55.

[823] Sannami, A. 1989. A Topological classification of the periodic orbits of the Hénon family. Japan J. Appl. Math. **6**: 291–300.

[824] Sannami, A. 1994. On the structure of the parameter space of the Hénon map. In: Towards the Harnessing of Chaos. (1994) 289–303. Amsterdam, Elsevier.

[825] Sano, M., Sato, S. and Sawada, Y. 1986. Global spectral characterization of chaotic dynamics. Prog.Theor. Phys. **76**(4): 945–948.

[826] Sataev, E. A. 1992. Invariant measures for hyperbolic maps with singularities. Russian. Math. Surveys. **47**(1): 192–251.

[827] Sataev, E. A. 2005. Non-existence of stable trajectories in non-autonomous perturbations of systems of Lorenz type. Sbornik. Mathematics. **196**(4): 561–594.

[828] Sauer, T., Yorke, J. A. and Casdagli, M. 1991. Embedology. J. Stat. Phys. **65**(3-4): 579–616.

[829] Sawyer, A. 1984. The dynamics of piecewise linear mappings of the plane. Ann Arbor. Michigan. U.S.A.

[830] Schuster, H. 1984. Deterministic Chaos. Physik-Verlag GmbH, Weinheim (F.R.G.).

[831] Sebesta, V. 1999. Predictability of chaotic signals. J. Elec. Engineering. **50**(9-10): 302–304.

[832] Silva, C. P. 2003. Shi'lnikov theorem—a tutorial. IEEE Trans. Circuits Syst.-I . **40**: 675–682.

[833] Simitses, G. J. and Hodges, D. H. 2006. Fundamentals of Structural Stability. Elsevier.

[834] Simon, R. 1972. A 3-dimensional Abraham–Smale example. Proc. Amer. Math. Soc. **34**: 629–630.

[835] Sinai, Y. 1963. On the Foundations of the ergodic hypothesis for a dynamical system of statistical mechanics, Dokl. Acad. Nauk. 153 (6) (1963) (in English, Sov. Math Dokl. **4**: 1818–1822).

[836] Sinai, Y. 1968a. Markov partitions and C-diffeomorphisms. Func. Anal. and its Appl. **2**(1): 64–89.

[837] Sinai, Y. 1968b. Construction of Markov partitions. Func. Anal. and its Appl. **2**(2): 70–80.

[838] Sinai, Y. 1970. Dynamical systems with elastic collisions. Russian Math Surveys. **25**: 141–92.

[839] Sinai, Y. 1972. Gibbs measures in ergodic theory. Uspehi. Mat. Nauk. **27**(4) (1972) 21–64. English translation. Russian. Math. Surveys. **27**(4): 21–69.

[840] Sinai, Y. 1972. Gibbs measure in ergodic theory. Russian Math. Surveys. **27**: 21–69.

[841] Sinai, Y. 1979. Stochasticity of Dynamical Systems. In: Nonlinear Waves, edited by Gaponov-Grekhov, A.V, Nauka, Moscow, P. 192 (in Russian).

[842] Sivak, A. 1997. On the periodicity of recursive sequences, in: S. Elaydi, I. Gayori, G. Ladas (Eds.), Proc. of the Second ICDEA, pp. 559–566.

[843] Sharkovsky, A. N. and Chua, L. O. 1993. Chaos in some 1–D discontinuous maps that appear in the analysis of electrical circuits. IEEE Trans. Circuits & Systems–I. **40**: 722–731.

[844] Shaw, R. S. 1984. The dripping faucet as a model chaotic system, Aerial Press.

[845] Shibayama, K. 1989. Connections of periodic orbits in the parameter space of the Lozi family. World Scientific Advanced Series in Dynamical Systems. **7**, The Study of Dynamical Systems, Kyoto, Nobuo Aoki Ed., London, 10–25.

[846] Shilnikov, L. P. 1965. A case of the existence of a countable number of periodic motions. Sov. Math. Docklady. **6**: 163–166 (translated by S. Puckette).

[847] Shi'lnikov, L. P. 1970. A contribution of the problem of the structure of an extended neighborhood of rough equilibrium state of saddle-focus type. *Math*. U.S.S.R. Shornik. **10**: 91–102 (translated by F. A. Cezus).

[848] Shilnikov, L. P. 1981. The bifurcation theory and quasi-hyperbiloc attractors. Uspehi Mat. Nauk. **36**: 240–241.

[849] Shilnikov, L. P. 1991. Bifurcations and chaos in the Shimizu-Marioka system (In Russian) in Methods and qualitative theory of differential equations, Gorky State University, (1986) 180–193. (English translation in Selecta Mathematica Sovietica **10**: 105–117).

[850] Shil'nikov, L. P. 1993. Strange attractors and dynamical models. J. Circuits Syst. Comput. **3**(1): 1–10.

[851] Shil'nikov, L. P. 1993. Chua's circuit: rigorous results and future problems. IEEE Trans. Circuits Syst.-IFund. Th. Appl. **40**(10): 784–786.

[852] Shil'nikov, A. L., Shil'nikov, L. P. and Turaev, D. V. 1993. Normal forms and Lorenz attractors. Int. J. Bifurcation Chaos. **3**(5): 1123–1139.

[853] Shil'nikov, L. P. 1994. Chua's circuit: rigorous results and future problems. Inter. J. Bifur. Chaos. **4**(3): 489–519.

[854] Shil'nikov, L. P. and Turaev, D. V. 1997. Simple bifurcations leading to hyperbolic attractors. Comput. Math. Appl. **34**(2–4): 173–193.

[855] Shilnikov, L. P. 2002. Bifurcations and Strange Attractors, ICM, Vol. III, 1-3.

[856] Shilov, G. E. and Gurevich, B. L. 1978. Integral, measure, and derivative: A unified approach, Richard A. Silverman, trans. Dover Publications.

[857] Shimada, I. and Nagashima, T. 1979. A numerical approach to ergodic problem of dissipative dynamical systems. Prog. Theor. Phys. **61**: 1605–1616.

[858] Shub, M. 1971. Topological Transitive Diffeomorphism *on* $\mathbb{T}^4$. Lecture Notes in Math. **206** Springer-Verlag, New York.

[859] Shub, M. and Sullivan, D. 1985. Expanding endomorphisms of the circle revisted. Ergod. Th. Dynam. Syst. **5**: 285–289.

[860] Shub, M. 1987. Global stability of dynamical systems, Springer-Verlag.

[861] Shub, M. and Wilkinson, A. 2000. Pathological foliations and removable zero exponents. Inv. Math. **139**: 495–508.

[862] Slodkowski, Z. 1991. Holomorphic motions and polynomial hulls. Proc. A. M. S. **111**: 347–355.

[863] Smale, S. 1963. Stable manifolds for differential equations and diffeomorphisms. Ann. Scuola Norm. Sup. Pisa. **17**: 97–116.

[864] Smale, S. 1965. diffeomorphisms with many periodic points, In Differential and Combinatorial Topology: A Symposium in Honor of Marston Morse, 63–70 S.S. Cairns (ed). Princeton University Press, Princeton, NJ.

[865] Smale, S. 1965. Diffeomorphisms with many periodic points, in Differential and Combinatorial Topology. A Symp. In Honor of Marston Morse, 63–80 Princeton Univ. Press, Princeton, N.J.

[866] Smale, S. 1967. Differentiable dynamical systems. Bull. Amer. Math.l Soc. **73**: 747–817.

[867] Smale, S. 1968. The 2126-stability theorem. Proc. Sympos. Pure.Math., Vol. XIV, Berkeley, Calif., 289–298 Amer. Math. Soc., Providence, R.I.4.

[868] Smale, S. 1998. Mathematical Problems for the Next Century. Math. Intelligencer. **20**(2): 7–15.

[869] Smale, S. 2000. Mathematical Problems for the Next Century. Mathematics, Frontiers and Perspectives (2000) (Ed. Arnold,V., Atiyah, M., Lax, P., and Mazur, B.,), Providence, RIAmer. Math. Soc.

[870] Smillie, J. 1997. Complex dynamics in several variables. In: Flavors of geometry., **31** of Math. Sci. Res. Inst. Publ., 117–150. Cambridge University Press, Cambridge, (1997). With notes by Gregery, T. Buzzard.

[871] Spany, V. and Pivka, L. 1990. Boundary surfaces in sequential circuits. Inl. J. on Ckl. Th. and Appl. **18**(4): 349–360.

[872] Sparrow, C. 1982. The Lorenz equations, bifurcations, chaos, and strange attractors. Springer-Verlag, New York.

[873] Spivak, M. 1999. A comprehensive introduction to differential geometry. Publish of Perish, Inc. Houston, Texas.

[874] Sprott, J. C. 1993a. Automatic generation of strange attractors. Comput. & Graphics. **17**(3): 325–332.

[875] Sprott, J. C. 1993b. Strange Attractors: Creating Patterns in Chaos. M &T Books, New York.

[876] Sprott, J. C. 1993c. How common is chaos?. Phys. Lett. A. **173**: 21–24.

[877] Sprott, J. C. 1994a. Some simple chaotic flows. Phys. Rev. E. **50**(2): R647–650.

[878] Sprott, J. C. 2003d. Chaos and Time-Series Analysis. Oxford University Press.

[879] Sprott, J. C. 1994b. Predicting the dimension of strange attractors. Phys. Lett. A. **192**: 355–360.

[880] Sprott, J. C. 2006. High-dimensional dynamics in the delayed Hénon map. Electronic journal of theoretical physics. **312**: 19–35.

[881] Sprott, J. C. 2007. Maximally complex simple attractors, Chaos. **17**: 033124-1–033128-6.

[882] Sprott, J. C. and Rowlands, G. 2001. Improved correlation dimension calculation. International Journal of Bifurcation and chaos. **11**(7): 1865–1880.

[883] Starrett, J. 2002. time-optimal chaos control by center manifold targeting. Phy. Rev. E. **66**: 046206-1–046206-6.

[884] Strelkova, G. and Anishchenko, V. 1997. Structure and properties of quasihyperbolic attractors. In Proc. of Int. Conf. of COC'97 (St. Petersburg, Russia, August 27–29,1997, **2**: 345–346.

[885] Sterling, D., Dullin, H. R. and Meiss, J. D. 1999. Homoclinic bifurcations for the Hénon map. Physica. D. **134**: 2153–2184.

[886] Stewart, I. 2000. The Lorenz attractor exists. Nature. **406**: 948–949.

[887] Stoffer, D. and Palmer, K. J. 1999. Rigorous verification of chaotic behaviour of maps using validated shadowing. Nonlinearity. **12**(6): 1683–1698.

[888] Stojanovska, L. F. 1989. Stochastic stability of Lozi mappings. PhD thesis. University of Arizona.

[889] Svitanovic, P. 1984. Universality in Chaos, Bristol Adam Hilger Ltd.

[890] Sucharev, Y. I. 2004. Fractal macroscopic forming of living oxyhydrate gels of rare metals. Proceedings of the Chelyabinsk Scientific Center.

[891] Sucharev, Y. I., Krupnova T. G. and Yudina, E. P. 2004a. Geometry of attractors of yttrium oxyhydrate gels which have undergone influence of the shearing. Proceedings of the Chelyabinsk Scientific Center.

[892] Sucharev, Y. I., Krupnova T. G. and Yudina, E. P. 2004b. Geometry of strange attractors in oxyhydrate gel systems. Proceedings of the Chelyabinsk Scientific Center.

[893] Sucharev, Y. I., Kostrjukova, A. M. and Marcov, B. A. 2005. Experimental phase diagrams of currental features of zirconium oxyhydrate gel systems. Proceedings of the Chelyabinsk Scientific Center.

[894] Sullivan, D. 1991. The universalities of Milnor, Feigenbaum and Bers, in Topological Methods in Modern Mathematics. SUNY at Stony Brook. Proc. Symp. held in honor of John Milnor's 60th birthday, 14–21.

[895] Sullivan, D. 1992. Bounds, quadratic differentials, and renormalization conjecture. A.M. S. Centennial Publ. **2**: 417–466.

[896] Sutherland, B. and Kohmoto, M. 1987. Resistance of a one-dimensional quasicrystal: Power-law growth. Phys. Rev. B. **36**: 5877–5886.

[897] Szpilrajn, E. 1937. La dimension et la mesure. Fundamenta Mathematica. **28**: 81–89.

[898] Szustalewicz, A. 2008. Minimal coverage of investigated object when seeking for its fractal dimension. In: Advances in Information Processing and Protection, Springer.

[899] Tahzibi, A. 2004. Stably ergodic systems which are not partially hyperbolic. Isr. Journal of Math. **142**: 315–344.

[900] Taixiang, S. and Hongjian, X. 2007. On the basin of attraction of the two cycle of the difference equation. J. Difference Equations and Applications. **13**(10): 945–952.

[901] Takens, F. 2005. Multiplications in solenoids as hyperbolic attractors. Topology and its Applications. **152**: 219–225.

[902] Tang, T. W., Allison, A. and Abbott, D. 2004. Parrondo's games with chaotic switching. Proc. SPIE. 5471, **520**, doi:10.1117/12.561307.

[903] Tapan, M. and Gerhard, S. 1999. On the existence of chaotic policy functions in dynamic optimization. Japanese Economic Review. **50**(4): 470–484.

[904] Tedeschini-Lalli, L. and Yorke, J. A. 1985. How often do simple dynamical prosses have infinitely many coexisting sinks?. Commun. Math. Phys. **106**(4): 635–657.

[905] Tél. T. 1982a. On the construction of stable and unstable manifolds of twodimensional maps. Z. Phys. B. **49**: 157–160.

[906] Tél. T. 1982b. On the construction of invariant curves of period-two points in two-dimensional maps, Phys. Lett. A. **94**: 334–336.

[907] Tél, T. 1983a. Invariant curves, attractors and phase diagram of a piecewise linear map with chaos. J. Stat. Phys. **33**(1): 195–221.

[908] Tél, T. 1983b. Fractal dimension of the strange attractor in a piecewise linear two-dimensional map. Physics Letters. **97A**(6) 219–223.

[909] Tibor, C., Barnabás, G. and Balázs, B. 2006. A verified optimization technique to Locate chaotic regions of Hénon systems. J. Global Optimization. **35**(1): 145–160.

[910] Theiler, J., Eubank, S., Longtin, A., Galdrikian, B. and Farmer, J. D. 1992. Testing for nonlinearity in time seriesthe method of surrogate data. Physica. D. **58**: 77–94.

[911] Thomas, G. B. and Finney, R. L. 1992. Maxima, Minima, and Saddle Points, §12.8 in Calculus and Analytic Geometry. 8th ed. Reading, MAAddison-Wesley, 881–891.

[912] Tovbis, A., Tsuchiya, M. and Jaffé, C. 1998. Exponential asymptotic expansions and approximations of the unstable and stable manifolds of singularly perturbed systems with the Hénon map as an example. Chaos. **8**: 665–681.

[913] Tresser, C. 1983. Une theor`eme de Shilnikov en $C^{1,1}$. C.R. Acad. Paris. **296**(1): 545–548.

[914] Tresser, C. 1982. The topological conjugacy for generalized Hénon mappings: Some negative results. Bol. Soc. Bras. Mat. **13**: 115–130.

[915] Tsai, J. S. H, Yu, J. M., Canelon, J. I. and Shieh, L. S. 2005. Extended-Kalman-filter-based chaotic communication. IMA Journal of Mathematical Control and Information. **22**(1): 58–79.

[916] Tse, C. K. and Chan, W. C. Y. 1997. Experimental verification of bifurcations in current-programmed dc/dc boost converters. in Proc. European Conf. Circuit Theory Design, Budapest, Hungary, Sept. 1274–1279.

[917] Tsuji, R. and Ido. S. 2002. Computation of Poincaré map of chaotic torus magnetic field line using parallel computation of data table and its interpolation. Parallel Computing in Electrical Engineering, PARELEC apos; 02. Proc.s. Inter. Conference on Volume , Issue, 386–390.

[918] Tucker, W. 1996a. Rigorous models for the Lorenz equations. U.U.D.M Report (1996) **26**, ISSN 1101–3591.

[919] Tucker, W. 1996b. Transitivity of Lorenz-like maps and the tired baker's map. Preprint.

[920] Tucker, W. 1999. The Lorenz attractor exists. C. R. Acad. Sci. Ser. I. Math. **328**(12): 1197–1202.

[921] Tucker,W. 2002a. A Rigorous ODE Solver and Smale's 14th Problem. Found. Comput. Math. **2**: 153–117.

[922] Tucker, W. 2002b. Computing accurate Poincaré maps. Physica. D. **171**(3): 127–137.

[923] Tufillaro, N. 1992. Abbott, T. and Reilly, J., An Experimental Approach to Nonlinear Dynamics and Chaos. Addison-Wesley, Redwood City, CA.

[924] Turaev, D. V. and Shilnikov, L. P. 1995. On a blue sky catastrophy. Soviet. Math. Dokl. **342**(5): 596–599.

[925] Turaev, D. V. and Shilnikov, L. P. 1996. An example of a wild strange attractor. Sbornik. Mathematics. **189**(2): 291–314.

[926] Turaev, D. V. and Shil'nikov, L. P. 2008. Pseudohyperbolicity and the problem on periodic perturbations of Lorenz-type attractors. Doklady Mathematics. **77**(1): 17–21.

[927] Ueta, T., Kawabe, T., Chen, G. and Kawakami, H. 2004. Calculation and control of unstable periodic orbits in piecewise smooth dynamical systems. Lecture Notes in Control and Information Sciences, Chaos Control. 691–693.

[928] Ustinov, Y. 1987. Algebraic invariants of the topological conjugacy classes of solenoids. Mat. Zametki. **421** (1987) 132–144, English transl. Math. Notes. **42**: 583–590.

[929] van Dantzig, D. 1930. Über topologisch homogene Kontinua. Fund. Math. **15**: 102–125.

[930] van Strien, S. 1981. In Dynamical Systems and Turbulence, eds. Rand, D.A. and Young, L.S., LNM 898 Springer Verlag, New York.

[931] Viana. M. 1993. Strange attractors in higher dimensions. Bull. Braz. Math. Soc. **24**: 13–62.

[932] Viana, M. 1998. Dynamics: A probabilistic and geometric perspective. In Proceedings of the International Congress of Mathematicians, Vol. I (Berlin, (1998)) volume 557–578.

[933] Viana, M. 2002. Dynamics beyond uniform hyperbolicity. Lecture at Collége de France, March.

[934] Viana, R. L., Barbosa, J. R. R. and Grebogi, C. 2004. Unstable dimension variability and codimension-one bifurcations of two-dimensional maps. Phys. Lett. A. **321**(4): 244–251.

[935] Vietoris, L. 1927. Über den hoheren Zusammenhang kompakter Raume und eine Klasse von zusammenhangstreuen Abbildungen. Math. Ann. **97**: 454–472.

[936] von Neumann, J. 1932a. Proof of the Quasi-ergodic Hypothesis. Proc. Natl. Acad. Sci. U. S. A. **18**: 70–82.

[937] von Neumann, J. 1932b. Physical Applications of the Ergodic Hypothesis. Proc. Natl. Acad. Sci. U. S. A. **18**: 263–266.

[938] Voronov, S. S., Kolpalrova, I. V. and Kuznetsov, V. A. 1996. Measurement methods using the properties of nonlinear dynamic systems. Izmeritel'naya Tekhnika. **39**(12): 16–18.

[939] Voronov, S. S., Kolpakova, L. V. and Kuznetsov, V. A. 2000. Chaotic oscillator method approaches to diagnosing the parameters of nonlinear chaotic systems. Izmeritel'naya Tekhnika. **43**(4): 19–21.

[940] Walczak, P. G., Langevin, R., Hurder, S. and Tsuboi, T. 2005. Foliations. Proceedings of the International Conference Lodz, Poland. 13–24.

[941] Walters, P. 1982. An introduction to ergodic theory. Springer, New York.

[942] Wang, L. and Smith, K. 1998. On chaotic simulated annealing. IEEE Trans. Neural Networks. **9**: 716–718.

[943] Wang, Y., Yang, L. and Xie, H. M. 1999, Complexity of unimodal maps with aperiodic kneading sequences. Nonlinearity. **12**: 1151–1176.

[944] Wang, Q. and Young, L. S. 2001. Strange attractors with one direction of instability. Comm. Math. Phys. **218**(1): 1–97.

[945] Wang, Y. and Xie, H. M. 1994. Grammatical complexity of unimodal maps with eventually periodic kneading sequences. Nonlinearity. **7**: 1419–36.

[946] Wang, X. M. and Fang, Z. J. 2006. The properties of borderlines in discontinuous conservative systems. The European Physical Journal D. **37**(2): 247–253.

[947] Watkins, W. T. 1982. Homeomorphic classification of certain inverse limit spaces with open bonding maps. Pacific J. Math. **103**: 589–601.

[948] Weisstein, E. 2002. Homoclinic tangle. In: Smale Horseshoe. Homoclinic Point. Eric Weisstein's World of Mathematics., http//mathworld.wolfram.com.

[949] Wen, L. 2004. Generic diffeomorphisms away from homoclinic tangencies and heterodimensional cycles. Bulll. Braz. Math. Soc, New Series. **35**: 419–452.

[950] Wen, L. 2008. The selecting lemma of Liao. Discrete Contin. Dyn. Syst. **20**: 159–175.

[951] Williams, R. F. 1955. A note on unstable homeomorphisms. Proc. Amer. Math. Soc. **6**: 308–309.

[952] Williams, F. R. 1970. Classification of one-dimensional attractors. Proc. Symp. in Pure Math. 361–393.

[953] Williams, R. F. 1974. Expanding attractors. IHES Publication, Math. **43**: 169–203.

[954] Williams, R. F. 1979. The Structure of Lorenz Attractors. Publ. Math. IHES. **50**: 321–347.

[955] Wolf, D. M., Varghese, M. and Sanders, S. R. 1994. Bifurcation of power electronic circuits. J. Franklin Inst. **331B**(6): 957–999.

[956] Wolf, C. 2006. Generalized Physical and SRB measures for hyperbolic diffeomorphisms. J. Statistical Physics. **122**(6): 1111–1138.

[957] Wolfram, S. 1984. Computation theory of cellular automata. Commun. Math. Phys. **96**: 15–57.

[958] Xie, H. and Chen, X., Question about chaotic pieces of some nonsmooth mapping. No date.

[959] Xie, H.M. 1993. On formal languages in one-dimensional dynamical systems. Nonlinearity. **6**: 997–1007.

[960] Xie, H. M. 1996. Grammatical complexity and one-dimensional dynamical system, Singapore: World Scientific.

[961] Xiang, X., Dong, C. and Lin, S. 1996. An extension of Smale's transverse homoclinic theorem and chaotic phenomena of the Lozi map (Chines, Chinese summary), Acta math. Sci. (Chinese) **16**(4): 385–390.

[962] Xu, M., Chen, G. and Tian, Y. T. 2001. Identifying chaotic systems using Wiener and Hammerstein cascade models. Math. & Computer Modelling. **33**(4): 483–493.

[963] Yan, Y. and Qian, M. 1985. The transversal heteroclinic cycles and it's application to Hénon map. Kexue Tongbao (in Chinese) **30**: 961–965.

[964] Yang, T. and Chua, L. O. 2000. Piecewise-linear chaotic systems with a single equilibrium point. Inter. J. Bifur.Chaos. **10**(9): 2015–2060.

[965] Yang, T. and Chua, L. O. 2001. Testing for local activity and edge of chaos. Inter. J. Bifur. Chaos. **11**(6): 1495–1591.

[966] Yang, X. S. 2003a. A proof for a theorem on intertwining property of attraction basin boundaries in planar dynamical systems. Chaos, Solitons and Fractals. **15**(4): 655–657.

[967] Yang, X. S. 2003b. Structure of basin boundaries of attractors of ODE's on S2. Chaos, Solitons and Fractals. **16**(1): 147–150.

[968] Yang, X. S. and Tang, Y. 2004. Horseshoes in piecewise continuous maps. Chaos, Solitons and Fractals. **19**: 841–845.

[969] Yang, J. 2007. Newhouse phenomenon and homoclinic classes. Arxiv preprint arXiv0712. 0513.

[970] Ying-Cheng, L., Grebogi, C., Yorke, J. A. and Kan, I. 1993. How often are chaotic saddles nonhyperbolic?. Nonlinearity. **6**: 779–797.

[971] Yang, T., Chun-Mei, Y. C. -M. and Yang, L. -B. 1998. A detailed study of adaptive control of chaotic systems with unknown parameters. Dynamics and Control. **8**(3): 255–267.

[972] Yildiz, I. B. 2011. Discontinuity of Topological Entropy for the Lozi Maps. Ergodic Theory and Dynamical Systems. DOI: 10.1017/S0143385711000411: 1–18.

[973] Yildiz, I. B. 2011. Monotonicity of the Lozi family and the zero entropy locus, Nonlinearity **24**: 1613–1628.

[974] Yoccoz, J. C., Polynômes quadratiques et attracteur de Hénon. Séminaire Bourbaki., 33 Exposé N°. **734**(1990)–(1991)).

[975] Yoccoz, J.C. 1995. Introduction to Hyperbolic Dynamics. Real and complex dynamical systems. Proceedings of the NATO Advanced Study Institute held in Hillerod, June 20-July 2 (1993) 265–291. Edited by Bodil Branner and Paul Hjorth. NATO Advanced Science Institutes Series CMathematical and Physical Sciences, **464**. Kluwer Academic Publishers, Dordrecht.

[976] Yorke, J. A. and Alligood, K. T. 1983. Cascades of period-doubling bifurcations: a prerequisite for horseshoes. Bull. Amer. Math. Soc. (N.S.) **9**(3): 319–322.

[977] Yorke, J. A., Grebogi, C., Ott, E. and Tedeschini-Lalli, L. 1984. Scaling behavior of windows in dissipative dynamical systems. Phys. Rev. Lett. **54**(11): 1095–1098.

[978] You, Z. P., Kostelich, E. and Yorke, J. A. 1991. Calculating stable and unstable manifolds. Int. J. Bifurcation. Chaos. **1**: 605–623.

[979] Young, L. S. 1982. Dimension, entropy, and Lyapunov exponents. Ergod. Theor. Dynam. Syst. **2**: 109–124.

[980] Young, L. S. 1985. A Bowen-ruelle measure for certain piecewise hyperbolic maps. Trans. Amer. Math. Soc. **287**: 41–48.

[981] Young, L. S. 1986. Stochastic stability of hyberbolic attractors. Ergodic Theory and Dynamical Systems. **6**: 311–319.

[982] Young, L. S. 1992. Decay of correlations for certain quadratic maps. Comm. Math. Phys. **146**: 123–138.

[983] Young, L. S. 1998a. Developments in chaotic dynamics. Notices Amer. Math. Soc. **45**(10): 1318–1328.

[984] Young, L.S. 1998b. Statistical properties of dynamical systems with some hyperbolicity. Ann. Math. **147**(3): 585–650.

[985] Young, L. S. 1999. Recurrence times and rates of mixing. Israel J. Math. **110**: 153–188.

[986] Yuan, G. H. 1997. Shipboard crane control, simulated data generation and border-collision bifurcations. Ph. D. dissertation, Univ of Maryland, College Park, USA.

[987] Yuan, G. H., Banerjee, S., Ott, E. and Yorke, J. A. 1998. Border collision bifurcations in the buck converter. IEEE Transactions on Circuits and Systems-I, **45**(7): 707–716.

[988] Yulmetyev, R., Emelyanova, N., Demin, S., Gafarov, F., Hänggi, P. and Yulmetyeva, D., Fluctuations and noise in stochastic spread of respiratory infection epidemics in social networks, Unsolved problems of noise and fluctuations UPoN (2002) 3rd international conference, 408–421.

[989] Zalzala, A. M. and Fleming, P. (eds) 1997. Genetic algorithm in engineering systems. IEE Control Engineering, ser. **55**. The Institution of Engineering and Technology.

[990] Zhang J.-Y. 1984. The Smale horseshoe in Hénon mapping, Kexue Tongbao (in Chinese). **29**: 1478–1480.

[991] Zakrzhevsky, M. 2008. New concepts of nonlinear dynamics: Complete bifurcation groups, protuberances, unstable periodic in nitiums and rare attractors. J. Vibroengineering (JVE). **10**: 421–441.

[992] Zeraoulia, E. and Hamri, N. 2005. A generalized model of some Lorenztype and quasi-attractors type strange attractors in three-dimensional dynamical systems. International Journal of Pure & Applied Mathematical Sciences. **2**(1): 67–76.

[993] Zeraoulia, E. 2005. A new chaotic attractor from 2-D discrete mapping via border-collision period doubling scenario. Discrete dynamics in nature and society. **Volume 2005** 235–238.

[994] Zeraoulia, E. 2007. On the dynamics of a n-D piecewise linear map. Electronic Journal of Theoretical Physics. **4**(14): 1–7.

[995] Zeraoulia, E. 2008. On the occurrence of chaos via different routes to chaos: period doubling and border-collision bifurcations. Translated from Sovremennaya Matematika i Ee Prilozheniya (Contemporary Mathematics and its Applications). **61** Optimal Control.

[996] Zeraoulia, E. and Sprott, J. C. 2008. A two-dimensional discrete mapping with C^∞-multifold chaotic attractors. Electronic journal of theorical physics. **5**(17): 111–124.

[997] Zeraoulia, E. and Sprott, J. C. 2008. On the robustness of chaos in dynamical systems: Theories and applications, Front. Phys. China, **3**: 195–204.

[998] Zeraoulia, E. and Sprott, J. C. 2009. The discrete hyperchaotic double scroll. International Journal of Bifurcations & Chaos. **19**(3): 1023–1027.

[999] Zeraoulia, E. and Sprott, J. C. 2010a. A new simple 2-D piecewise linear map. Journal of Systems Science and Complexity. **23**(2): 379–389.

[1000] Zeraoulia, E. and Sprott, J. C. 2010b. 2-D quadratic maps and 3-D ODE's systems: A rigorous introduction, World Scientific Series on Nonlinear Science Series A, **73**.

[1001] Zeraoulia, E. and Sprott, J. C. 2011. Robust choas and its applications. World Scientific Series on Nonlinear Science Series A, **79**.

[1002] Zgliczynski, P. 1996. Fixed point index for iterations, topological horseshoe and chaos. Topological Methods in Nonlinear Analysis. **8**(1): 169–177.

[1003] Zgliczynski, P. 1997. Computer assisted proof of the horseshoe dynamics in the Hénon map. Random & Computational Dynamics. **5**: 1–17.

[1004] Zheng,W. M. and Hao, B. L. 1990. In: Experimenta [Study and characcferization of chaos, edited by Hao, B. L., World Scientific, Singapore.

[1005] Zheng, W. M. 1991. Symbolic dynamics for the Lozi map. Chaos, Solitons & Fractals. **1**(3): 243–248.

[1006] Zheng, W. M. 1992. Admissibility conditions for symbolic sequences of the Lozi map. Chaos, Solitons and Fractals. **2**(5): 461–470.

[1007] Zhirov, A. Yu. 1995. Hyperbolic attractors of diffeomorphisms of oriented surfaces. I. Coding, classification, and coverings. Mat. Sb. 1856 (1994(a)) 3–50, English transl. Russian Acad. Sci. Sb. Math. **82**: 135–174.

[1008] Zhirov, A. Yu. 1996. Hyperbolic attractors of diffeomorphisms of oriented surfaces. II. Enumeration and application to pseudo-Anosov diffeomorphisms. Mat. Sb. 1859 (1994(b)) 29–80, English transl. Russian Acad. Sci. Sb. Math. **83**: 23–65.

[1009] Zhirov, A. Yu. 1995. Hyperbolic attractors of diffeomorphisms of oriented surfaces. III. A classification algorithm. Mat. Sb. 1862 69–82, English transl. Sb. Math. **186**: 221–244.

[1010] Zhirov, A. Yu. 2000. Complete combinatorial invariants of conjugacy of hyperbolic attractors of diffeomorphisms of surfaces. J. Dynam. Control Systems. **6**: 397–430.

[1011] Zhou, Z. and Liu, Z. 1988. Structure of attracting set of a piecewise linear Hénon mapping. Appl. Math and Mech. **9**(9): 771–836.

[1012] Zhou, X. and Zhang, X. 2002. Control of discrete-chaotic dynamic system based on improved relativity. TENCON'02. Proceedings. 2002 IEEE Region 10 Conference on Computers Communication, Control and Power Engineering, 28–31 Oct. 2002. **3**: 1393–1396.

[1013] Zuppa, C. 1979. Regularisation C^∞ des champs vectoriels qui préservent l'élément de volume. Bulletin of the Brazilian Mathematical Society. **10**(2): 51–56.

Index

For Product Safety Concerns and Information please contact our EU
representative GPSR@taylorandfrancis.com
Taylor & Francis Verlag GmbH, Kaufingerstraße 24, 80331 München, Germany

www.ingramcontent.com/pod-product-compliance
Ingram Content Group UK Ltd.
Pitfield, Milton Keynes, MK11 3LW, UK
UKHW021116180425
457613UK00005B/113